George H. Dadd, Fairman Rogers Collection

The Anatomy and Physiology of the Horse

with anatomical and questional illustrations - containing, also, a series of

examinations on equine anatomy and physiology, with instructions in reference to

dissection

George H. Dadd, Fairman Rogers Collection

The Anatomy and Physiology of the Horse
with anatomical and questional illustrations - containing, also, a series of examinations on equine anatomy and physiology, with instructions in reference to dissection

ISBN/EAN: 9783337256135

Printed in Europe, USA, Canada, Australia, Japan

Cover: Foto ©berggeist007 / pixelio.de

More available books at **www.hansebooks.com**

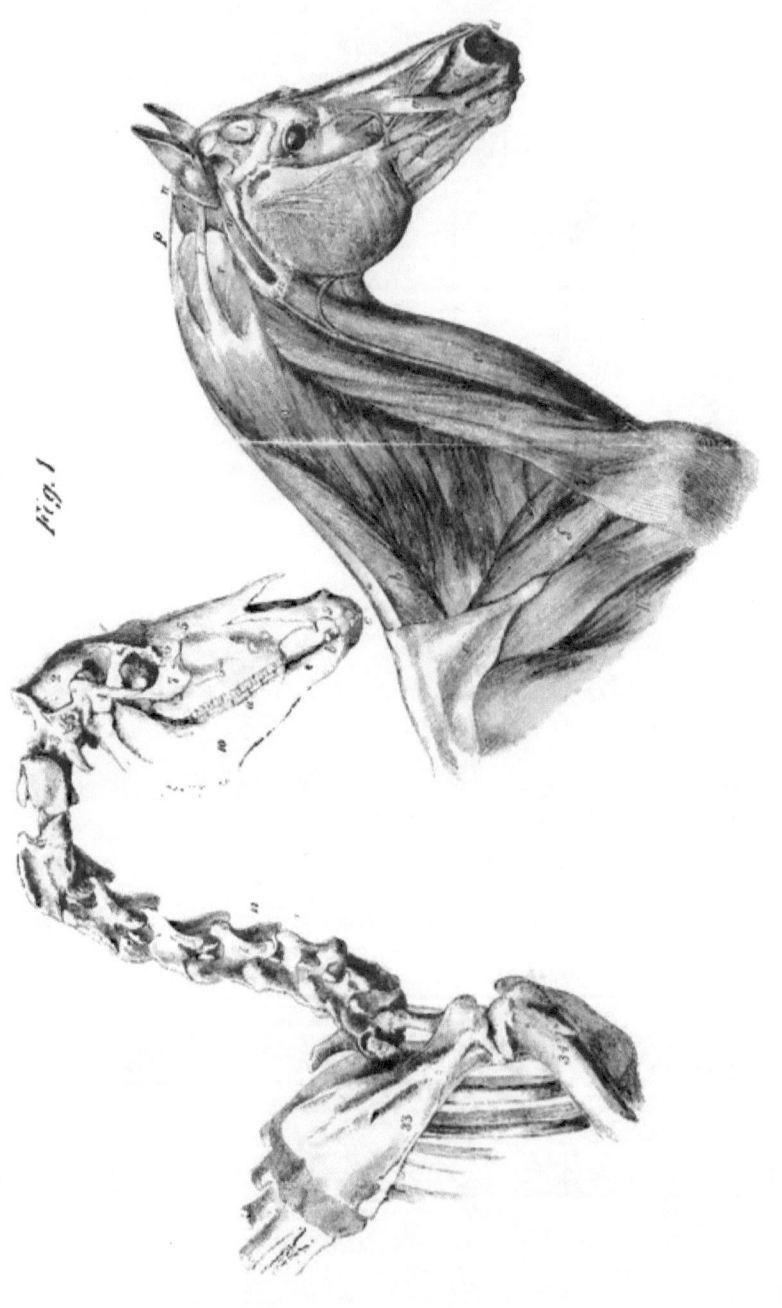

Fig. 1

EXPLANATION OF FIGURE I.

OSSEOUS STRUCTURE.

1. Frontal bone.
2. Parietal.
3. Occipital.
4. Temporal.
5. Nasal.
6. Lachrymal.
7. Malar.
8. Superior maxillary.
9. Anterior "
10. Inferior "
11. Cervical vertebræ.
33. Scapula.
34. Humerus.

From 4 to 7 is the zygomatic arch.
  c. Alveolar processes.

MUSCLES OF THE HEAD, NECK, AND SHOULDER.

*. Ligamentum colli.
a". Trapezius.
b". Rhomboideus longus.
c". Scalenus.
e". Sterno scapularis — pectoralis transversus.
f". Antea spinatus.
g". Postea spinatus.
h". Teres.
c. Dilator naris lateralis.
d. " " anterior.
e. Orbicularis oris.
f. Nasalis longus.
g. Levator labii superioris.
h. Buccinator.
i. Zygomaticus.
j. Retractor labii inferioris.
k. Masseter.
l. Abducens aurem.
m. Attolentes et adducens aurem.
n. Retrahentes aurem.
o. Adducens vel deprimens aurem.
p. r. Tendon of the splenius and complexus major.
q. Obliquus capitis superior.
s. Splenius.
t. Obliquus capitis inferior.
u. Levator humeri.
v. Sterno maxillaris.
x. Subscapulo-hyoideus.

VEINS.

1. Temporal vein.
2. Facial vein.
3. Jugular vein.
10. Parotid gland.

THE
ANATOMY AND PHYSIOLOGY
OF
THE HORSE:

WITH

ANATOMICAL AND QUESTIONAL ILLUSTRATIONS.

CONTAINING, ALSO,

A SERIES OF EXAMINATIONS ON EQUINE ANATOMY AND PHYSIOLOGY, WITH INSTRUCTIONS IN REFERENCE TO DISSECTION, AND THE MODE OF MAKING ANATOMICAL PREPARATIONS.

TO WHICH IS ADDED,

Glossary of Veterinary Technicalities, Toxicological Chart, and Dictionary of Veterinary Science.

BY

GEORGE H. DADD, M. D., V. S.,

AUTHOR OF "THE MODERN HORSE DOCTOR," "CATTLE DOCTOR," ETC., ETC.

NEW YORK:
PUBLISHED BY C. M. SAXTON,
AGRICULTURAL BOOK PUBLISHER.
1863.

Entered according to Act of Congress, in the year 1855, by
JOHN P. JEWETT AND COMPANY,
In the Clerk's Office of the District Court of the District of Massachusetts.

LITHOTYPED BY THE AMERICAN STEREOTYPE COMPANY,
PHŒNIX BUILDING, BOSTON.

PREFACE.

AMERICAN VETERINARY LITERATURE has hitherto possessed no work devoted to the anatomy and physiology of the Horse; consequently such subjects are either discussed theoretically and imperfectly, or else fail to be noticed. But a new era is now dawning upon Veterinary Science; a spirit of inquiry is abroad; and the people of this Republic find themselves in possession of some of the most magnificent specimens of "live stock" to be found in the world. The natural inquiry is, "How shall we protect our property?" And the conclusion arrived at is, "Veterinary science offers the only probable and practicable security against the numerous casualties incidental to the *habitats* of the *barn* and *stable.*"

Hitherto, much indifference has been manifested regarding this science, in consequence of the difficulty encountered in its study, for want of proper text-books and teachers; and its unsatisfactory results when tested by men unacquainted with its fundamental principles. The well-known works of English and French authors furnish all the necessary information, yet their cost is beyond the means of many, and, therefore, their circulation is very limited.

In view, therefore, of supplying the above deficiency, which is disclosed in the barrenness of our anatomical and physiological knowledge, and for the purpose of furnishing a work that shall come within the reach and financial means of all men, the author has undertaken the double task; and it is hoped that the effort will not be thought untimely.

There are a vast number of highly educated physicians in this country who are often urged by their employers to give advice in the management, medical and surgical treatment, of the *inferior* orders of creation; yet decline to do so, in consequence of a lack of authoritative knowledge regarding anatomy, physiology, therapeutics, and pathology. To such, whose sympathies lean in the right direction, and who are willing to give counsel, and lend a helping hand in the restoration of a sick or dying animal, this work is offered, and the author, therefore, submits it to their candid perusal and criticism.

The work, however, is principally intended for veterinary surgeons, teachers of the art, and students of veterinary medicine, whose wants the author professes to have some knowledge of; and he has endeavored, to the best of his ability, to cater to the same.

The necessity for such a work, at the present time, is evident from the facts, that three veterinary colleges have lately come into legislative existence, and it is very natural to suppose that, ere long, many candidates for the honors of these institutions will knock at the door of science, and seek admittance; they must then need *text*-books; and, in view of furnishing a part of what the author foresees every teacher and student must necessarily need, he offers this, not as a work pregnant with his own ideas, for that were presumptuous, when anatomy and physiology are the texts; but, as a work carefully prepared from the writings of our best authorities, the work may be considered as the legitimate offspring of *scientific* observation and experience.

Another argument in favor of the necessity that will soon exist for a text-book of anatomy and physiology is founded on the fact, that agricultural colleges will soon be endowed in every State of the Union; many already exist; and each will, probably, endow a professorship of veterinary science. With such, and among the young and aged men that may seek for knowledge, the author hopes that his work may find favor; and, if such should be the result, he will have the satisfaction of knowing that he has not labored in vain.

There are other classes of men that need a work of this description; namely, the husbandman, the horse-owner, and the horse-lover, as well as the purely scientific man. The three first, incited by laudable sentiments, or pecuniary motives, will read the following pages, and study the anatomical illustrations; some with veneration of that wonderful piece of mechanism, a horse's structure; others for the purpose of making themselves acquainted with the form, action, and capacities of the same.

The *purely scientific* man, who desires to inform himself how veterinary science is to be studied, — what are its legitimate objects, and its appropriate sphere, — will read these pages with considerable profit.

GEORGE H. DADD.

Boston, January, 1857.

REMARKS IN REGARD TO THE COMPOSITION OF THIS WORK.

The plan of the author, in the commencement of this work, was to prepare a *complete* manual of examinations on the Anatomy and Physiology of the Horse; but, finding that he could not bring the *matter* within the prescribed limits, the plan was speedily abandoned. The examinations, excepting those descriptive of the osseous structure, are intended, either to elicit some physiological fact, or to introduce topics that have not been treated of in the body of the work.

In attempting to furnish the public with a systematic treatise on Anatomy and Physiology, it will be obvious that the author must necessarily avail himself of the labor of others; for, as regards the science of anatomy, no one has anything new to offer. The industrious anatomists and dissectors of early times have borne off all the laurels, and there remains but little, if anything, for future discovery. As regards physiology, also, there are very few facts to discover; we now allude, however, to practical physiology — that science which teaches us the functions of the animal body, or the uses of its parts. The author has, however, occasionally stepped beyond the details of practical physiology, and has endeavored to throw some light on the complex combinations in which vital phenomena present themselves, and the nature of their dependencies one upon another. Matter of this kind he has thought best to introduce in the form of examinations.

In preparing this work, the author has endeavored to select the most recent and reliable information. The following list of authors consulted and compiled from, together with the foot notes and the writers' names appended, will serve to indicate the principal sources on which the author has relied for information:

Mr. Percivall's *Anatomy of the Horse* has been freely employed in composing the anatomical part of the work. The description of the abdominal viscera is from the pen of Mr. Gamgee, and was written as a *prize essay*, and published in the London *Veterinarian*. Carpenter's *Physiology*, general and comparative, is also quoted. Liebig's *Chemistry*, Hooper's *Dictionary*, Percivall's *Hippopathology*, Roget's *Outlines of Physiology*, have also been consulted, and extracts made from the same. The illustrations, not otherwise indicated, are by Girard; explanations translated by the author. For the loan of the French plates, the author is indebted to C. C. Grice, V. S., of New York City.

The *plan* of the *examinations* was suggested to the author by Ludlow's *Manual of Examinations,*—a work which he formerly, while studying medicine, had occasion to use. The subject matter, in this work, of course differs from that of the former.

In preparing the "Definitions of Veterinary Technicalities," and "Dictionary," the author has availed himself of the works of Cooper, Hooper, Cleveland, Blaine, Mahew, and White; and, regarding the method of making anatomical preparations, etc., the works of Parsons, Pope, and Swan, have been consulted.   G. H. D.

CONTENTS.

	PAGE
PREFACE,	3
Remarks regarding the composition of the work,	5
Remarks on the osseous, cartilaginous, and ligamentous structures,	11
TEGUMENTARY SYSTEM. — On the hair of horses; examinations on the common integument; physiology of the skin, of the cellular membrane, of the adipose tissue; examinations on the same,	14–17
OF THE EXTERNAL PARTS. — The hoof; its form, spread, color, magnitude; the wall; its situation and relation, connection, figure, division, solar border, laminæ, quarters, heels, coronary border and bars,	17–23
THE SOLE. — Situation and connection, figure, arch, division, surfaces, and thickness,	23–25
THE FROG. — Its situation and connection, figure, division, surfaces; the cleft of the frog, its superior surface, the sides, the commissures, toe, heels or bulb, coronary frog band; development of hoof; structure of the hoof; production of the hoof; properties of horn,	25–30
INTERNAL PARTS OF THE HOOF. — The coronary substance; its situation, connection, structure, and organization,	30–31
THE CARTILAGES. — Their situation, attachment, and form; the false cartilages, and their use; the sensitive laminæ; division of the same; elasticity, number, dimensions, and organization; the sensitive sole; its structure, connection, thickness, and organization; the sensitive frog; its situation, division, structure, and organization,	31–34
A tabular view of the bones of the horse,	35
Anatomy of the skeleton, introduced in the form of questions and answers, embracing a complete system of osteology,	36–54
Remarks on the changes which horses' teeth undergo, with examinations on the same,	54–56
MYOLOGY. — A table of the names and number of muscles, divided into regions,	57–60
A tabular synopsis of the number, name, region, situation, insertion, and action of all the muscles,	61–78
ON DISSECTION. — Dissecting instruments; subjects suitable for dissection; rules in reference to dissection of muscles,	79–80
ANATOMICAL PREPARATIONS. — Injecting instruments; directions for using the syringe.	80–81
ON THE DIFFERENT KINDS OF INJECTIONS. — Formulæ for coarse warm injections; fine injections; minute do; plaster injection; cold injection; as regards the course of injections; quicksilver injections; mode of injecting the lymphatics with quicksilver; method of injecting the lacteals, and parotid gland; wet preparations; preparations by distension; method of preparing and distending the lungs; menstrua for preserving specimens; method of preserving the brain and lungs; method of macerating and cleaning bones; to render bones flexible and transparent; method of cleaning and separating the bones of the cranium; exposition of Mr. Swan's new method of making dry anatomical preparations,	81–87
DIGESTIVE SYSTEM. — The mouth, lips, cheeks, gums, palate, tongue, salivary glands, pharynx, œsophagus, and nasal fossa; cavity of the cranium; the orbits and cavities of the nose; the mouth, peritoneum, stomach, intestines; the vessels, nerves, and lymphatics of the intestines; the spleen, liver, pancreas, kidneys, supra-renal capsules, ureters, bladder, urethra,	87–119
GENERATIVE ORGANS OF THE MALE. — Vasa deferentia, vesiculæ seminales, ejaculatory ducts, prostrate gland, Cowper's glands,	119–121
ORGANS OF GENERATION CONTINUED. — Testicles and scrotum, spermatic cord, epididymis, penis, and urethra,	121–125
FEMALE ORGANS OF GENERATION,	125–128
Physiological considerations on the reproduction of organized beings,	128–136
Examination on the digestive system,	136–138
Remarks and examinations on the eye,	139–143

(vii)

RESPIRATORY SYSTEM. — Observation on the same; the larynx, glottis, epiglottis, trachea, bronchial tubes, pleura, mediastinum, lungs, bronchial glands, - - - - - - - 144–154
CIRCULATORY SYSTEM. — Remarks on the blood; examinations resumed on the blood, pericardium, and heart, - - - - - - - - - - - - - 155–157
ARTERIAL SYSTEM. — Distribution of the arteries, - - - - - - - 158–163
A table showing the mode of the distribution of the arteries, - - - - - 164–166
Distribution of the veins, - - - - - - - - - - - 166–168
A table showing the mode of distribution of the veins, - - - - - - 169–170
The brain and its appendages; the nervous system, - - - - - - 171–177
Examinations on neurology, - - - - - - - - - - 177–180
Distribution of the lymphatics, - - - - - - - - - - 181–184
A glossary of veterinary technicalities, - - - - - - - - 185–193
Toxicological chart, - - - - - - - - - - - 195–209
A dictionary of veterinary science, - - - - - - - - - 211–287
APPENDIX. — Ligamentary mechanism of articulations and joints, - - - - 289–291

INDEX OF THE ILLUSTRATIONS.

	PAGE
FIGURE I. — Presents two views: *one* of a portion of the osseous structure, showing the head, neck, and shoulders; and the *other* is composed of the superficial muscles, covering the above parts; precedes the title page.	
FIGURE II. — Is a section of the osseous structure, giving a side of the spinal column, ribs, and a part of the near, anterior, and posterior extremities.	10
FIGURE III. — Is a representation of the superficial muscles of the body, of a part of the neck, and of the extremities.	20
FIGURE IV. — Has four illustrations of the hind extremities, as follows: No. 1 is a side view of the bones of the off-hind leg; No. 2 shows the muscles and tendons of the off-hind leg; No. 3 is a front view of the bones of the same; No. 4 shows the muscles and tendons in the anterior region, or front part, of the off-hind extremity.	30
FIGURE V. — Presents two illustrations: the first shows the superficial muscles in the region of the head, neck, and shoulders, on the near side; and the other is a corresponding section of the osseous structure, on which the insertions of the ligamentum colli into the occiput, cervical vertebræ, and dorsal spines, are shown.	40
FIGURE VI. — Presents four views of the forward extremities: No. 1 shows the bones which enter into the composition of the near fore-leg; No. 2 is a side view of the muscles and tendons of the near fore-leg; No. 3 is an anterior view of No. 1; No. 4 is an anterior view of No. 2.	50
FIGURE VII. — Presents four views of the near fore-extremity: Nos. 1 and 3 are side and posterior views of the bones of the near fore-limb; Nos. 2 and 4 show the muscles and tendons belonging to the above regions.	60
FIGURE VIII. — Has four views of the off-hind extremity: Nos. 1 and 3 are side and posterior views of the bones entering into the composition of the limb; Nos. 2 and 4 show the muscles and tendons of the same.	70
FIGURE IX. — Presents two views: one, of the bones; the other shows the superficial muscles of the head, neck, shoulders, and breast, viewed in an anterior direction.	80
FIGURE X. — Has two cuts: one of which shows a portion of the osseous framework; the other shows the superficial muscles belonging thereto.	90
FIGURE XI. — Is illustrated by two cuts: one of which shows a portion of the muscles of the body, neck, and limbs; it is a sort of anterior side view; the second cut shows the bones which enter into the composition of these parts.	100
FIGURE XII. — Has two illustrations, which are the counterpart of Fig. XI., as seen from the opposite or posterior direction.	110
FIGURE XIII. — Presents a side view of the deep-seated muscles: it is taken from Mr. Blaine's "Outlines," and is one of the most magnificent plates ever presented to the public.	120
FIGURE XIV. — Is a view of the muscles and tendons of the fore and hind extremities.	140
FIGURE XV. — Is illustrated with five views of the off and near fore extremity: Nos. 1, 2, 3 show very distinctly the *action* of the flexors of the limb, as well as their location, and that of the extensor tendons and muscles. The triceps extensor brachii, and pectoral muscles, are also quite prominent and easily recognized; No. 4 is the same limb divested of the soft parts; No. 5 is an interior view of the near fore-leg, and shows some of the tendons and muscles which are not seen in the other cuts.	150
FIGURE XVI. — Presents five views of the hind extremities, in which the use and action of several very important muscles and tendons are accurately delineated: Nos. 1, 2, 3, 4 compose the bones, muscles, and tendons of the near-hind extremity; No. 5 shows the muscles and tendons on the inside of the near-hind leg.	160

FIGURE XVII. — Presents two views (as seen from a posterior direction): one contains a great portion of the superficial muscles of the body and limbs, and the other shows the basis of their superstructure.   170

FIGURE XVIII. — Is the skeleton of a horse, for which the author is indebted to Blaine's "Outlines of the Veterinary Art."   180

FIGURE XIX. — Is a counterpart of Fig. XVII., as seen from an opposite direction.   175

FIGURE XX. — Is an excellent representation of the muscles of one side of the head, neck, body, and limbs.   211

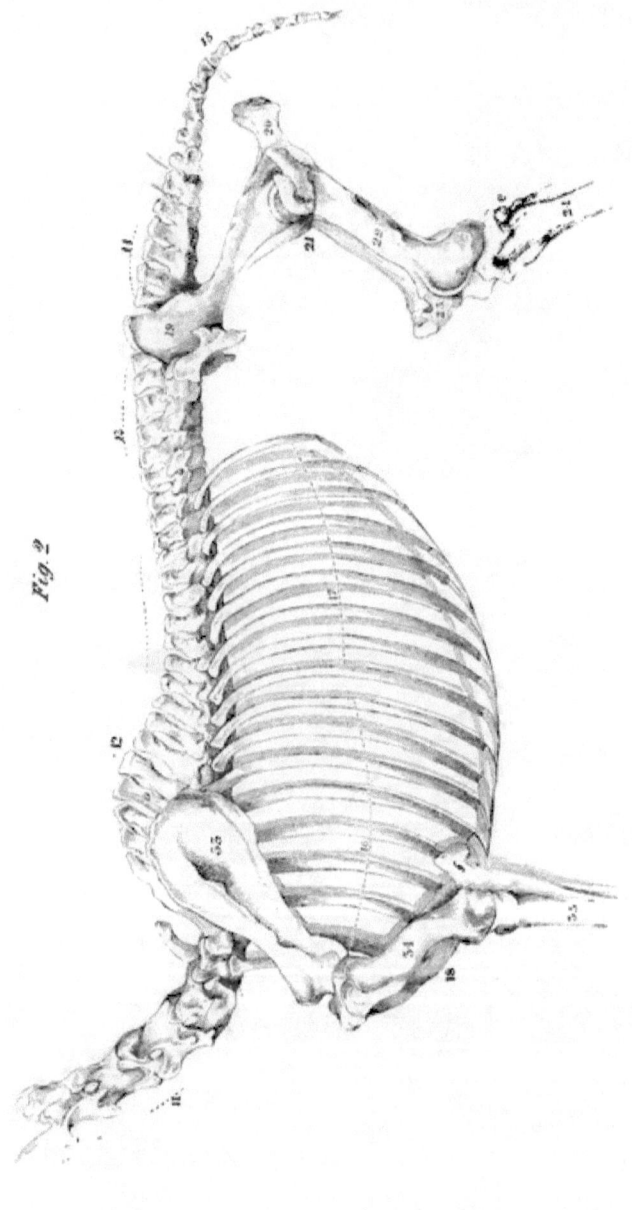

EXPLANATION OF FIGURE II.

OSSEOUS STRUCTURE.

11. Cervical vertebræ.
12. Dorsal   "
13. Lumbar   "
14. Sacrum.
15. Coccygeal bones.
16. True ribs.
17. False "
18. Sternum.
19. Pelvis.
20. Posterior part of the pelvis, or ischiatic spines.
21. Inferior, or pubic region.
22. Femur.
23. Patella.
24. Tibia.
33. Scapula.
34. Humerus.
35. Radius.
 e. Fibula.
 f. Ulnar.

REMARKS

ON THE

OSSEOUS, CARTILAGINOUS, AND LIGAMENTOUS STRUCTURES.

The bones are the solid framework which gives stability to the whole fabric, and afford fixed bearings upon which the powers regulating the varied movements operate. The bones, then, are considered as the most dense and solid structures of the animal frame; affording support, and in many parts protection, to some of the softer parts; at the same time, the leverage which regulates the action of a limb is derived from the osseous structure.

On making an examination of a bone, we find that its external surface is the hardest part, and it differs very much in thickness in different bones, and in different animals. The long bones (or cylindrical) of the horse contain less marrow, and are more cancellated within, than the bones of the human subject; in many of the former the whole areum is occupied by cancelli. The bones of the ribs have an osseous plating differing in thickness in various subjects, and within is a cellular structure which may be termed diploe.

The marrow, as it is termed, is a soft substance of an oleaginous character, contained in an infinite number of sacs, deposited and suspended in the cavities of bones and in the cancelli. The marrow sacs are composed of a delicate vascular membrane, which isolates them from each other, and prevents the marrow from gravitating or passing into the osseous structure.

Bones present the appearance of lamellæ yet they are fibrous; the fibres of the cylindrical bones are longitudinal; in the flat bones they have a radiated appearance, and in the short and peculiar shaped bones, the fibrous arrangement is more irregular and difficult to trace.

The basis of the osseous structure is nearly the same as the membranous parts,* being composed of fibrous laminæ or plates, which are connected together so as to form, by their intersection, a series of cells analogous to those of the cellular structure. This theory has been disputed by some distinguished physiologists; the moderns contend that the osseous fabric is cellular.†

Bones are invested, on their exterior, except those parts plated with cartilage, with a membrane termed periosteum. Through this medium an arterial and venous com-

---

* "The analysis of a bone into its two constituent parts is easily effected by the agency either of acids or of heat. By macerating a full-grown bone for a sufficient time in diluted muriatic acid, the earthy portion of the bone, amounting to nearly one-third of its weight, is dissolved by the acid; the animal portion only remaining. This animal basis retains the bulk and shape of the original bone, but is soft, flexible, and elastic; possessing, in a word, all the properties of membranous parts, and corresponding in its chemical character to condensed albumen. A portion of this solid animal substance affords gelatin by long boiling in water, especially under the pressure, admitting of a high temperature, to which it may be subjected in Papin's digester. On the other hand, by subjecting a bone to the action of fire, the animal part alone will be consumed, and the earth left untouched, preserving, as before, the form of the bone, but having lost the material which united the particles, presenting a fragile mass which easily crumbles into powder. This earthy basis, when chemically examined, is found to consist principally of phosphate of lime, which composes eighty-two hundredths of its weight; and to contain also, according to Berzelius, minute portions of fluate and carbonate of lime, together with the phosphates of magnesia and of soda."—*Rept.*

† The best authority in support of the cellular theory is Scarpa. Percivall advocates the laminated and fibrous theories.

munication is established between the dense and soft parts. The periosteum is analogous to the fibrous textures, being composed of numerous inelastic fibres of great strength and density.

The inner surface of the periosteum is connected with the bone by the vessels passing from the one to the other, and also by numerous prolongations, which pervade the osseous substance.

The blood-vessels of the periosteum are numerous, and are easily demonstrated by injection.

CARTILAGE.*

The structure which appears most intimately connected with the osseous is cartilage. It is a firm and dense substance, apparently homogeneous in its texture, semipellucid, and of a milk-white or pearly color.

The surface of cartilage is smooth and uniform, presenting neither eminences nor cavities, pores nor inequalities. It has, however, minute capillary vessels, the diameters of which are too small for ocular demonstration. Notwithstanding its density, it has a minute circulating apparatus, which is demonstrated in diseases known as spavin and ringbone, in which absorption of cartilage occurs.

Cartilaginous structures are chiefly composed of gelatin, albumen, and phosphate of calcium.

Cartilage occurs in two forms, temporary and permanent. The former prevails previous to adult life; the latter are identical with the permanent structures after the animal has migrated from coldhood.

There are three or four different forms of cartilages, viz: the membriform, interosseal, articular, and inter-articular.

The membriform are fibro-cartilaginous; they furnish a basis of support to the softer parts, supply the place of bone, and give form, shape, and firmness, to parts unossified. By their elasticity, they admit of considerable variation of figure and form, yield to external pressure, and recover their proper shape as soon as pressure is removed. This kind of cartilage is found in the nostrils, ears, larynx, and trachea.

The interosseal cartilages pass from one bone to another, adhering firmly by their extremities to each. They permit of an increase of extent or motion, as observed between the ribs; when macerated, they are divisible into laminæ of an oval shape, which are united by fibres passing obliquely between them.

The articular cartilages are those plates of articular substance which adhere firmly and inseparably to the surfaces of bones which are opposed to each other in the joints, or over which tendons and ligaments play. The elastic resistance of this cartilage has a powerful tendency to lessen the shocks incident to sudden and violent actions.

The inter-articular cartilages do not differ in composition from the preceding. They are attached to the inside of the capsular ligament, by which they are rendered somewhat movable; and, being interposed between the bones of the knee and hock, allow them a greater latitude of motion, while at the same time they contribute to adapt their surfaces more perfectly to each other. The structure of these cartilages is laminated.

---

* "The mechanical property which particularly distinguishes cartilage is elasticity, a quality which it possesses in a greater degree than any other animal structure, and which adapts it to many useful purposes in the economy. Hence it forms the basis of many parts where, contrary to the purposes answered by the bones, pliancy and resistance as well as firmness are required; and hence cartilage is employed when a certain shape is to be preserved, together with a capability of yielding to an external force. The flexibility of cartilage, however, does not extend beyond certain limits; if these be exceeded, fracture takes place. Great density bestowed upon an animal structure, indeed, appears to be in all cases attended with a proportionate degree of brittleness. These mechanical properties of cartilages, as well as their intimate structure, although nearly homogeneous in all, are subject to modification in different kinds of cartilage. Cartilages are covered with a fine membrane, termed the *perichondrium*, analogous in its structure and office to the periosteum, which we have already had occasion to point out among the fibrous membranes, as investing the bones." — *Roget.*

FIBRO-CARTILAGINOUS STRUCTURES.

Fibro-cartilage appears to be of an intermediate nature between ligament and cartilage. Having a fibrous texture united to a cartilaginous basis, it combines the characteristic properties of both of the above textures.

Fibro-cartilaginous structures are found to unite the bodies of the bones of the vertebræ; they then get the name of inter-vertebral substance. They impart great elasticity to the spine, and also diminish the effects of concussion.

LIGAMENTOUS STRUCTURES.

The ligamentous structures are dense; possess a considerable degree of solidity in some parts, while in others they are modifications of fibrous membrane. The ligamentous system includes a number of parts which have received different names, such as ligaments, tendons, fasciæ, aponeurosis, capsules or bursæ mucosæ; and fibres of ligamentous matter also enter into the composition of other organs, imparting to them different degrees of mechanical strength. The ligamentous structures vary; we find that in some places they are expanded into fasciæ, etc., at others they collect into dense, elongated cords. The first division includes fibrous membranes, fibrous capsules, tendinous sheaths, and aponeuroses.

Fibrous membranes: these resemble ordinary membranes, only that their fibres are denser. The periosteum is a membrane of this description, and the dura mater has a similar structure.

Fibrous capsules are presented in the form of sacs, which surround various tendons and joints. These capsules are also lined by a synovial membrane, which secretes the synovia.

Tendinous sheaths are formed by fibrous membranes which surround the tendons, in those parts that are subjected to friction, or liable to displacement, during the action of the muscles which move the joint.

Aponeuroses are those extended sheets of fibrous texture which in some instances form coverings of parts, while in others they constitute points of attachment to muscles. In the former case they are termed fasciæ, and either surround the muscles of a limb, forming a sheath for it, or else invest or confine some particular muscle.

In the latter case the aponeurosis presents broad or narrow surfaces and fibres which give attachment to particular portions of muscle.

ANATOMY AND PHYSIOLOGY.

TEGUMENTARY SYSTEM.

ON THE HAIR OF HORSES.

Hair is a peculiar tegumentary appendage, characteristic of the horse and other mammals. It is developed on the interior of follicles which are formed by a depression of the true skin. These follicles are lined by a continuation of the epidermis, the cells of which are developed in peculiar abundance from a spot at its deepest portion; the dense exterior of the cluster thus formed being known as the bulb of the hair, while the softer interior is termed its pulp. The elementary parts of hair are: a cortical or investing substance of a dense horny texture; and a medullary or pith-like substance, of a much softer character, occupying the interior. The cortical envelope of hairs is a continuation of the outer scaly layers of the epidermis that lines the follicle; whilst the medullary is derived from the deeper stratum whose cells are produced in unusual abundance at the cæcal extremity; and it is by the constant development of new cells at this point, that the continual growth of the hair is kept up.

An excoriation or moulting of the hair, which falls off, is replaced by a new growth, which as it comes to maturity assumes the original color. This change in the covering with which nature has so wisely clothed the horse, usually takes place either in spring or autumn, or at both periods. The hair of the mane and tail, however, is not subjected to these periodical changes; hence, it acquires considerable calibre and length.

By analysis, the hair yields carbon, hydrogen, nitrogen, oxygen, and sulphur, and its variation in color is due to the presence of different shades of matter which infiltrates the cortical substance.

EXAMINATIONS ON THE COMMON INTEGUMENTS.

Q. Of how many parts do the common integuments consist? — *A.* Three: cuticle, cutis, and rete-mucosum.

Q. Describe the cuticle or epidermis? — *A.* It is a thin, transparent, tough, and elastic porous membrane, serving as an envelope to the cutis, or true skin. It is composed of flexible lamellæ, so arranged as to bear some analogy to the scales of fish; it pervades the whole body, and insinuates itself into porous structures and follicular passages, inlets, and outlets of the system; it is supposed to be continuous from the mouth to the anus.

Q. Describe the pores or perforations. — *A.* There are three. First, those surrounding the hair. Secondly, exhalant pores. Thirdly, absorbent pores.

Q. How is the cutis designated? — *A.* As the cutis vera, or true skin.

Q. What is the structure of the cutis? — *A.* It is of a fibrous texture, tough, elastic, vascular, and highly sensitive, and is what we commonly denominate leather.

Q. What are the attachments of the cutis? — *A.* The cutis is attached to the subjacent parts by cellular membrane, in some places so tensely that little or no motion is admitted of; in others so loosely that it admits of being thrown into folds. About the forehead, upon the back, around the tail, and upon the pasterns, it can scarcely be pinched up; but upon the sides of the face and neck, upon the ribs, along the flanks, and upon the anus and thighs, it will easily admit of duplication.

Q. What varieties are there in the density of the cutis? — *A.* It varies in density, not only where it covers different parts in the same animal, but in horses of different breeds and temperaments, it varies very essentially.

Q. What is the organization of the rete mucosum? — *A.* It is composed of a fine, delicate, laminated tissue, interposed between the cuticle and cutis, and serves as their connecting medium, and is supposed to secrete the coloring matter of the external surface and hair.

PHYSIOLOGY OF THE SKIN.

The skin is highly sensitive; yet those persons who are in the habit of making free use of the whip scarcely ever realize the fact. The author has an impression that the skin of a horse is more sensitive than that of man; for example, let a small quantity of turpentine be applied to a horse's back,—very soon he evinces signs of pain, which cannot be elicited when a *man* becomes the subject of the same experiment. Every horse-owner, also, must have noticed the uneasiness a horse manifests when a common fly, or gad-fly, alights on him; and in a variety of other ways the highly sensitive state of a horse's skin admits of demonstration.

The skin is one of the principal emunctories of the body, from the surface of which passes off a large quantity of morbid fluid in the form of perspiration, sensible or insensible, as the case may be. The skin is the great external outlet; and, should the kidneys or any other organ fail to play their part in eliminating useless fluids, the skin opens its flood-gates, and thus purifies the body. The amount of fluid exhaled from the external surface has been the subject of some very interesting experiments, and the results are truly astonishing.

OF THE CELLULAR MEMBRANE BENEATH THE SKIN.

This tissue abounds in almost every part of the body; thus, says Carpenter, "it binds together the ultimate fibres of the muscles into minute fasciculi, unites these fasciculi into larger ones, these again into still larger, which are obvious to the eye, and these into the entire muscle; and also forms the membranous divisions between distinct muscles. In like manner it unites the elements of nerves, glands, etc., binds together the fat-cells into minute masses, these into larger ones, and so on; and in this manner penetrates and forms a considerable part of all the softer tissues of the body. It also serves as the bed in which blood-vessels, nerves, and lymphatics may be carried into the substance of the different organs."

This tissue consists of a net-work of minute fibres and bands, which are interwoven in every direction, so as to leave innumerable arcolæ or spaces, which communicate freely with one another.

Of these fibres, some are of the yellow or elastic kind, but the majority are composed of the white fibrous tissue; and, as in the other form of elementary structure, they frequently present the form of broad flattened bands, or membranous shreds, in which no distinct fibrous arrangement is visible. The proportion of the two forms varies, according to the amount of elasticity or simple resisting power which the endowments of the part require. The interstices or arcolæ are filled, during life, with a fluid which resembles very dilute serum of the blood; consisting chiefly of water, but containing a sensible quantity of common salt and albumen. It is the undue accumulation of this fluid which constitutes dropsical effusion, the influence of gravity upon the seat of which, shows the free communication that exists among the interstices. This freedom of communication is still more shown, however, by the fact, that either air or water may be made to pass, by a moderate continued pressure, into almost every part of the body containing cellular or arcolar tissue, although introduced only at a single point. In this manner it is the habit of butchers to inflate veal; and impostors have thus blown up the scalps and faces of their children, in order to excite commiseration. The whole body has been thus spontaneously distended with air by emphysema in the lungs; the air having escaped from the air-cells into the surrounding arcolar tissue, and thence, by the continuity of this tissue with that of the body in general at the root or apex of the lungs, into the entire fabric.

The structure of the serous and synovial membranes is essentially the same as the above. The true cellular membrane is sometimes termed reticular, while that containing fat is called adipose.

ADIPOSE OR FATTY TISSUE.

The adipose tissue is composed of isolated cells, which appropriate fatty matter from the blood after the same manner as the secreting cells appropriate the elements of bile, urine, and milk. "The portion of fatty matter separated from the circulating fluid to form adipose tissue, is only that which can be spared from the other purposes to which they have to be applied; and hence the production of this tissue depends, in part, upon the amount of fatty matter taken in as food.* This is not entirely the case, however, as some have maintained; for there is sufficient evidence that animals may produce fatty matter by a process of chemical transformation, from the starch or sugar of their food, when there is an unusual deficiency of it in the aliment." Liebig writes: " Whatever views we may entertain regarding the origin of the fatty constituents of the body, this much, at least, is undeniable, that the herbs and roots consumed by the cow contain no butter; that, in the hay or other fodder of oxen, no beef-suet exists; that no hog's-lard can be found in the potato refuse given to swine; and that the food of geese or fowls contains no goose nor capon fat. The masses of fat found in the bodies of these animals are formed in their organism; and, when the full value of this fact is recognized, it entitles us to conclude, that a certain quantity of oxygen, in some form or other, separates from the constituents of their food, for without such a separation of oxygen, no fat could possibly be formed from any one of these substances."

The chemical analysis of the constituents of the food of the graminivora shows in the clearest manner that they contain carbon and oxygen in certain proportions; which, when reduced to equivalents, yield the following series:

"In vegetable fibrine, albumen, and caseine, there are contained, for—

	120 eq. carbon,		36 eq. oxygen.	
In starch,	120 "	"	100 "	"
" cane sugar,	120 "	"	110 "	"
" gum,	120 "	"	110 "	"
" sugar of milk,	120 "	"	120 "	"
" grape sugar,	120 "	"	140 "	"

* "*Exposition.*— In almost all animals that are healthy, copious food of a nutritive kind, combined with little labor, will increase the deposition of fat; but in the human subject, and, indeed in many quadrupeds, the animal spirits appear to have very considerable influence over this secretion. We see numberless examples of people, who appear to enjoy the best bodily health, and yet are constantly meagre, though their food and habits of life tend to an opposite state; and we may occasionally observe horses and dogs, particularly circumstanced, in which, from their natural leanness, or *poorness upon the rib*, something of a mental nature would appear to be operating; indeed, it is a well known truth, that if you separate a horse of an irritable disposition from others with whom he is accustomed to be stalled, he will fall away in condition, in consequence of (to use the vulgar expression) *fretting from being alone;* and so much does this act of segregation affect some, that I have known them even refuse their food. Those horses are commonly the fattest that are fed on easily digestible food — such as bruised or scalded corn, roots of a nutritive kind, chopped hay, etc., and that have little or no exercise: a fact well appreciated by the horse-dealer, whose horses are *fine* and *fit for sale*, but incapable of fatigue.

Absorption.— Constitutional diseases, generally speaking, extenuate the body, and more particularly such as are of the acute or painful description; hence, the irritation caused by a simple puncture in the foot, will, if it be of long duration, induce a state of emaciation: under which circumstances, the absorbents are supposed to act with more than ordinary effect, and to take up the adeps from the interior of its cells."— *Percivall*

EXAMINATIONS RESUMED.

CELLULAR MEMBRANE.

Q. What is the principal use of cellular membrane? — *A.* It is employed in uniting, covering, and defending various parts of the body.

Q. Does cellular differ from serous or nervous membranes? — *A.* No, they are all resolvable into the same constituents.

Q. How does the periosteum differ from the above? — *A.* It presents itself in a more condensed form.

Q. How do capsules of joints differ from common cellular membrane? — *A.* They are a modification of it, under a condensed form.

Q. In what part of the animal does cellular membrane exist in greatest abundance? — *A.* Immediately beneath the skin; upon the ribs, and about the breast, and under the jaws, in the scrotum, on the inside of the elbow and thigh.

"Now in all fatty bodies there are contained, on an average:

"For 120 eq. carbon, only 10 eq. oxygen.

"Since the carbon of the fatty constituents of the animal body is derived from the food, seeing that there is no other source from whence it can be derived, it is obvious, if we suppose fat to be formed from albumen, fibrine, or casein, that for every 120 equivalents of carbon deposited as fat, 26 equivalents of oxygen must be separated from the elements of these substances; and, further, if we conceive fat to be formed from starch, sugar, or sugar of milk, that for the same amount of carbon there must be separated 90, 100, and 110 equivalents of oxygen from these compounds respectively.

"There is therefore but one way in which the formation of fat in the animal body is possible, and this is absolutely the same in which its formation in plants takes place; it is a separation of oxygen from the elements of food."

OF THE EXTERNAL PARTS.

THE HOOF.*

"The hoof is the horny case or covering nature has provided for the protection of the sensitive parts of the foot. It may be said of itself to constitute such a shoe or defence, as enables the animal in his wild state to travel about in quest of food, not only without injury to the structures underneath it, but with a degree of elasticity that preserves his whole frame from concussion. Were one forced into any comparison of the sort, it must be admitted that the hoofs of animals bear some anatomical affinity to the human nails, or claws of other animals; though they are vastly superior in physiological importance to any such appendages as these.

* Percivall's Anatomy.

EXAMINATIONS RESUMED.

ADIPOSE TISSUE.

Q. What is the fatty matter contained in the adipose cells composed of? — *A.* Stearine, margarine, and oleine.

Q. How do they appear when isolated? — *A.* The two former are solid, and the latter is fluid.

Q. How are they preserved in a fluid state in the animal body? — *A.* By the ordinary temperature of the body.

Q. What are the observable differences in color occurring in different parts of the body, and in animals of diverse temperaments? — *A.* In some parts of the body it is white, in others it has a yellow tinge; in animals of lymphatic and nervous temperaments it is white; in the sanguine it has somewhat of a red tinge; in the bilious it presents a yellow appearance.

Q. The fat at the ordinary temperature of the living body being fluid, how is it retained in the fat cells without transudation? — *A.* The intervals of the fat cells are traversed by a minute net-work of blood vessels, from which they derive their secretion; and it is probably by the constant moistening of their walls with a watery fluid, that their contents are retained.

Q. What are uses of the adipose tissue? — *A.* It is intended to fill up spaces; forms a sort of cushion or pad for the support of movable parts. It also acts as a non-conductor of heat, thus preserving the animal temperature; it serves as a reservoir of combustible matter, at the expense of which the respiration may be maintained when other materials are deficient.

Q. Suppose you desired to fatten a horse or an ox, what method should you adopt?* — *A.* I should keep the animal at rest, and furnish him with an abundance of nitrogenized food.

Q. In what vegetable constituents does nitrogen abound? — *A.* In vegetable fibrine, albumen, and caseine.

* Experience teaches us that, in poultry, the maximum of fat is obtained by tying the feet, and by a medium temperature. These animals in such circumstances may be compared to a plant possessing in the highest degree the power of converting all the food into parts of its own structure. The excess of the constituents of blood forms flesh and other organized tissues, while that of starch, sugar, etc., is converted into fat. When animals are fattened on food destitute of nitrogen, only certain parts of their structure increase in size. Thus, in a goose, fattened in the method above alluded to, the liver becomes three or four times larger than in the same animal, when well fed with free motion, while we cannot say that the organized structure of the liver is thereby increased. The liver of a goose fed in the ordinary way is firm and elastic; that of the imprisoned animal is soft and spongy. The difference consists in a greater or less expansion of its cells which are filled with fat.

In some diseases, the starch, sugar, etc. of the food obviously do not undergo the changes which enable them to assist in respiration, and consequently to be converted into fat. Thus, in diabetes mellitus, the starch is only converted into grape sugar, which is expelled from the body without further change.

In other diseases, as for example in inflammation of the liver, we find the blood loaded with fat and oil; and in the composition of the bile there is nothing at all inconsistent with the supposition that some of its constituents may be transformed into fat.

"*Form.*—Sainbel viewed the foot as 'the segment of an oval, opened at the back, and nearly round in front.' To a common observer, the hoof exhibits a conoid form; the part resting upon the ground being the basis, the vacuity above, the obtruncated apex. Mr. Bracy Clark asserts that this view is incorrect, and that the general figure of the hoof is a *cylinder*, very obliquely truncated upon its ground surface. This he demonstrates in two ways; either by rolling up a piece of paper into the shape of a cylinder, and afterwards cutting one of its ends in a very slanting direction; or by taking a carpenter's square, and placing one limb beneath the foot across the quarters, then sloping the other backward against the side of the quarters, parallel to the front, when the edge of the iron will be found parallel to the wall of the hoof. This corrected view of its figure will serve to account for the general equiformity manifest in the hoof, and also for the undeviating correspondence found to exist between its slope or slant, as well in front as behind, which in an ordinary or healthy foot may be estimated at an angle of 45°. Around the coronet, where the hoof unites with the skin, the cylinder is cut directly across its perpendicular—at right angles with it: it is the oblique truncation of its ground-surface that occasions the slant, which latter we may consequently increase at pleasure by any means that augment the former, viz.: by lowering the heels; by cutting away a prominent frog; or by putting on thin-heeled shoes. At the same time that we increase the slant of the hoof, we increase the obliquity of the pasterns, and likewise proportionately augment the ground-surface of the hoof, from heel to toe, the breadth remaining unaltered; and in the same ratio, consequently, extend the surface of tread.*

"*Spread.*—By the *spread*, is meant the inclination the hoof manifests, when left unshod, around the toe and sides, to bulge or protrude at bottom, whereby its ground-surface becomes augmented, particularly around the outer quarter. To a certain extent this is worthy of observation; although, in my opinion, it is to be regarded rather as an effect of *pressure* than one of abstract growth. The surface of inclination upon which the horn is produced has no such spread, nor can the hoof itself be said, *from growth alone*, to have any such *natural* tendency; but, as it continues to grow and shoot beyond the inner foot that produced it, and to which it was so intimately united, it yields to the pressure of the animal's weight, and bulges or spreads out, and more at the outer side than the inner, in consequence of the pressure tending more in that direction. If we examine a number of hoofs of neglected growth, and consequent exuberance and deformity, of various descriptions, we may discover that, in them all, the spread seems to have been the first or *incipient* deviation from that line of growth viewed as consistent with the health and well-doing of the foot. It is only in the unshod hoof that any spread is found: as soon as the ground-surface comes to be confined by a shoe, pressure can no longer exert its influence to produce such consequences.

"Mr. Goodwin aptly observes, that 'to take the form of the hoof correctly, we must strip it of its exuberant or superfluous parts, the same as one would pare the superabundant growth off our own nails. The neglect of this necessary preparative has led to a considerable difference of opinion about the natural, healthy, or true form of the ground-surface of the foot. Mr. Bracy Clark, I conceive, has inclined to the side of error in this particular; though, in the substitution of the cylindrical for the conical figure of the entire hoof, he has certainly the advantage of other writers. His natural foot is one with great spread to it, much of which the smith would find it necessary to deprive it of, *even on the first shoeing;* and the protuberance of the outer quarter (which Mr. C. points out as an attribute

---

* For further elucidation on the cylindrical form of the foot, consult Mr. Bracy Clark's works on the Foot of the Horse.

of health) being *wholly owing to the spread*, will, of course, disappear with the annihilation of the spread.' *

"Although Mr. Goodwin has not here explained what he conceives to be the origin or cause of the spread, it is evident we both concur in viewing it rather as a *deviation* from health or nature than a circumstance worthy of the consideration it has been accounted of by Mr. Clark.

"*Color.* — Hoofs are black or white, or some intermediate shade, or they may exhibit a black and white striped or marbly aspect. It is an old observation, and one that passes current among us at the present day, that black or dark-shaded hoofs possess greater strength and durability, and indicate less proneness in the feet to disease, than such as are composed of white or striped horn. The rationale of which appears to be, that white horn (the same as white hair) is the product of parts weaker by nature than such as produce dark or black horn, and, being weaker, consequently are more liable to disease, less able to resist those impressions that tend to disorder. White hoofs are more porous than black ones, and consequently absorb moisture and lose it again by evaporation with more facility: a fact that may probably aid us in accounting for the failures attributed to them.

"*Magnitude.* — It requires no veterinary skill to discover any very material disproportion in the magnitude of the foot: it will strike us at once as being *large* or *small*, in comparison to the limb or the size of the animal. A foot of any description that is out of proportion is to the horse possessing it more or less objectionable: but, for all that, these out-of-proportion feet, abstractedly considered, have their advantages as well as their disadvantages. Sainbel tells us, that a large wide hoof, by extending the surface of tread, ' will increase the stability and firmness of the fabric;' but then, he adds, ' this partial advantage grows into an evil when it becomes applied to a body capable of translation, and considered in a state of actual motion; because, then, the mass and weight of the foot overburthen the muscles of the extremity.' And because, I would add, the surfaces of contact being greater, the attraction of cohesion becomes greater, and so much the more muscular force is required to raise the foot (particularly in moist ground) from the earth. Besides which, a large foot is apt to become objectionable from its striking, during action, the opposite leg. On the other hand, it is contended, that a large foot will not sink so deep into soft ground as a small one, and consequently will not demand so great an effort of strength to draw it out. This *is* an argument, however, that can only hold good under the supposition, that in both cases the muscular strength is equal, which we know but rarely to happen, — in general, broad or flat-footed horses possessing superior strength ; small, narrow-footed ones, superior speed. There cannot be a doubt about a large foot being unfavorable for speed, a small one for stability; neither one nor the other can be indiscriminately found fault with ; both within certain limits possess their respective advantages ; though to turn out as such, they each of them require to be combined with suitable conformation and action.

"Large bulky hoofs are found to be mechanically weaker than others, in consequence of being composed of a thin, soft, porous description of horn. Sainbel ascribes all this to ' a relaxation of the fibres composing the hoof: in which case, the diameters of the vessels are increased, the porosities are multiplied, and the fluids abound in them in too great quantities ; consequently this kind of foot is soft, tender, and sensible.' Small feet, on the contrary, in general possess a close-woven horn, thick in substance, and consequently prove strong: they are rather oval than circular in figure, with great depth of substance, and are found to be of a durable nature. ' In feet of this description,' says Sainbel, ' from the too close union and too close tension of

---

* Goodwin's New System of Shoeing, edit. second, page 33.

their fibres, the vessels destined to conduct the nutritious fluid are contracted and obliterated; whence proceeds that dryness of the part which renders the horn brittle and liable to split.'*

"*Division.*— To the common observer the hoof appears to consist of one entire or indivisible case; but the anatomist finds, by subjecting it to maceration, or coction, or even to putrefaction, that it resolves itself into three separate pieces: still, so long as the hoof maintains its integrity, such is the force of cohesion existing between these three parts, that we as easily rend it in any other place as dissever one of its jointures. These constituent parts are the *wall*, the *sole*, and the *frog*.

THE WALL.

"The wall or crust is the part of the hoof which is visible while the foot stands upon the ground. It forms a circular boundary-wall or fence inclosing the internal structures. On taking up the foot, we find the wall prominent all round beyond the other parts, making the first impression upon the ground, and evidently taking the largest share of bearing. It is the part to which the shoe is nailed. It is, in fact, the most important division of the hoof; appearing to form (in the words of Mr. Clark) 'the basis or first principle in the mechanism of the hoof, the other parts being all subordinate to this.'

"*Situations and Relations.*— The wall takes its beginning at the coronet, from the terminating circular border of the skin, with which it is intimately united; their line of union being concealed by a row of overhanging hairs. From the coronet the wall descends in an oblique direction to the bottom of the foot, where it embraces the sole, and terminates in a circular projecting border. The anterior and lateral parts of the hoof are formed entirely by the wall; but at the posterior part, instead of the heels of the wall being continued one into the other so as to complete the circle, they become inflected, first downward, afterwards forward and inward, and are elongated in the latter direction until they reach the centre of the bottom of the foot, where they terminate: these inflections or processes of the wall constitute the *bars*. Altogether, the wall may be said to form about two-thirds of the entire hoof.

"*Connection.*— Superiorly, around the coronet, the wall is united with the skin; inferiorly, within its circumferent border, with the sole; posteriorly, between its heels, with the heels of the frog; inferiorly, between the bars, with the sides of the frog; and internally, with the sensitive laminæ. Let us now consider the wall in its detached or separate state.

"*Figure.*— That of a hollow cylinder, having the sides presented to the ground cut much aslant, and whose circle exhibits a *hiatus* or deficiency behind, from the lateral boundaries of which issue two narrow processes or appendages. Taking a lateral view, the wall assumes a conical shape, being broad and deep in front, and gradually narrowing as it stretches backward.

"*Division.*— For facility of reference, and in aid of our descriptions, we distinguish in the wall, First, the *toe*; secondly, the *quarters*; thirdly, the *heels*; fourthly, the *superior or coronary border*; fifthly, the *inferior or solar border*; sixthly, the *laminæ or lamellæ*; lastly, the *bars* or *appendages*.

"*The Toe* forms the bow or front of the hoof, and comprehends about two-thirds of the superficies of the wall. It is the deepest, broadest, and thickest part of the wall; for reasons that will appear hereafter. It exhibits a degree of slant about equal, naturally, to an angle of forty-five degrees; though there are variations from this which (as was explained before) will be found, in a measure, to be dependent upon the oblique truncation of the cylinder. When we come to understand the physiology of this part, however, a more operative and efficient cause for this variation will be found in the weight the wall has to sustain, and in its own mechanical strength or force of resistance: on which principle it is that light horses, thorough-breds, and ponies, as well

---

* Saintbel's Lectures on the Elements of Farriery.

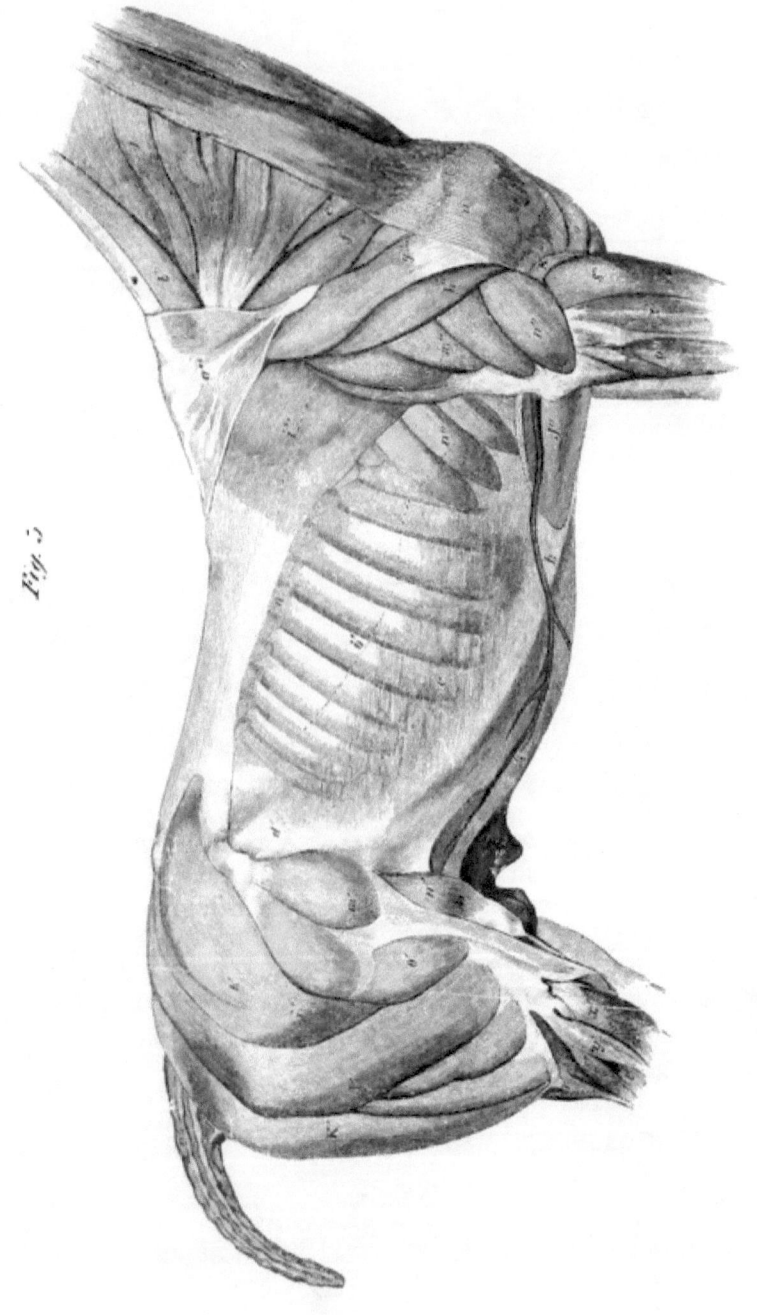

Fig. 5

EXPLANATION OF FIGURE III.

MUSCULAR STRUCTURE.

a''. Trapezius.
b. Rhomboideus longus.
c''. Scalenus; and J. Splenius.
e''. Pectoralis transversalis.
f'''. Antea spinatus.
g''. Postea spinatus.
h. Serratus magnus.
b'. Intercostales.
c'. Obliquus externus abdominis.
d'. Obliquus internus abdominis.
e'. Erector coccygis.
f'. Depressor coccygis.
g'. Compressor coccygis.
h'. i'. Glutcii muscles. J. Triceps abductor.
K. Biceps abductor femoris.
m'. Tensor vaginæ.
n'. Rectus.
o'. Vastus externus.
r', s'. Gastrocnemius externus, and plantaris.
v'. Flexor pedis accessorius; its fleshy belly.
y'. Peroneus.
x'. Extensor pedis.
h''. Teres major.
i''. Latissimus dorsi.
J. Pectoralis magnus.
K. Humero cubital. (Pectoral region.)
l''. Scapulo-ulnaris.
m''. n''. Triceps extensor brachii.
o''. Pectoralis transversalis.
p''. Flexor metacarpi externus.
u. v. Levator humeri, and Sterno maxillaris.
s''. Extensor metacarpi magnus.
z''. Extensor pedis.
    At the inferior part of the abdomen the letter h. occurs; it should be 4. 4 The subcutaneous thoracic vein $h.$, however, serves to indicate the region of the rectus abdominalis.
7. The sheath.
*. Ligamentum colli.
A. (Region of the back.) Levatores costarum.

as mules and asses, have *upright* or *strong feet* (i. e.), walls but moderately sloped; whereas heavy horses, cart-horses, and coach-horses, have commonly *flat* or *weak feet* (i. e.), walls that slant immoderately. And (as was before observed) upon the degree of obliquity of the wall must very much depend that of the pasterns. In estimating the slant or slope of the wall, it is proper to distinguish between that which is consequent on the detruncation of the hoof, and such as is the effect of a burthen under which the wall succumbs. The depth of horn in front of the toe, measuring from the termination of the skin to the most prominent point below (and supposing the hoof to be cut and ready to receive the shoe), may be rated at about three and a half inches. The bow or degree of convexity of the toe in front must depend upon its obliquity as well as upon the circularity of the foot. The thickness of the horn composing the toe may be estimated at three-eighths of an inch, or from that to half an inch, and this substance is the same from immediately beneath the coronary circle to the junction of the wall with the sole; at which part there is an accession of horny matter to block up the interstices between the laminæ, and also to fill the angular vacuity that would otherwise exist here between the wall and sole. In the fore-feet, the toe is thicker in substance than either the quarters or heels; but (we have it from Sainbel) 'in the hind, on the contrary, the heels and quarters are generally thicker than the toe.'

"*The Quarters* are the portions of the wall intermediate between the toe and the heels. They are commonly described as standing *upright*, and, according to a carpenter's square set against the wall, so they appear to do; this is not, however, the view the anatomist ought to take of their position: to him the oblique course of their component fibres, together with the slant of their laminæ, demonstrate that they slope in the same manner and degree as the toe does. The quarters do not run in straight lines from before backward, but by their prominence describe gentle curves, the outer making a wider sweep than the inner. This gives the hoof altogether a sort of *twisted* appearance, and makes the inner part of the toe look more projecting than the outer; a deviation that seems principally to have originated in the *spread*, and one, methinks, that has had more attention given it than any consequences attachable to it render it deserving of. The quarters range in depth from two to three inches; and measure in thickness from one-fourth to three-eighths of an inch.

"*The Heels* are the two protuberant portions of the wall by which it is terminated posteriorly. They are the shallowest, and thinnest, and (*in connection*) only flexible parts of the wall. Though their surfaces recede from the perpendicular, they maintain the same slope as the toe and quarters. At their angles of inflection, from which are continued the bars, they form (in conjunction with the heels of the sole) pouches or sockets into which are received the heels of the sensitive foot. In depth they range from one and a half to two inches. In substance they do not exceed a quarter of an inch, the outer heel being rather thicker than the inner.

"*The Superior* or *Coronary Border* is the circular, attenuated, concavo-convex part entering into the composition of the coronet. Its extent is marked exteriorly by the whitish aspect it exhibits, and also by some partial separation and eversion of the outer flakes of horn around its junction with the wall below. Externally, it assumes the same character as the wall below it; but its internal surface is altogether different. Instead of possessing laminæ, the surface is smooth and uniformly excavated, being moulded to the form of the sensitive coronet, and everywhere presenting numerous pores for the purpose of receiving the secreting *villi*. Superiorly, the coronary border presents two edges, having a groove between them for the reception of the terminating border of the cutis. It is this groove that marks the reception of the coronary border into two parts: the internal edge belonging to the inner part, which is the beginning of the wall itself; the external edge to the white band by which the other is embraced, and to which Mr. Clark has in particular drawn our attention, under the appellation of *Coronary Frog-band*. This covers the proper or veritable coronary border of the hoof; having, through its fibres, which are very fine, a sort of dove-tailed connection with it. As it recedes backward, it grows broader to that degree, that its breadth at length becomes doubled; being about half an inch broad in front, and one inch behind. It is thickest around its middle parts; its inferior edge, like the superior, becoming attenuated, until it grows so fine as to end in imperceptible union with the substance of the wall, giving it its beautifully polished surface: from the heat, however, to which the hoof is artificially exposed, the thin part below the coronet often grows arid, splits from the crust, and becomes everted; turning, at the same time, in consequence of dryness, of a whitish complexion. Posteriorly, we find it continued round the heels of the wall and frog, and from thence across the back of the cleft, forming altogether a complete circle, and everywhere showing itself to be the *medium of connection* between the skin and the hoof. It has been already stated that the cutis terminates in a circular border, let into a groove around the summit of the wall: the cuticle, however, does not end here — it is continued down; in fact, we trace it to the horny band we have been describing, the one being continuous in substance with the other. Indeed, the only detectible differences in them are, that one is thicker than the other, and grows hard, and dry, and white, from the effects of heat upon it from without, and the want of moisture from within. This cuticular origin and assimilation may be demonstrated in the putrefied foot; or, better still, in the foot of the fœtus. The band is broader at the heels than elsewhere, in consequence of the greater breadth of exposed cutis at those parts. In its texture it is fibrous, and its fibres pursue the same direction as those of the wall, from which they differ only in being of a finer texture. Mr. Bracy Clark appears to entertain some singular notions in regard to the structure, but more particularly the *uses*, of this part; which, in the respect I bear for their author, I shall consider, when the time may arrive for me to treat of the physiology of the foot.

" *The Inferior* or *Solar Border* offers but little worthy of observation. It constitutes the ground or wearing surface of the wall, and is the part to which we nail the shoe. It grows thicker and more exuberant around the toe than in other places, and, from its projecting beyond the sole, presents a convenient and suitable hold for the nails of the shoe. Around the anterior and lateral parts, it embraces the sole; behind, it joins the bars, which two points of union form two principal bearing places for the shoe. The inferior border possesses a larger circumference than the superior, in consequence of the oblique detruncation of the hoof.

" This is a part that requires paring down every time the horse is shod. Such is its exuberating nature, that (like the human nail), were it not continually kept worn down, or broken, or cut off, it would elongate very considerably, and gradually turn up, exhibiting forms not only of the most unsightly but even grotesque description, and proving incommodious to a degree to be almost entirely destructive of progression.

" *The Laminæ* (better named *lamellæ*) consist of numerous narrow thin plates or processes, arranged with the nicest order and mathematical precision upon the internal surface of the wall. They extend, in uniform parallels, in a perpendicular direction from the lower edge of the superior border down to the line of junction of the wall with the sole; and are so thickly set that no part of the superficies remains unoccupied by them. They are likewise continued upon the surfaces of the *bars*. In the recent subject they are found soft, yielding, and elastic; but from exposure they become dry and rigid.

" Every lamellæ exhibits two *edges* and

two *surfaces*. By one edge it grows to the wall; the other, which is somewhat attenuated, hangs loose and floating within the cavity of the hoof. The surfaces, which are two lateral, are smooth, and, considering the magnitude of the lamella itself, of enormous extent; so much so that it might be said almost to be constituted entirely of superficies. And this leads us to the contemplation of the great and magnificent design which Nature evidently had in view in their formation, viz., the production of ample surface within a small space, an end that has been obtained through the means of multiplication. Mr. Bracy Clark procured from the late Thos. Evans, L.L. D., mathematical teacher of Christ's Hospital, a calculation of what their united superficies amounted to; and it appeared to afford an increase of actual surface more than the simple internal area of the hoof would give of about twelve times, or about 212 square inches, or nearly one square foot and a half.

"The lamellæ exhibit no differences but in their dimensions. In length they correspond to the respective depths of the wall; being longest, and likewise broadest, around the toe, and gradually decreasing towards the hinder parts.

"In composition they are horny. Viewed through a microscope, Mr. Clark discovered in their substance two planes of fibres, 'the one running in parallel lines to the axis of the hoof, the other obliquely intersecting these.' When stretched, they exhibit signs of elasticity; but this appears greater in the transverse than in the perpendicular direction.

"By means of its lamellæ, the wall presents a superficies of extraordinary amplitude for the attachment of the coffin-bone. A structure consisting of similarly formed lamellæ envelops the bone, and these are dovetailed in such a manner with the horny lamellæ, as to complete a union which, for concentrated strength, combining elasticity, may vie with any piece of animal mechanism at present known to us.

"*The Bars* are processes of the wall, inflected from its heels obliquely across the bottom of the foot. For a long time, by farriers, they were confounded with the substance of the sole, an error that owed its origin and perpetuation to the malpractice they exercised in paring the foot — in cutting both bars and sole down, without any distinction, to a common level. In the natural healthy foot the bars appear, externally, as elongated sharpened prominences, extending from the bases of the heels into the centre of the foot, between the sole and the frog; posteriorly, they are continuous in substance with the wall, with which they form acute angles; anteriorly, they stretch as far as the point of the frog, constituting two *inner* walls or lateral fences between that body and the sole. Sainbel conceives, from their position, that they offer resistance to the contraction of the heels. Their internal surfaces exhibit rows of lamellæ, continued from those lining the wall, but which are here **short**, and in their direction *transverse*, two circumstances referable to the narrowness and inflection of the bar. Towards the extremity of the bar they gradually grow shorter, and less distinctly marked, until we at length lose all vestige of any more of them. While the prominence of the bars is such as to give them a secondary bearing upon the ground, their sharpened forms will sink them more or less deeply into every impressible surface.

"THE SOLE.

"The sole is the arched plate entering into the formation (as its name implies) of the bottom of the hoof: or, to adopt Sainbel's definition, 'it is that part which covers the whole inferior surface of the foot, excepting the frog.' It is a very just practical observation of Mr. Coleman's, that although a knowledge of every part of the foot is indispensably necessary to render us scientific overseers of the farrier's art, no individual part requires such undivided attention, as regards shoeing, as the *sole*, since the success of this mechanical operation mainly depends on the paring and defence of this arched horny plate.

"*Situation and Connection.* — It fills up the interspaces between the outer and inner walls (or bars) of the crust. I differ in opinion from those who describe it to surround the toe of the frog. I hold its circumferent support and connection to be the wall of the hoof, to which it is firmly cemented by an interstitial horny matter, filling the crevices between the laminæ.

"*Figure.* — The circumferent outline of the sole measures about two-thirds of a circle, the remaining third being omitted to form a triangular-shaped *hiatus* or opening for the reception of the frog and bars. This circular form, however, is by no means true, or even invariably the same, in its dimensions. Generally, the longitudinal exceeds the transverse diameter. Its greatest diameter is shown by a line extended from either heel across its middle to the opposite point of the toe.

"*Arch.* — Commonly, the sole presents an arch of more or less concavity inferiorly, and convexity superiorly. But it is not a regular or uniform arch, being one that rather waves or undulates, so as to bear a comparison, made of it by Mr. Clark, 'to the mouth of a bell extremely extended or flattened.' Like that of the bell, the arch is highest in the middle, from which it slopes, laterally, down to a flat, subsequently to rise again around its border, in order to present a dilated surface for attachment towards the wall. There is, however, vast variety in the degree of arch of the sole: in some feet it is of surprising depth; in others, the arch is converted into a flattened surface; and yet both seem to perform equally well. In the hind feet the sole is generally more arched than in the fore, and approaches in figure nearer to the oval than the circle.

"*Division.* — In the sole we distinguish an *anterior part* or *toe*; a *middle* or *central part*; two *points* or *heels*; and two *surfaces*. These divisions are not very well defined; but they prove serviceable in aid of our descriptions. The toe of the sole is the part encircled by the toe of the wall, against which it abuts, and to which it is intimately united by horny matter, the two together forming a stout bulwark of defence to those parts of the internal foot included between them. The points or heels are the two posterior salient angles received into the angular intervals between the outer and inner walls or bars. Although naturally the least exposed, these are the parts most subject to injury or pressure from the shoe, being the seat of that disease mistakenly called *corn*. The middle or centre of the sole is the portion more immediately surrounding the fore parts of the frog, and would (were the sole a regular arch) be the most elevated part; but, in general, we find the sole flattened hereabouts; the highest parts of the arch being the angles alongside of the bars; the lowermost, those around the toe.

"*Surfaces.* — Of the surfaces, the superior (as was mentioned before) is unevenly convex; the inferior, correspondingly concave. The former is everywhere pitted, particularly about the heels, with numerous circular pores, running in an oblique direction, the marks of which remain evident upon the inferior surface likewise. These pores are the impressions made in the soft horn by the *villi* of the sensitive sole, from whose orifices the horny matter is produced. They also form the bond of union between the horny and the sensitive soles: which is of a nature so strong and resisting, that it requires the whole strength of a man's arm to effect their separation — an operation of a cruel description that was wont to be practised in times past, under the fallacious notion that 'drawing the sole' was extirpating the malady.

"*Thickness.* — The natural thickness of the sole may be estimated at about one-sixth of an inch. There will be found, however, variations from this standard in different horses; and it will also very much depend on the part selected for measurement. The portion of the sole most elevated from the ground — that which forms a union with the bars — is nearly double the thickness of the central or circumferent parts; and next to this, in substance, comes the heel. I do not find that the sole 'grows thinner from the circumference to the centre,' as has been stated by an author of celebrity.

"THE FROG.

"The frog is the prominent, triangular, spongy body, occupying the chasm left by the inflection of the bars.

"*Situation and Connection.*— The frog is fitted into the interval between the bars; the three, altogether, filling up the vacuity in the sole, and thereby completing the circle, and establishing the solidungulous character of the foot. The frog extends forward, towards the toe, about two-thirds of the longitudinal diameter of the ground-surface of the hoof, terminating a little beyond the central point (or what would be the central point) of the sole—or rather shooting directly through it, so as to annihilate the spot. Posteriorly, it is embraced by the heels of the wall; laterally, it possesses firm and solid junctions with the bars, and through their medium with the sole: and these unions are effected not by simple apposition and cohesion of surface, but by a *lamellated structure*, apparent on the sides both of the frog and bars, by which the parts are reciprocally dovetailed into each other. Lamellæ are discoverable upon its sides, even all round the toe of the frog; and this is a circumstance that confirms me in my belief that the bars reach thus far.

"*Figure.*— The frog may be called pyramidal, or cuneiform, or triangular in figure; its outline forming the geometrical figure denominated an isosceles triangle. I know of no comparison so familiarly apt as that of resembling it to a ploughshare: not only do they both correspond, as near as such comparisons can be expected to do, in outline and make, but they likewise exhibit a singular coincidence in function; the frog, like the ploughshare, being intended by its point to plough or divide the surface of the earth, and in that manner serve as a stay or stop to the foot.

"*Division.*— We distinguish in the frog two *surfaces*, an *inferior* and a *superior*; two *sides*; a *point* or *toe*; and two *bulbs* or *heels*.

"*Surfaces.*— Both surfaces of the frog manifest striking irregularities, and these are respectively reversed, making one surface the exact counterpart of the other. In other respects, the only difference they exhibit, is, that the superior exceeds the inferior both in length and breadth.

"The inferior surface presents to our view a remarkable cavity, broad, deep, and triangular in its shape, bounded on the sides by two sloping prominences, which divaricate from the convexity forming the toe of the frog, and terminate, after a short divergent course, at the heels. This cavity or hollow is denominated

"*The Cleft of the Frog*: with seeming reference to the relationship existing, through its presence, between the horse's foot and the cloven one of the ox, deer, sheep, etc. In consequence of its sides sloping inward, the cleft at bottom gapes wide open; but along the top is roofed by a simple linear mark running from before backward. The horn is kept continually soft and pliant within the cleft by a peculiar secretion from the sensitive parts it covers, the odor of which is notorious.

"The solid wedge-like portion of horn in front of the cleft, extending from it to the point of the toe, has been observed by Mr. Clark to exhibit, in the natural foot at its full growth, 'a considerable bulbous enlargement,' which, by way of distinction, he calls the *cushion of the frog*. On making a perpendicular section of the foot, Mr. C. finds this part is situated 'nearly opposite or under the navicular bone.' And it would appear (according to this author) that this 'rotundity, or swell of the frog,' is never reproduced after it has once been annihilated by the knife of the smith.

"The superior surface of the frog, everywhere continuous, uniform, and porous, being the counterpart in form of the inferior, presents us with nothing but reverses: where the one is hollow or depressed the other rises into swells and eminences, and *vice versa*. This accounts for our finding the part opposite to the cleft elevated into a conspicuous eminence, bounded on its sides by two deep channels, and a hollow of broader but shallower dimensions in the front. To this central conical elevation Mr. Clark has given the name of *frog-stay*, from some novel notions he entertains of its physiology. Such a bold promontory of horn rising in the middle of broad and deep channels is well calculated to form that dovetailed sort of connection with the sensitive foot, which greatly augments their surfaces of apposition, and establishes their union beyond all risk or possibility of dislocation. It is a part which (as far as my observations on it have extended) grows and becomes developed together with other parts of the foot; and one that is apt to vary in its relative volume in different feet. In front of the frog-stay, the lateral borders, bounding the hollow in the middle, describe a waving line, which, near half-way to the point of the toe, exhibits a dip or impression: this marks the impression of the navicular bone, and is the part immediately opposite to the 'cushion of the frog,'—a coincidence important to be borne in mind, as tending to throw some light on the nature of this new-christened structure.*

"*The Sides* are the parts by which the frog establishes its union with the borders of the triangular vacuity in the hoof into which it is admitted. Along their superior borders they are transversely lamellated, or rather indentated, in order that they may be fitted to the internal surfaces of the bars, which exhibit a similar structure.

"*The Commissures* are the two deep triangular-shaped hollows between the bars and the sides of the frog. It being only the *superior* borders of these parts that are engaged in their union, their broad, unattached parts, below, form the boundary walls of the commissures. Looking into the interior of the hoof, we discover that the commissures, internally, are converted into rounded promontories, similar in appearance and texture to the one in the middle—the frog-stay—on the sides of which they are rising. In the natural state, the commissures must unavoidably get plugged with dirt, or whatever the animal may happen to tread upon; a circumstance from which some far-fetched notions have been extracted concerning their use.

"*The Toe* or point of the frog is the anterior, undivided, elongated portion; that which forms the apex of the pyramid or wedge — the acute or extended angle of the triangle — the only part displaying that prominent or rounded form that would warrant us in using the epithet 'conical' to the frog. It possesses solidity of substance, firmness of texture, and luxuriance of growth in an eminent degree; facts well known to the farrier, who, in paring the foot, seldom fails to make more free with this than any other part of the frog.

"*The Heels* or *bulbs* of the frog are the posterior protuberant parts embraced by the heels of the wall, and separated from each other by the cleft, forming, together, the base of the wedge or triangle. They present greater depth of substance than the toe, but are of a softer, more spongy texture, and are less resisting and stable, in consequence of being deprived of mutual support by the division of the cleft. Anteriorly, the heels unite with the lateral prominences bounding the cleft; inferiorly, they present two surfaces of tread to the ground, evidently designed to take a share in the bearing of the foot; posteriorly and superiorly, they exhibit a bulbous fulness, in consequence of receiving at this part a supplementary covering from a production which has been (in the description of the wall) adverted to, under the appellation given it by Mr. Clark, of

"*Coronary Frog-band.*—It was there stated, that the coronary groove (the groove or canal in the coronary border of the cutis) broadened considerably as it descended to and turned round upon the heels; in like manner does the horny band produced by it broaden, and not only grow broader but thicker in substance, and consequently in

---

* In fact, the cushion of the frog appears to be nothing more than a bulge of the part produced by the superincumbent pressure of the navicular bone.

the same degree augments the substance of the heels, occasioning that swell of them which has suggested the appellation 'bulb.' The horny band itself is everywhere lamellated upon its internal surface; but these broadened parts of it display lamellæ of a much bolder character, and consequently render their union with the heels so much the more intimate and enduring. The inferior edge of the band is denticulated, and the denticulations become so interlaced with the lamellated fibres of the wall, that their union is rendered, in the ordinary state of the hoof, altogether imperceptible. For drawing our attention to this part, we are indebted to Mr. Clark; and, insomuch as he considers it to be a production of the cutis (not having any connection with the glandular circle that secretes the wall), and to serve the purpose of 'uniting the sensible parts with the insensible,' I agree with him. I find something very similar to this growing upon the human nail, issuing from the superior edge of the terminating border of the cutis, and continued from the cuticle, which proceeds for some way upon the nail, uniting it more closely and firmly with the cutis, and protecting the latter from external injury. This production is no more the beginning of the nail itself than is the so-called *frog*-band the commencement of the wall: they are both distinct parts, though but supplementary ones, and seem to be of a nature partaking both of horn and cuticle. It has no more important relation to the frog, in my opinion, than it has to the wall: it serves the same purpose to both, —that of strapping up the heels of the frog and binding them in closer and more intimate connection with the neighboring parts. Were I asked what other use it appeared to have, I should say, that it was formed *to cover and protect from injury the new-formed horn* of the hoof, guarding it in its passage downward, until it has acquired substance and hardness sufficient to resist external impressions of itself.

"DEVELOPMENT OF THE HOOF.

"During the early months of fœtality, no horn or hoof is to be found. The foot is covered with a substance, white, firm, and elastic, resembling cartilage in its appearance, but proving more of the nature of cuticle on examination, which supplies the place of hoof. At the coronet this substance takes its origin from the cutis, being found to be continuous with the cuticle; but that which covers the bottom of the foot is a production from the sensitive sole and frog. Altogether, it possesses the general form and appearance of the hoof, differing however in these particulars—that the substitute for the wall is comparatively thin in its substance; while that which grows from the bottom of the foot is enormously thick, and, instead of being shaped into sole and frog, exuberates to a degree to constitute club-footedness. About the same period at which the pastern and coffin-bones take on ossification, horn makes its appearance underneath this cuticular wall, in the form of plates descending from the coronet, exhibiting with peculiar distinctness the lamellated structure. The horny wall becomes considerably advanced before we perceive any change in the bottom of the foot. At length, horn is detected forming underneath the cuticular substance, which, increasing in thickness, gradually represents sole and frog. Not, however, in an undeveloped state; for even at birth these parts are yet concealed by the exuberant cuticular covering, now become loose in its texture, and shaggy and ragged, in consequence of not receiving any further supply from the parts that produced it, and of being near its decadence; for it not long after falls off, disclosing sole and frog both ready formed.

"STRUCTURE OF THE HOOF.

"Horn is found to differ in its texture or quality, not only in the many animals in which it is met with, but in different parts, and even in the same part of the body of the same animal. That which composes the hoof of the horse is a remarkable example of this. How different is the horn of the frog from the horn of the wall; and yet neither of them agree in texture with the sole. The horny substance of the wall is resolvable into fibres, bearing a resemblance to thick or coarse hairs, which in the entire hoof are so intimately matted and glued together, as to have the appearance and strength of solidity. By close and accurate inspection these fibres may be seen descending in parallel lines, taking the obliquity of the wall, from the coronet to the inferior or solar border; they do not run promiscuously, but are arranged in rows, forming sorts of beds or *strata*, lying one upon another—a disposition made manifest in the foot of the fœtus. A clean-cut transverse section of the wall exhibits upon its surface numerous minute, circular, whitish spots, which grow larger and more distinct towards the internal part, and through a glass appear to be hollow or tubular. These spots I take to be produced by sections of the horny tubes, apparently containing a whitish matter, a sort of pith, or pulp, or gelatinous instillation which pervades them from their origin from the *villi* of the coronary circle; the same as hairs derive their unctuous matter from the bulbs producing them, and (as this matter does the hair) renders the horny fibre tough and elastic—in fact, imbues it with the peculiar attributes so well known to smiths by the appellation of *living horn;* the epithet "living" being here used to denote the obvious differences the hoof of a living animal evinces from one that has been long detached from the body, or that is dead. We are too apt to believe that the various agents known to act upon the dead hoof or horn must take similar effect on the living; and upon this erroneous belief we employ hot and cold water, etc., etc., in treating disease of the feet, forgetting that we have opposed to our remedies the resisting or self-preserving properties of living horn.

"The sole, as well as the wall, is fibrous in its structure; but its fibres appear to be of a finer quality, and, in course, are very much shorter: they, however, take an oblique direction, from behind forwards, following the same degree of slope as those of the wall. They issue from the *villi* penetrating the superior surface. To the fineness of its fibres, combined with the relative magnitude of the tubular canals, and consequent proportions of horny and gelatinous substances, may be ascribed the comparative softness and elasticity of the sole.

"The frog, however, displays these qualities in such a remarkable degree as to appear, in fact, to be composed of quite another kind of horn; though, on examination, we find it to evince the same fibrous structure, the only perceivable differences being the comparative fineness of the fibres and their proportionably greater tubularity: their direction is oblique, correspondent with those of the wall.

"PRODUCTION OF THE HOOF.

"The wall is produced by the *coronary substance*, a sensitive and glandular part we shall have occasion soon to examine. Its *villi*, by some peculiar, mysterious, secretory process, convert the blood circulating through them into a soft pulpy gelatinous matter, which by exposure becomes hard horn, descending from the villous point that produced it, in the form of a tubular fibre, down to the sole. The fibres are united together at their very origin, but their tubes or canals diminish, the lower they descend; which accounts for the porous or honey-comb-like structure of the interior of the coronary border and the comparative solidity of the parts below. The outer layers or strata of fibres are found to be more compact and of closer texture than the inner; which arises, in part, from the villi producing them being removed to a greater distance, and to the comparative smallness of their canals, and which, consequently, the sooner become obliterated. The use of Mr. Clark's coronary frog-band becomes now more apparent, serving, as it evidently does, to cover and protect these external fibres until they grow sufficiently firm and solid of themselves to bear exposure and resist casualties.

"The sensitive laminæ make no addition to the substance or thickness of the wall:

they simply produce the horny lamellæ arranged along its interior; as one proof of which, the wall measures as much in thickness at the place where it quits the coronet as it does at any point lower down. Other demonstrations of this fact come every day before such practitioners as have to treat canker, quittor, sandcrack, and other diseases of the feet.

"The horny sole is a production from the villi of the sensitive sole; after the same process as that by which the horny frog is secreted from the villi of the sensitive frog.

"In a state of health of the foot, the secretion of horn is unceasingly going on. Disease or injury of the glandular parts may diminish or altogether suspend the process; disease, under certain other forms, appears also to have the effect of increasing it; but whether we have any artificial means of effecting this, seems questionable. The wall grows from above downwards. If a mark be made in any part of the wall, it will remain until it grows down and becomes cut off below, at the inferior border; and by observations made on the gradual descent and disappearance of these marks, calculations may be formed of the period of time required for the renewal or restoration of the wall.

"PROPERTIES OF HORN.

"Horn is a tough, flexible, elastic substance, consisting of tubular fibres, more or less intimately connected together, taking the direction from the surface of the body on which it grows. Its property of toughness or resistance much depends on its condition in regard to moisture; for if it is exposed to a degree of heat sufficient to abstract much of its natural juice or imbibed moisture, it loses its flexibility and toughness, and becomes brittle. On the other hand, saturated with moisture, it is converted into a soft and highly flexible substance, but at the same time becomes weak and unresisting. This known effect aids us to account for the flat-footedness of horses reared in low, fenny, or marshy situations; the hoof being constantly in a state of saturation with moisture, the wall and sole yield to the superincumbent burthen of the body, and the latter grows flat (instead of remaining concave or arched), and even in some instances bulges. If oily or unctuous applications have any effect in softening the hoof, they appear to do so by filling the crevices and interstices between the fibres on the surface, and in this manner checking or suppressing evaporation. Horn takes a high and beautiful polish. Although much inferior in transparency to tortoise-shell, it may be worked up to bear so near a resemblance to it as to be often, in manufactures, substituted for it, as in combs, etc. The hoof admits of an elegant polish; and in that altered and improved state has been manufactured into articles no less useful than valuable and ornamental;* even the hoofs of the living animal may, by being kept clean, and when dry rubbed with linseed oil, be numbered among the ornamental beauties Nature has bestowed upon quadrupeds.

"By chemical analysis horn has been found to consist of membranous substance, having the properties of coagulated albumen, and of some gelatine. The horns of some animals, the deer species, from containing bone, become exceptions to this. Mr. Hatchett burnt five hundred grains of ox's horn, and the residuum proved only one and a half grain, not half of which was phosphate of lime.

"Shavings of hoof thrown into nitric acid become soft, and speedily melt into a yellow mass, which in about eight hours disappear in complete solution.

"The same thrown into sulphuric acid turn black, in becoming soft, and require thrice the time for their solution. Muriatic acid also turns horn black, and corrodes it, but has so little effect towards its solution, that after ten days a piece of hoof soaked in it was found to have become only more brittle or rotten. Common vinegar will turn horn dark-colored, but does not

* The Eclipse hoof, presented by his Majesty at Ascot Races, as the reward of the best horse on the turf, forms a notable illustration of this.

appear to have any power in impairing its texture, or, at least, in dissolving it. Liquor potassæ will not only turn it black, but will corrode the horn of the hoof. Ammonia does not change its color, but slowly destroys its texture, rendering it brittle and rotten.

"INTERNAL PARTS OF THE FOOT.

"The internal, sensitive, organic parts of the foot, comprise the *bones, ligaments, tendons, coronary substance, cartilages, sensitive laminæ, sensitive sole,* and *sensitive frog.*

"The *bones* entering into the composition of the foot are the *coffin* and *navicular bones:* to which may be added (as forming part of the coffin-joint, and consequently having intimate relation to them), the *coronet bone.*

"The *tendons* immediately connected with the foot are those of the *extensor pedis* and the *flexor pedis perforans:* the former being inserted into the coronal process; the latter into the posterior concavity of the coffin-bone.

"THE CORONARY SUBSTANCE.

"A less inappropriate name for the part commonly called the *coronary ligament.**

"To revert, for the sake of elucidation here, to former description—after the hoof has been detached by a process of maceration or putrefaction, in a perfectly entire, uninjured condition, it presents around its summit a circular groove, bounded in front by a soft whitish substance, having a thin edge, and being of a nature between horn and cuticle; and behind, by an attenuated margin, more horny in its character, whose thin edging is denticulated or serrated. Into this circular groove or canal is received the terminating margin of the cutis: the cuticulo-horny layer of the hoof, in front of it, having every appearance of being a continuation of the cuticle.

"*Situation — Dimension.* — The coronary substance occupies the concavity formed upon the inside of the superior or coronary border of the wall of the hoof: it is the part constituting the basis of the circular prominence commonly distinguished in the living animal as the *coronet.* It is broadest around the toe of the wall, diminishing in breadth towards the quarters and heels, and being somewhat broader around the outer than the inner side. It is thickest in substance around its middle and most prominent parts, growing gradually thinner both above and below.

"*Connection.*— Externally, the coronary substance is connected with the hoof; and the connection appears to be principally, if not entirely, of a vascular nature: the surface of the wall presenting a porous honeycomb-like texture, and the *rilli* or vessels issuing from the coronary substance entering the pores, and thus establishing an intimate and extensive vascular union between these organic and inorganic parts. Internally, the coronary substance is connected with the coffin-bone, the extensor tendon, and the cartilages, by a fine, dense, copious cellular tissue, which at the same time forms a bed for the assemblage and ramification of the blood-vessels concerned in the secretion of the wall of the hoof. Superiorly, its union with the skin is so intimate and complete, that one has been thought to be a continuation of the other; and, so far as meets the eye of a common observer, they might be taken as such; but, when we come to examine them by anatomical tests, we not only find a line of external demarcation between them, but discover such difference of internal structure as forbids the adoption of this delusive notion. As it descends upon the coffin-bone, the coronary substance not only grows thinner, but in growing attenuated becomes imperceptibly gathered or puckered into numerous points, from which issue a like number of plaits or folds, which afterwards form the sensitive laminæ. It is worthy of remark, that the part of the bone upon which this transformation takes place is smaller in circumference than the coronet; consequently the

---

* Averse as I am to changing or altering names, nothing less than a palpable contradiction, in regard both to structure and *function,* would have induced me to do so in the present instance.

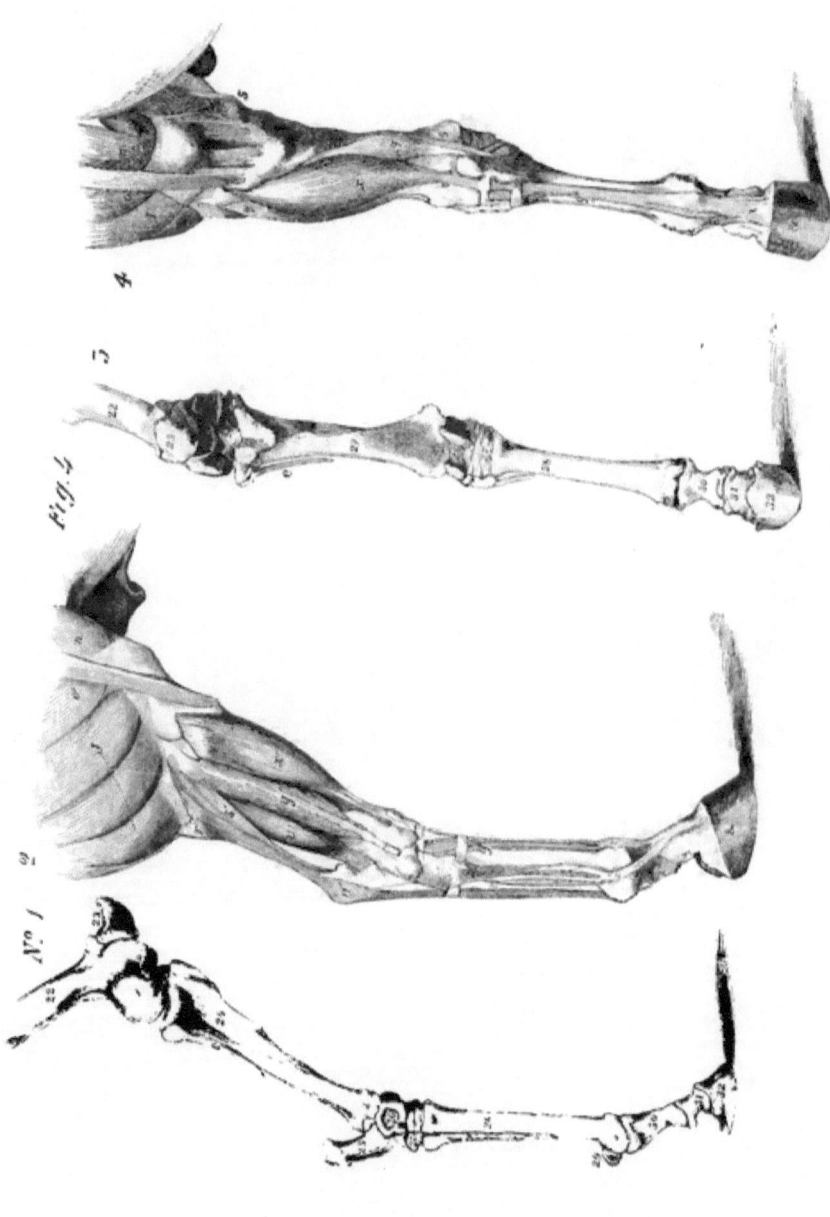

EXPLANATION OF FIGURE IV.

NO. 1.—OSSEOUS STRUCTURE.
OFF-HIND EXTREMITY.

22. Femur or thigh bone.
23. Patella.
24. Tibia.
25. Os calcis.
26. Astragalus.
27. One of the tarsal bones.
28. Metatarsi magnum.
29. The sessamoids.
30. Os suffraginis.
31. Os coronæ.
32. Os pedis.
 e. The fibula.

The above explanation will answer for "No. 3," of this plate.

NO. 2.—MUSCULAR STRUCTURE.
SIDE VIEW OF THE OFF-HIND EXTREMITY.

j. Triceps.
n. Rectus.
o'. Vastus.
v. s'. Gastrocnemii and perforans.
v. v. Flexor pedis accessorius.
u'. (At the hock.) The insertion of the gastrocnemius into the point of the hock.
x'. Extensor pedis.
y. y'. Peroneus.
v. (Beneath the pastern.) Flexor perforatus and perforans.
z. Bifurcation of the suspensory ligament.
$\&'$. The hoof.

NO. 4. MUSCULAR STRUCTURE.
ANTERIOR VIEW OF THE OFF HIND EXTREMITY.

n. Rectus.
o'. Vastus externus.
J. Triceps abductor tibialis.
y. y'. Peroneus.
x'. x'. Extensor pedis.
y. Flexor pedis accessorius.
8. Bifurcation of the suspensory ligament.
5. Saphena vein.
$\&'$. The hoof.

same measure of coronary substance which but tensely and smoothly covered the latter, admitted of being disposed in gathers or folds so soon as it reached the former. Posteriorly, the coronary substance forms a junction, indeed becomes continuous in substance, with the heels of the sensitive frog.

"*Structure.* — The coronary substance discloses three different parts in its composition: 1. *A fibro-cartilaginous circling band*, forming the substratum and basis of the entire structure. 2. *A cuticular covering*, so called from its resemblance in texture to the cutis. 3. *A network of blood-vessels*, reposing upon the former, and covered by the latter. The cartilaginous structure, freed from its vascular connections, is found to be wrought in the form of a coarse, open, irregular network, and appears designed mainly for the purpose of affording a bed for the lodgment and ramification of the blood-vessels destined to produce the wall. The looseness of its connection, added to its own elasticity, renders this substance peculiarly adapted to accommodate itself to the motions of the coffin-joint, and thus preventing those movements from operating prejudicially to the superimposed glandular structure.

"*Organization.* — The coronary substance may be ranked among the most vascular parts of the body: no gland even possesses, for its magnitude, a greater abundance of blood-vessels, or of blood-vessels (taking them generally) of larger size; nor does there exist any part in which greater care appears to have been taken to arrange its vessels so as to insure an uninterrupted supply of blood. These vessels it is that produce the wall: and there is every reason to believe that they perform this office without any assistance from the vessels of the laminæ.

"THE CARTILAGES

"Are two broad, scabrous, concavo-convex, cartilaginous plates, erected upon the sides and wings of the coffin-bone. Professor Coleman calls them 'the *lateral* cartilages,' in contradistinction to two others he has named 'the *inferior* cartilages.'

"*Situation.* — The cartilages form the postero-lateral parts of the sensitive foot, extending the surface considerably in both these directions.

"*Attachment.* — The cartilages are fixed into fossæ excavated in the supero-lateral borders of the coffin-bone. Their anterior parts become united, on each side, with descending lateral expansions from the extensor tendon, and are also attached to the coronet bone by cellular membrane. Their posterior parts surmount the *alæ* or wings of the bone, to which they are firmly fixed, and from which they project backwards, beyond the bone, giving form and substance to the heel. Supposing one of the cartilages to be divided into two equal parts by a line drawn horizontally across its middle, the superior half, which extends as high as the pastern-joint, is covered by skin only; and on that account is quite perceptible to the feel, and (in form) to the sight, as the animal stands with his side towards us. The lower half is covered, superiorly, by the encircling coronary substance; inferiorly, by sensitive laminæ: consequently, over all by the hoof, which envelopes both the coronary substance and the laminæ. The extreme posterior ends of the cartilages incurvate downward and backward; but, being overreached by the heels of the sensitive frog, any abrupt or exposed termination of them is prevented. Around these points also the coronary substance makes its inflections upon the sensitive frog, thereby giving them additional substance and support.

"*Form.* — Considered in the detached state, the cartilage in its general figure describes an irregular quadrangle, of which the supero-anterior and infero-posterior angles are the most projecting; the latter at the same time being incurvated inwards. Externally, the cartilage is pretty regularly convex; internally, it is unevenly concave, the surrounding border turning inwards into the substance of the sensitive frog. The posterior part of the cartilage is somewhat thinner than the anterior, and has several foramina through it — three or four of large size — which transmit vessels to the frog.

"*The False Cartilages.*— From the inferior and posterior sides of the *true* cartilages, proceed in a direction forward — towards the heels of the coffin-bone — two fibro-cartilaginous productions, to which Mr. Coleman has given the name of '*inferior* cartilages.' If they are to be considered as *cartilages* at all, I prefer denominating them *false;* they being, as well in structure as in use, different from the *true* or lateral cartilages. They spread inwards upon the surface of the tendo perforans; become united at their inner sides with the superior margin of the sensitive frog; are covered inferiorly by the sensitive sole; and at the same time assist in the support of the sensitive frog. They are triangular in their figure, and are arched in the same manner as the sole.

"*Use.*— Their use appears to me to be, to fill up the triangular vacant spaces left between the tendo perforans and heels of the coffin-bone, thereby completing the surface of support for the sensitive frog, and extending that for the expansion of the sensitive sole. Bone in these places must have proved inconvenient by more or less impeding the impression upon, and consequent reaction of, the sensitive frog.

"THE SENSITIVE LAMINÆ OR LAMELLÆ.

"So is denominated the laminated, membranous, vascular structure clothing the wall of the coffin-bone.

"*Production.*— The sensitive laminæ appear to be derived from the coronary substance — the one, in fact, seems to be a continuation from the other; for if, in a foot in a putrid condition, we attempt to part them by force, we may make an artificial rent somewhere, but can find no natural separation between them. The cuticular covering of the coronary substance having descended upon the coffin-bone, the circumference of which is less than that of the coronet, because thereupon gathered into numerous little plaits or folds, which proceed in parallel slanting lines down the wall of the bone: a transformation it may be difficult to explain, since the laminæ unfolded would occupy a much larger surface than the coronet; at the same time, it is one that has its parallels in the animal constitution, and a remarkable one in the instance of the ciliary processes.

"*Division.*— According to this mode of derivation, every lamina consists of one entire plait or duplication of substance, having its *inward* sides intimately and inseparably united; its *outward* sides being the surfaces of attachment for the horny laminæ. It has also two *borders:* one opposed to the coffin-bone, the other to the hoof; and two *ends* or *extremities*, one issuing out of the coronary substance, the other vanishing in the sensitive sole.

"*Structure.*— The substance of the laminæ when held to the light evinces a degree of transparency; although its nature is extremely dense, and it possesses extraordinary toughness and tenacity. Veterinary writers and lecturers have endowed the laminæ with a high degree of elasticity: but it appears to me that the property is referable to their *connections*, and not one that is inherent in their own substance.

"*Elastic Structure.*— This is a *substratum* of a fibrous periosteum-like texture, attaching the laminæ to the coffin-bone, in which it is that the property of elasticity resides to that remarkable extent usually ascribed to the laminæ themselves: indeed, so elastic is it found to be, that it can be made to stretch and recede the same as a piece of India rubber. Its fibres take a direction downward and backward. At the same time, it affords a commodious bed for the ramification of blood-vessels issuing from the substance of the bone, in which they are (particularly in the stretched condition of the substance) protected from injurious compression and consequent interruption to their circulation.

"*Number.*— In round numbers we may estimate the laminæ at about 500; not including those of the bars. They vary, however, in number: I have reckoned upwards of 600.

"*Dimensions.* — In length they decrease from around the toe towards the sides and heels in a corresponding ratio with the wall; those in front, the longest, being rather more than two inches in extent; the shortest, those at the heels, being rather less than one inch. In breadth there is no variation: all measure alike, one-tenth of an inch.

"*Organization.* — The laminæ are highly organized, though they are not equally so with either the sensitive sole or sensitive frog; nor are they so red as those parts: and the obvious explanation of this is, that (over and above what is requisite for their own nutrition) all the blood they have occasion for is only that which is sufficient for the secretion of the horny laminæ.

"THE SENSITIVE SOLE.

"The sensitive sole, or (as Sainbel calls it) the fleshy sole, is the fibro-vascular substance covering the arched concave, or ground surface, of the coffin-bone; in fact, is the part corresponding to the horny sole.

"*Structure.* — The same kind of elastic fibrous structure that sustains the laminæ is found constituting the groundwork of the sensitive sole; only that in the latter case it is closer, denser, and firmer in its texture. Upon this is spread a remarkably beautiful venous network. And the whole is enveloped in an outer cuticular covering, derived from the heels and frog, from which are sent villous processes, loaded with the points of arteries into the porosities of the horny sole: not, however, perpendicularly downward, but in an oblique direction — downward and forward — the same in which the horny fibres grow.

"*Connection.* — Around the circumference of the coffin-bone, the sensitive sole is connected with the fibrous substance descending from the wall, together with the tapering, vanishing points of the laminæ. In the centre, it is united with the bars and frog. But its principal attachment consists in its being firmly rooted into the sole of the coffin-bone; a connection that receives considerable addition from the blood-vessels issuing out of the substance of the bone.

"*Thickness.* — The sensitive sole varies in thickness at different places. On an average, it may be said to measure one-eighth of an inch in thickness. In the vicinity of the frog, it is something less than this. At the heels, it possesses double that thickness.

"*Organization.* — This is one of the most vascular and sensitive parts in the body. Independently of the much admired venous network expanded over the fibrous substance of the sole, arteries enter it issuing from the substance of the bone, and penetrate its villi, which, by taking this course, elude all compression and obstruction: there are also others — the nutrient arteries; but these have an external origin, from the inferior coronary artery. The chief assemblage of arteries takes place within the villi, upon the cuticular surface — those issuing out of the interior of the bone simply passing through (without ramifying within) the fibrous substance: so that, if the substance of the sole is laid open by transverse section, the incised edge, near the surface, exhibits a deep red tint; while the interior, nearer the bone, has a pinkish or pale red aspect.

"THE SENSITIVE FROG.

"Under this head is included the cleft, cuneiform body, projecting from the bottom of the foot, together with the substance continued from it and filling the interval between the cartilages. Sainbel calls it 'the fleshy frog.'

"*Division.* — We distinguish, in the sensitive as in the horny frog, an *apex* or *toe*; two *heels*, separated by the *cleft*; and a portion intermediate between these, which is the *body*.

"*Situation and Connection.* — The sensitive frog occupies the posterior and central parts of the bottom of the foot, forming in the tread a firm and secure *point d'appui*

Being in the hoofless foot equally prominent with the projecting edge of the coffin-bone, one might be led to infer that the horny frog should take the same line of bearing with the crust. The frog, altogether, is lodged in a capacious irregular space, bounded superiorly by the tendo-perforans and common skin, laterally by the cartilages, and inferiorly by the horny frog; with all which parts it has connections; besides being continuous with the sensitive bars and sole, and at the heels with the coronary substance. On its sides are two shallow, ill-defined hollows, corresponding to the commissures of the horny frog, into which are received the horny prominences opposed to them.

"*Structure.* — Entering into the composition of this body we distinguish four parts: An exterior or cuticular covering; a *congeries* or network of blood-vessels; a fibro-cartilaginous texture; and an elastic interstitial matter.

"*The exterior* or *cuticular covering* invests the prominent bulbous portion of the frog, and also gives a lining to the cleft. Superiorly, it is continuous with the skin descending upon the heels; anteriorly, with the cuticular covering of the coronet; inferiorly, with that of the sole. Numerous villous processes sprout from its surface, and enter the porosities in the interior of the horny frog, taking a direction downward and forward, the same as that in which the fibres of the horn grow.

"*The vascular covering* succeeds the cuticular, lying immediately underneath it. It consists of a network of blood-vessels, principally veins, but which are not so thickly set as upon the sole.

"*The fibro-cartilaginous case* comes next. We find it spread over those parts most subjected to pressure, and to be, in many places, one-fourth of an inch in thickness. From its interior are sent off numerous processes, pervading the elastic matter of the frog, forming so many *septa* intercrossing one another, and dividing it without any notable regularity into many unequal compartments. In the posterior and bulbous parts, the septa exist in greater numbers, and are closer arranged than in the middle parts. The fibres of this vaginal substance run obliquely downward and forward, and become intermixed around the borders with those of the bars and sole.

"*The elastic interstitial matter*, however, composes the bulk of the sensitive frog. It consists of a pale yellowish soft substance, which has been mistaken for fat or oil, and hence has been named 'the fatty frog.' When cut deeply into, it exhibits a granulated appearance, and the fibrous intersecting chords become apparent, putting on the ramous arrangement of a shrub or tree. Altogether, the sensitive frog forms a peculiar, spongy, elastic body, for which we lack some more appropriate name."

A TABULAR VIEW OF THE BONES OF THE HORSE.

BONES OF THE CRANIUM.

	Number.
Frontal,	1
Parietal,	2
Temporal, two pairs,	4
Occipital,	1
Ethmoid,	1
Sphenoid,	1

BONES OF THE FACE.

Nasal,	2
Superior and anterior maxillary,	4
Malar,	2
Lacrymal,	2
Palatine,	2
Superior and inferior turbinated,	4
Vomer,	1
Lower jaw,	1

TEETH.

Incisors,	12
Canine,	4
Molars,	24

BONE OF THE TONGUE.

Os Hyoideus,	1

BONES OF THE EAR.

Malleus,	2
Incus,	2
Stapes,	2
Orbiculare,	2

BONES OF THE SPINE.

Cervical,	7
Dorsal,	18
Lumbar (sometimes 6 are found),	5

BONES OF THE SACRUM AND TAIL.

Sacral,	1
Coccygeal (tail), about	15

BONES OF THE CHEST.

Ribs, on each side 18,	36
Sternum,	1

PELVIS.

Innominata (or bones without a name),	2

BONES OF THE SHOULDER.

Scapular,	2

BONES OF THE ARM.

Humeral,	2

BONES OF THE FORE-ARM.

Radial and ulnar. The ulnar being, in the adults, connected with the radius, we shall consider them as one bone. Radial, . . 2

BONES OF THE KNEE.

The carpal bones are thus named:

First Row	Second Row
Scaphoid,	Pisiform,
Lunar,	Trapezoid,
Cuneiform,	Magnum,
Trapezium.	Unciform.

Eight bones to each knee, . . . 16

BONES BELOW THE KNEE.

Metacarpal,	2
Splents,	4
Pastern,	2
Coronet,	2
Sessamoid,	4
Navicular,	2
Pedal or foot bones,	2

BONES OF THE HIND EXTREMITY.

Femur,	2
Stifle,	2

BONES OF THE LEG.

Tibia and fibula. These we shall consider as one to each extremity, . . . 2

BONES OF THE HOCK.

Astragalus,	2
Os Calcis,	2
Cuboid,	2
Cuneiform,	6

BONES OF THE LEG.

Two cannons and four splents,	6

BONES BENEATH THE CANNON.

Pastern,	2
Coronet,	2
Sessamoids,	4
Navicular,	2
Pedal or foot bones,	2

Total number of bones, . . 238

The correct technical nomenclature of the above bones will be found in "Osteology," which see.

ANATOMY OF THE SKELETON.—OSTEOLOGY.

OSSEOUS SYSTEM OF THE HORSE.

In the form of answers to a series of questions, the student will become acquainted with the name, location, form, use, and general peculiarities of the various bones composing the horse's skeleton.

Q. What is understood by the *natural skeleton?*—A. The term is applied when the whole bones are held together by their natural attachments: ligaments, cartilages, and synovial membranes.

Q. Why is the term, *artificial*, sometimes applied to the skeleton?—A. Because the bones, having been divested, by maceration or otherwise, of their connecting ligaments, etc., are united *artificially*, by wire and plates of metal.

BONES OF THE CRANIUM.

Q. Enumerate the cranial bones.—A. Frontal, two parietal, occipital, four temporal, ethmoid, sphenoid: ten.

FRONTAL BONE (OS FRONTIS).

Q. Describe the situation of the frontal bone.—A. It occupies the antero-superior part of the cranium in the region known as the forehead.

Q. What are its peculiarities?—A. In form it is irregular, having two surfaces and four borders. Its surfaces are flat externally, concave internally. Its internal surface is divided by a septum into anterior and posterior concavities. The posterior one is occupied by a portion of the anterior lobe of the cerebrum; the anterior constitutes the frontal sinuses, they being separated from each other by the nasal spine. The concavity is further divided into shallow chambers by imperfect septa.

Q. Describe the borders of the os frontis.—A. They are denticulated and squamous. The posterior is arched, describing segments of two circles. The anterior or nasal is waving, inclines backwards and outwards. The frontal border is straight, anteriorly broad and triangular. The ethmoidal or outer border is irregular, and unites with the lachrymal, sphenoid, and ethmoid bones.

PARIETAL BONES (OSSA PARIETALIA).

Supposing the horse to be an adult, we shall consider these bones as one.

Q. What is the situation of the parietal bone?—A. It occupies the mesio-superior part of the cranium.

Q. Describe the same.—A. Its form is quadrilateral: vaulted, concave internally, and convex externally. It has two surfaces and four borders, denticulated and squamous.

Q. What is observable on the convex surface?—A. A longitudinal messian crest, bifurcating anteriorly; which indicates the location of the sutures, now obliterated by age. Between the bifurcatures arises an eminence above the cranial surface.

Q. Describe the appearance of the internal surface?—A. It is indented by, and receives, the lobular eminences of the cerebrum, and it is also furrowed by arterial ramifications which supply the dura mater.

TEMPORAL BONES (OSSA TEMPORUM).

Q. What portion of the cranium do the ossa temporum occupy?—A. Its sides and base.

Q. How do these bones differ from those in man?—A. In man they are divided into three portions, squamous, petrous, mastoid; yet in reality they are united. In the horse they constitute four distinct bones, two on each side.

Q. Name them. — A. Two ossa temporum, pars squamosa, pars petrosa: four.

Q. Describe their appearance. — One pair is composed of laminæ, vaulted; form ovoid, surmounted by irregular projections; the other pair are solid and convex.

OCCIPITAL BONE (OS OCCIPITIS).

Q. What is the situation of the os occipitis? — A. It is located in the postero-superior and inferior parts of the cranium.

Q. What is its form? — A. Convex externally, irregular, having an occipital tuberosity and condyles.

Q. What are the connections of this bone? — A. It unites, superiorly, with the parietal bones; inferiorly and anteriorly with the sphenoid; laterally, with the temporal, and it articulates posteriorly with the atlas.

Q. What is the fœtal state of the bone? A. It is easily separable into four portions.

Q. State its use. — A. It forms the posterior and inferior parts of the cranium, protects this portion of the brain, and gives exit to the spinal cord.

SPHENOID BONE (OS SPHENOIDES).

Q. What is the situation of the os sphenoides? — A. It passes from one temporal region to the other, across the antero-inferior part of the brain.

Q. What are its general divisions? — A. It is divided into body, situated in the middle, alæ or wings, on each side, and two pterygoid processes, considered as legs.

Q. To what bones is it connected? — A. Occipital, ethmoid, squamous-temporal, palate, and vomer.

ETHMOID BONE (OS ÆTHMOIDES).

Q. What part of the cranium does the os æthmoides occupy? — A. Anterior to the sphenoid, and is the boundary of the cranial, and commencement of the nasal, cavities.

Q. Describe its form. — A. The posterior portion bears resemblance to a bird with its wings extended, having no legs, but a long erected neck and a small round head; the anterior part consists of a slim, brittle, porous, spongy structure of considerable volume.

Q. What are its connections? — A. With the sphenoid, frontal, vomer, and superior turbinated bones; and with the cartilaginous septum of the nose.

BONES OF THE FACE.

Under this head we shall consider the

Ossa nasi,	2
" maxillaria superiora,	2
" maxillaria anteriora,	2
" malarum,	2
" lacrymalia,	2
" palati,	2
" turbinata, superiora et inferiora,	4
" vomer,	1
Os maxillare inferius (lower jaw,)	1
Total,	18

We shall now consider these bones in the above order.

NASAL BONES (OSSA NASI).

Q. How many nasal bones are there? — A. Two.

Q. Where are they situated? — A. In the superior part of the face.

Q. Describe their form? — A. They resemble the form of a pear; are broad posteriorly, pointed anteriorly; they are convex externally and concave internally.

Q. To what bones are they connected? — A. To the frontal, superior and anterior maxillaria, and lachrymal.

Q. What is their use? — A. To defend the nares, and retain in position the septum nasi.

SUPERIOR MAXILLARY BONES (OSSA MAXILLARIA SUPERIORA).

Q. Where are they situated? — A. In the supero-lateral parts of the face.

Q. Describe their form? — A. They are somewhat irregular — tri-lateral; from the centre (which is thickest) they taper, the anterior part being much thinner than the posterior.

Q. How are they divided? — A. Each

bone has a facial, palatine, and nasal surface. It has also nasal, alveolar, and palatine borders, and two extremities: posterior, which forms the maxillary tuberosity; anterior, or dental extremity.

Q. What are the connections of the ossa maxilaria?—A. With the squamous temporal, nasal, anterior maxillary, malar, lachrymal, palate, and inferior turbinated bones.

ANTERIOR MAXILLARY BONES (OSSA MAXILLARIA ANTERIORA).

Q. What is the situation of these bones? —A. They are placed in the supero-anterior and antero-lateral parts of the face.

Q. What is the general form of these bones?—A. Very irregular; consisting of a broad, thick base, turned forwards, from which is sent off a thin flexible plate; and a narrow, elongated, tapering portion turned backwards.

Q. How is each bone divided?—A. Into three surfaces and three borders.

Q. Describe the surfaces.—A. The superior or nasal surface is smooth, convex, and oblong. The inferior or palatine is vaulted, it contributing to the formation of the palate; within it, of an oval form, is the interdental space, which is occupied by two thin, flexible plates, the palatine processes, denticulating along the sides with each other. In the side of the bone is a deep hollow, for the reception of that portion of the superior maxillary bone which holds the tusk; and the remainder of the surface, posteriorly, is articulated with the same. The anterior or labial surface is broad, smooth, and convex, and gives attachment to the depressor labii superioris, and gums.

Q. Describe the borders.—A. The anterior border is broad and curved, and is composed of two laminae, formed apart and divided into septa for the insertion of six incisors. The posterior border is narrow and sloped, and denticulates with the nasal bone. The internal border is broad, quadrilateral, curved, and denticulates with its fellow, forming thereby the superior maxillary symphysis, through which runs the foramen incisivum, for the transmission of the palatine arteries.

Q. How is this bone connected?—A. It connects with the superior maxillary and nasal bones, and with its fellow.

MALAR BONES (OSSA MALARUM).

Q. What is the situation of the ossa malarum?—A. They occupy the antero-external part of the orbit.

Q. Describe their form.—A. Irregularly triangular, presenting a broad basis forwards.

Q. How is the bone divided?—A. Into three surfaces, three angles, a basiform and an apisform extremity.

Q. Name the surfaces.—A. Facial, maxillary, and orbital.

Q. Describe the same.—A. The facial surface is divided into two portions by the zygomatic spine; the upper division is smooth and nearly flat; the lower part is narrow and roughened, for the insertion of the masseter muscle. From this surface, posteriorly, arises the zygomatic process, which is very obliquely sloped off, and laminated for adaptation to the process of the same name, meeting it from the temporal bone, the two together forming the zygomatic arch. The maxillary surface is concave. The orbital surface has a smooth concavity which forms the infero-external part of the orbit.

Q. Describe the angles.—A. There are three, superior, inferior, and posterior. The superior constitutes the external portion of the orbital circumference. The inferior forms the zygomatic spine. The posterior is not so prominent nor defined, but forms an irregular link with the superior maxillary bone.

Q. What of the extremities?—A. The anterior extremity is broad, irregular, and denticulated, and articulates with the superior maxillary and lachrymal bones. The posterior or apisform extremity forms the zygomatic process.

Q. With what bones do the ossa malarum connect?—A. With the temporal, superior, maxillary, and lachrymal bones.

LACHRYMAL BONES (OSSA LACHRYMALIA).

There are two lachrymal bones: we shall describe but one, considering that they are both alike, as indeed are those already referred to, in a plural sense.

Q. What is the situation of the lachrymal bone? — A. It occupies the antero-external part of the orbit.

Q. How is it divided? — A. Into three surfaces and five borders.

Q. Name the surfaces. — A. Internal, external, and orbital.

Q. Name the borders. — A. External and internal facial, nasal, and external and internal orbital.

Q. What is observable in the orbital excavation of this bone? — A. The lachrymal fossa.

Q. What occupies this fossa or groove? — A. The lachrymal vessels, sac, and duct.

Q. With what bones is it connected? — A. With the frontal, nasal, malar, and superior maxillary bones.

PALATE BONES (OSSA PALATI).

Q. What is the situation of the palate bones? — A. They are placed in the inferior posterior part of the face, adjoining the base of the cranium.

Q. What does the palatine surface form? — A. The palatine arch or roof of the mouth.

Q. What of the nasal surface? — A. It forms the posterior surface of the nasal outlet.

Q. What other surfaces do these bones present? — A. Ethmoidal and orbital.

Q. To what part of the bone is the velum palati attached? — A. To the palatine.

Q. How are the palate bones united to the superior maxillary? — A. By their supero and infero lateral borders; each being denticulated.

Q. What other connections have the palate bones? — A. They are joined to the frontal, ethmoid, sphenoid, vomer, and inferior turbinated bones.

TURBINATED BONES (OSSA TURBINATA SUPERIORA ET INFERIORA).

Q. Where are the ossa turbinata located? — A. Within the nasal cavity: the superior above, and the inferior below.

Q. What is their form, and how are they divided? — A. In form they are oblong, thin, foliated, convoluted, scroll-like, and cavernous. They are divided into external and internal surfaces; superior and inferior extremities.

Q. How many bones are there? — A. Four.

Q. Describe the bones. — A. Their external surface is convex, and presents series of longitudinal grooves which mark the ramifications of small blood-vessels. The internal surface is cellular, being unequally divided by transverse septa. Their interior is capacious; they are open superiorly and closed anteriorly. They are porous and elastic. The superior bone exceeds in volume the inferior, and makes its convolution from below, its superior border being attached; whereas, the reverse is the case with the inferior one.

Q. What are their connections? — A. The turbinated bone is connected above with the ethmoid; and laterally, with the nasal bone.

VOMER.

Q. From what does the name of this bone arise? — A. From its resemblance to a ploughshare.

Q. What are its uses? — To divide the nasal chambers and permit the expansion of olfactory nerves.

Q. What is inserted into its superior groove? — A. The septum narium.

Q. What are its connections? — A. It unites with the ethmoid, sphenoid, superior and anterior maxillary, and palate bones.

LOWER JAW. — INFERIOR MAXILLARY BONE (OS MAXILLARE INFERIUS).

Q. What is the situation of this bone? — A. It composes the inferior and posterior parts of the face.

Q. What is the fœtal state of the bone? — A. In the fœtal state it is divided, at its inferior junction, by a connecting cartilage, hence the part has been called its symphysis.

Q. How is the bone divided? — A. Into

body, neck, sides, and branches; external and internal surfaces, and corresponding borders.

Q. What do you understand by these terms? — A. Body signifies the anterior part reaching posterior to the tusks; *neck* signifies the contracted part, immediately posterior to the *body*; *sides* are the parts comprehended between the neck and the branches; the *branches* are the parts posterior to the *neck*, which terminate in the condyles. As regards *surfaces*, the external is convex, rounded, rough, and porous, and affords attachment for muscle and gum. The internal surface is concave, rough, and porous, and answers for the attachment of muscles and gum, and as a channel for the tongue, and attachment for the frænum linguæ.

Q. What do you understand by borders? — A. Each superior *border* exhibits six alveolar cavities for the molar teeth; the septum is composed of osseous laminæ. The inferior border is thin and irregular. The posterior border is broad and roughened for the insertion of muscles.

GENERAL INQUIRIES.

It is now presumed that we understand the *location* and *names* of the different bones composing the cranium and face; and, before we proceed further, it may be profitable to make some *general* inquiries regarding the bony structure.

Q. Is not the number of bones greater during colthood than at mature life? — A. Yes, many of the bones separable at that period become united in the adult.

Q. How are bones divided? — A. They are divided into long or cylindrical, broad or flat, and thick.

Q. What do you understand by epiphysis of bones? — A. The region where cartilage is interposed between bones that finally become ossified.

Q. What is the structure of bones? — A. They consist of a cellular, reticular, and vascular parenchyma, and of osseous matter deposited in it: their base, therefore, is the same as that of the soft parts.

Q. Are bones vascular? — A. Yes.

Q. How can you demonstrate their vascularity? — A. By numerous small foramina and by the tinge they receive from the coloring matter of the food.

Q. Name the investing membrane of bones? — A. Periosteum.

Q. What is its organization? — A. Fibrous.

Q. Of what use is this periosteum? — A. It limits the growth of bones, is the medium of circulation and nutrition, and affords attachment for ligaments and muscles, and favors the free articulation of the latter.

Q. What does its internal surface secrete? — A. An oleaginous fluid, deposited in the cellular structure and cavity of bones.

Q. What are foramina? — A. Holes perforating the substance of bones.

Q. What are sinuses? — A. Occurring in bones, they are large cavities with small openings.

Q. What are sinuosities? — A. Superficial but broad irregular depressions.

Q. What are furrows? — A. Long, narrow, and superficial canals.

Q. What are notches? — A. Cavities in the margin of bones.

Q. What are fossæ? — A. Deep and large cavities on the surface of bones.

Q. What are glenoid cavities? — A. Cavities for articulation.

Q. What are tubercles? — A. Small eminences.

Q. What are tuberosities? — A. Rough elevations.

Q. What are spines? — A. Long projections upon a bone.

Q. What are heads? — A. The round tops of bones.

Q. What are necks? — The narrow portion of bones beneath their heads.

Q. What are processes? — A. Short projecting portions of bones.

THE TRUNK — (REMARKS ON THE SAME).

We shall now consider the peculiarities of the trunk; which comprehends the verte-

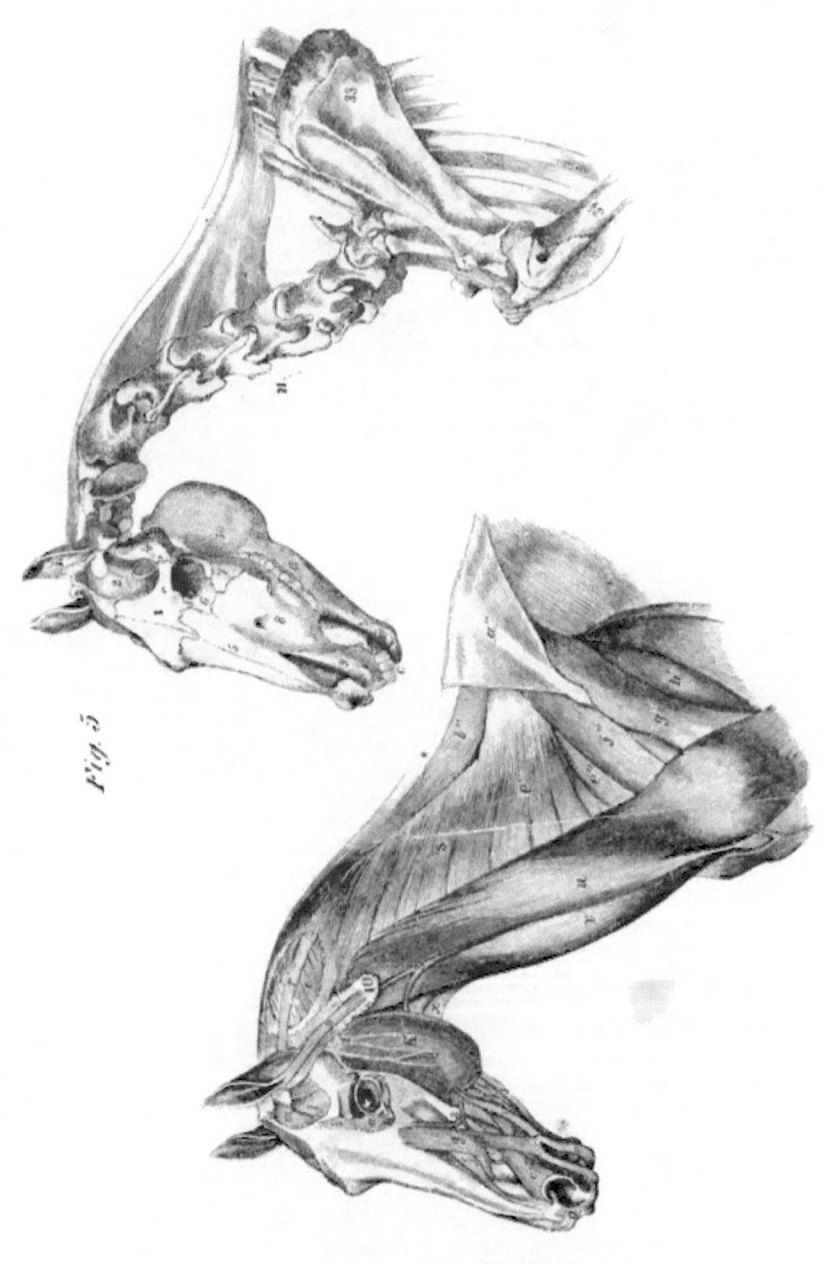

Fig. 5.

EXPLANATIONS OF FIGURE V.

MUSCULAR STRUCTURE.

LATERAL VIEW OF THE HEAD, NECK, AND SHOULDER. — THE HEAD.

- a. Orbicularis palpebrarum.
- b. Levator palpebræ.
- c. Dilator naris lateralis.
- d. "   " anterior.
- e. Orbicularis oris.
- f. Nasalis longus.
- g. Levator labii superioris.
- i. Zygomaticus.
- j. Retractor labii inferioris.
- k. Buccinator.
- K. Masseter.
- l. Temporalis.
- m. Attolentes et abducens aurem.
- f. Facial veins.

THE NECK.

- *. Ligamentum colli.
- b". Rhomboideus longus.
- s. A portion of the splenius.
- c". Scalenus.
- e". Pectoralis transversalis.
- o. Abducens vel deprimens aurem.
- r. Tendon of the splenius and complexus major.
- t. Obliquus capitis inferior.
- u. Levator humeri.
- v. Sterno maxillaris.
- x. Subscapulo hyoideus.
- 3. Jugular vein.

REGION OF THE SHOULDER.

- a". Trapezius.
- f". Antea spinatus.
- g". Postea spinatus.
- h". Teres major.

OSSEOUS STRUCTURE.

- *. Ligamentum colli, or subclavium.
- 1. Temporal bone.
- 2. Parietal bone.
- 4. Zygomatic arch.
- 5. Nasal bone.
- 6. Lachrymal bone.
- 7. Malar.
- 8. Superior maxilla.
- 9. Anterior   "
- 10. Inferior   "
- b. The neck of the same.
- 11. Cervical vertebræ.
- 33. Scapula.
- 34. Humerus.
- a. The molars.
- e. The incisors.
- h. The lining membrane of the ear.
- i. Nasal cartilage.

bral chain, thorax, and pelvis. It is generally called the spine, or back bone, and extends from the occipital bone to the sacrum. The spine is divided into three regions, denominated cervical, dorsal, and lumbar. The spine, as a whole, exhibits three surfaces and two extremities. The surfaces are named superior, inferior, and lateral. The superior surface is flat in the region of the neck; in the back and loins it offers a series of projections. The inferior surface is more uniform, and the lateral is very irregular.

CERVICAL VERTEBRÆ.

Q. How many cervical vertebræ are there?— A. Seven.

Q. What is the name of the first?— A. It is called atlas.

Q. How does it differ from the rest?— A. It has no superior spinous process nor body; the vertebral hole is larger than in the others, and its transverse processes are very broad. It has three pairs of foramina: one posteriorly, through which run the vertebral arteries; and two anteriorly.

Q. What is the name of the second cervical vertebra?— A. It is named dentata.

Q. How is it recognized from the rest?— A. By its anterior projection, which in the human subject resembles a tooth.

Q. With what does this tooth-like process articulate?— A. It articulates with the infero-posterior part of the ring of the atlas.

Q. Describe the *third, fourth,* and *fifth* cervical vertebræ.— A. They possess the genuine characters of *cervical vertebræ*, and closely resemble each other; the *third*, however, has commonly a more elevated superior spine than either of the others, and is narrower across the *mesio-superior* part of the *body* (measuring from the roots of the articular processes), which dimension increases in the fourth, but is greatest in the fifth.

Q. What of the sixth *vertebra*?— A. It has no inferior spine; and its transverse processes are trifid, consisting each of three eminences.

Q. What of the seventh?— A. It is the shortest, and in its general conformation resembles the first *dorsal*. Its body, posteriorly, presents two semilunar articular depressions, constituting a part of the socket for the first rib.

DORSAL VERTEBRÆ.

Q. How many *dorsal* vertebræ are there? — A. Eighteen.

Q. What is peculiar to the dorsal vertebræ?— A. They have each a *body, spinous process,* and *transverse process,* and are generally distinguished by the length, form, and direction of their spines.

Q. How is the first *dorsal vertebra* distinguished from the rest?— A. By the sharpness of its *spinous,* and singleness of transverse, processes, and by the breadth of its articulatory surfaces.

Q. How do the *articular depressions* for the insertion of the ribs differ in each bone? — A. They are less deeply marked, as we proceed *posteriorly.*

Q. How are the seventeenth and eighteenth distinguished from the rest?— A. They have perfect articulatory depressions on the bodies for the insertion of ribs.

LUMBAR VERTEBRÆ.

Q. How many lumbar vertebræ are there?— A. Five.

Q. How are the bodies of the *lumbar vertebræ* distinguished from the dorsal?— A. They are larger, contracted in the centre, and their edges are more prominent.

VERTEBRAL CANAL.

Q. What is the form of the vertebral canal?— A. In the *cervical* region it is capacious and semi-oval; through the *dorsal, transversely* oval and smaller. In the *lumbar* it is semi-circular, of less diameter than the *cervical* and greater than the *dorsal.*

Q. With what does the spinal canal connect?— A. Anteriorly, with the *cranial* cavity; posteriorly, with the *sacral* canal.

PELVIS, SACRUM, AND TAIL BONES.

We shall now consider the posterior boundary of the trunk.

OS SACRUM.

Q. What is the popular name of this bone?—*A.* It is called the "*rump bone.*"

Q. Where is it located?—*A.* At the superior part of the pelvis, between the *ossa ilia*.

Q. What is the popular name of the *ossa ilia?*—*A.* They are called the haunch bones.

Q. How many pieces enter into the composition of the sacral bone, in the foal?—*A.* It is composed of five pieces.

Q. How are they united?—*A.* By *fibro-cartilaginous* substance.

Q. What ultimate change takes place in this substance?—*A.* It becomes ossified, and hence the solid bone.

Q. How is the sacral bone divided?—*A.* Into three surfaces, two borders, base, and apex.

Q. Describe its surfaces.—*A.* They are named superior, inferior, and lateral. The superior is convex, very irregular; on its central line are five eminences, and laterally are superficial grooves pierced by the four sacral foramina. The inferior surface is smooth and slightly concave. The lateral surface is thick anteriorly, gradually tapering posteriorly; they are roughened for the reception of the sacro-iliac ligament.

Q. Describe, briefly, the base and apex. —*A.* The *base* is composed of a central and two lateral parts. The *apex* is oval, and articulates with the anterior bones of the tail.

BONES OF THE TAIL (OSSA COCCYGIS).

Q. What is the situation of the ossa coccygis?—*A.* Posterior to the sacrum.

Q. Of how many bones is the tail composed?—*A.* Fifteen.

PELVIS OR HAUNCH BONES (OSSA INNOMINATA).

Q. How do anatomists divide these bones, in the fœtal state?—*A.* Into *ilium*, ischium, and pubes.

Q. In the adult horse are there more than two bones?—*A.* They are considered as two, yet in reality they are united at the pubes so as to constitute but one bone. In this state, however, they are denominated *ossa innominata*—unnamed bones.

Q. What is the situation of the iliatic, ischiatic, and pubic portions?—*A.* They are in the anterior, superior, and lateral parts of the pelvic region. The ischiatic extends posteriorly and the pubic inferiorly.

Q. What are the connections of the ossa innominata?—*A.* They are connected, anteriorly and inferiorly, to the os sacrum; posteriorly and inferiorly, to each other, forming the symphysis pubis; laterally, with the thigh bones.

Q. What are the uses of the pelvis?— *A.* It affords an arch for supporting the posterior parts. It contains the urinary organs, rectum, etc., gives protection to bloodvessels and nerves, and origin and insertion to various muscles and ligaments.

CHEST OR THORAX.

The thorax or chest is formed by the dorsal vertebræ, superiorly; ribs, laterally; and sternum, inferiorly. It also affords protection to the principal organs of circulation and respiration.

Q. State the number of ribs and their arrangement?—*A.* Their number is generally thirty-six; eighteen on each side, eight of which are termed true, and the remainder false, ribs.

Q. Why are the anterior eight called true ribs?—*A.* Because they have a direct cartilaginous insertion into the breast bone or sternum.

Q. Why are the posterior ten termed false ribs?—*A.* Because they are *indirectly* connected with the sternum.

Q. What is the general conformation of a rib?—*A.* It is lengthy, curved; convex outwardly or laterally; terminating in a sharp border posteriorly, which forms a posterior convexity. On the inner surfaces it is concave, and of course the reverse of the external.

Q. What are the variations in ribs?— *A.* They vary in length, degree of curvature, and obliquity of direction.

Q. How shall we divide each rib?—*A.*

Into a body, external convexity, internal concavity, a superior and inferior termination, anterior and posterior edges.

Q. What do you understand when the term, *head*, is applied to a rib? — A. It signifies its protuberance — its superior portion; presenting a smooth convexity for articulation with the bodies of vertebræ.

Q. Where is the neck of a rib situated? A. Immediately below the head.

Q. What is the difference between the anterior and posterior edges of the ribs? — A. The anterior edge is circular and the posterior is sharp.

Q. Where is the tubercle of the rib situated? — A. Posterior to the head, at the root of the neck.

Q. How is the first rib distinguished from the rest? — A. It is the shortest and thickest, and is almost straight.

Q. How does the second rib differ from the first? — A. It is longer, less dense, and has a greater curvature in the region of its neck.

Q. How do the ribs differ from the second to the seventh? — A. They increase in breadth.

Q. How do they differ in length? — A. Up to the ninth.

Q. How do they differ in curvature? — A. Gradually up to the eighteenth, which is the most curved of all.

BREAST BONE (STERNUM).

Q. What is the situation of the *sternum*? — A. It occupies the anterior and inferior portion of the thorax.

Q. How does it differ from the human sternum? — A. In the human subject it is composed of three pieces; in the adult horse it is considered as a single bone. It is made up, however, of seven irregularly formed bones.

Q. What is the structure of the *sternum*? — A. It is composed of an *osseous* cellular substance and cartilages.

Q. Name the cartilages? — A. *Ensiform* and *cariniform*.

Q. What is the use of the *cariniform* cartilage? — A. It affords attachment to the sterno-maxillares and sterno-thyro-hyoidei muscles.

Q. To what part of the sternum is the *ensiform* cartilage inserted? — A. To its inferior and posterior part.

Having now considered the bones of the head (with the exception of the teeth), and hyoides (appendages), spine, thorax, and pelvis, we now commence on the bones composing the extremities. These are four in number, disposed in pairs, and known as the fore and hind extremities. Our examination will be conducted with reference only to *one* fore, and one hind, extremity; presuming that a description of the bones on one side will suffice for those on the other.

FORE EXTREMITIES.

Q. What is the situation of the fore extremities? — A. They occupy the antero-lateral parts of the trunk, from which they proceed *inferiorly*.

Q. How are the bones divided? — A. Into shoulder, arm, knee, leg, pastern, coronet, and foot.

Q. Name the bones composing each region? — A.

SHOULDER BONES.
Scapula, Humerus.

ARM BONES.
Radius, Ulnar.

BONES OF THE KNEE.

First Row.	Second Row.
Scaphoid,	Pisiform,
Lunar,	Trapezoid,
Cuneiform,	Os Magnum,
Trapezium.	Unciform.

BONES OF THE LEG.
Large Metacarpal.
Two small Metacarpal (splents).

IN THE REGION OF THE FETLOCK.
Two Sessamoid Bones.

PASTERN BONE.
Os Suffraginis.

CORONET BONE.
Os Coronæ.

FOOT BONES.
Navicular and Coffin-bones.

Q. How many bones compose one of the fore extremities? — A. Twenty-one.

OF THE SHOULDER.

Q. The shoulder being composed of the *scapula* and *humerus*, what portion of the thorax do they occupy? — A. They occupy its antero-lateral region.

SCAPULA, (SHOULDER BLADE).

Q. What is the position of the scapula? — A. It occupies the antero-lateral parts of the thorax.

Q. Describe the bone. — A. It is triangular, broad, and thin superiorly; narrower and thicker inferiorly; its external surface is unequally divided into two superficial concavities, named fossæ antea, et postea spinatæ. Its internal surface is smooth, yet excavated.

Q. Describe the borders. — A. The *superior* has a thin, roughened summit for the insertion of the cartilage of the scapula; the *anterior* is thin in its upper half, yet below it becomes rounded; the *posterior* is obtuse and rounded.

Q. How does the scapula terminate inferiorly? — A. By a glenoid cavity.

Q. What are the connections of the scapula? — A. It has a ligamentous connection with the spines of some of the dorsal vertebræ; to the thorax, it is connected by muscular fasciæ; and its inferior connection is by means of the glenoid cavity, to the head of the humerus; this latter forms the shoulder joint.

Q. Is there anything remarkable about the shoulder joint? — A. Its most remarkable feature is, the great disproportion in size between the head of the humerus and the glenoid cavity.

Q. How is this disproportion in magnitude compensated for? — A. By an extensive capsular membrane, which admits of extensive motion.

Q. What are the insertions of this capsular membrane? — A. It has a circular insertion into the rough margin of the glenoid cavity, and also around the neck of the humerus.

Q. How is this membrane protected internally and externally? — A. Internally it is clothed with a synovial membrane; externally by adherent muscles.

HUMERUS.

Q. What is the situation of the humerus? — A. It is situated beneath the scapula, occupying a diverse direction, viz., downwards and backwards, and is in contiguity with the lateral parts of the thorax.

Q. Describe the form of the humerus. — A. It is irregular, cylindroid, having a convoluted appearance, and its superior extremity is much larger than the inferior.

Q. How is this bone divided? — A. Into a body, superior and inferior extremities.

Q. Describe the *body*. — A. It is angular, with sides, contracted superiorly and flattened and rounded inferiorly. From its superior-anterior-lateral margin projects a roughened tuberosity, into which the levator humeri is inserted. The lateral part of the body is hollow or excavated. The inner side is somewhat roughened and prominent.

Q. Describe the *superior* extremity. — A. The superior extremity being much larger than the inferior, presents a head and several tubercles: it has a projecting, hemispherical surface, designed for extensive articulation. It presents a smooth surface, yet has an irregular, indented groove for the insertion of a capsular ligament.

Q. What is the use of the tubercles? — A. The anterior, three in number, serve as articulations for the flexor brachii to traverse. The fourth serves as a protection against dislocation.

Q. Describe the *inferior* extremity. — A. It consists of two heads or condyles, separated by deep ovoid fossæ into which is received the olecranon of the ulna.

Q. What are the connections of the humerus? — A. Superiorly, it connects with the scapula; inferiorly, it articulates with the radial and ulnar extremity of the *Os Brachii.*

OS BRACHII (ARM BONE).

Q. Describe the location of the os brachii. — A. It is located beneath the thorax, in the inferior region of the humerus.

Q. How does this bone differ from those of the human subject? — A. By being consolidated into a single bone.

Q. How is it divided? — A. Into radial and ulnar portions.

Q. Describe the *radial*. — A. It consists of a body, superior and inferior extremities. The body is lengthy, compared with other bones of the fore extremity; posteriorly it is excavated and roughened; anteriorly it projects with a smooth, cylindrical surface.

Q. Describe the superior extremity. — A. The superior extremity presents an interrupted articulatory surface, having a central eminence, with two cavities, which correspond to the articulations of the os humerii.

Q. Describe the *inferior* extremity. — A. It appears to consist of *three* articulatory surfaces, which correspond with those of the bones of the carpus.

Q. Describe the *ulnar* portion of the os brachii. — A. It presents a tapering triangular projection, firmly connected with the radius; at its junction with the same, it presents a semilunar concavity; this, with the articulatory surface of the radius, forms the humero-brachial articulation.

Q. Name the projection of the ulnar, commonly termed point of the elbow. — A. Olecranon.

Q. What muscle is inserted into the olecranon? — A. The *triceps extensor brachii*.

Q. What is the state of this bone in early colthood? — A. It is composed of two pieces named radius and ulnar, which afterwards become consolidated.

Q. With what bones does the inferior portion of the *os brachii* articulate? — A. With the scaphoid, lunar, and cuneiform bones.

BONES OF THE KNEE (CARPUS).

The bones of the knee correspond to the wrist, or carpus, of man.

Q. How are these bones arranged? — A. They are ranged in two rows, or tiers; one of the number, trapezium, is located in the posterior part of the carpus.

Q. Name the bones of the first row? — A. Scaphoid, lunar, cuneiforme, trapezium.

Q. Name the bones crossing the second row? — A. Pisiform, trapezoid, magnum, unciform.

Q. What is the general form and situation of each of the bones of the first row? — A. The os scaphoides is semi-ovoid in form, its superior surface is sigmoid and smooth, the inferior surface is somewhat oval, and rests upon the trapezoides and magnum of the second row. Its internal surface comes in contact with the os lunare. The os lunare is the second bone of the first row; it articulates superiorly with the brachii; inferiorly, with the ossa magnum and unciforme; its superior surface is triangular; inferior, oblong; on one side, internally, it articulates with the scaphoid, on the other with the cuneiforme. The cuneiforme is known as the external, yet smallest bone of the knee. Its superior surface is concave; inferior, smooth; its internal surface articulates with the os lunare, and posteriorly it unites with the trapezium.

Q. What is the general form and situation of each of the bones of the second row? — A. The os trapezoides is situated on the inner side of the knee, resting on the inner splent bone, and articulating with the os magnum; its form is that of an irregular, curvated, flattened cone; its superior surface is convex, and its inferior flat. The os magnum is the middle bone of the second row, and is known as the largest bone of the knee. Its superior surface presents two articulatory surfaces, one sigmoid and oblong for the os lunare, and the other ovoid and flat, to correspond with the surface of os scaphoides; its interior surface is flat, and articulates with large metacarpal bone. The os unciform is situated on the outer side of the second row, and in form resembles a blunt hook; its superior surface is convex; its inferior irregular, articulating with the outer splent and cannon. The os trapezium is situated in the posterior part of the carpus, and presents two smooth surfaces for articulation with the ossa cuneiforme and brachii. Its external, lateral surface is convex; its internal concave; its superior border gives attachment to the flexores metacarpi; and into the inferior is inserted a ligament. The ossa pisiformia — for sometimes there are two present — is situated posterior to the trapezoides; its form is orbicular or pea-shape.

METACARPAL BONES.

The metacarpal bones are three in number, viz: metacarpi magnum, 1; metacarpi parvum, 2. There seems, however, so great a disproportion between the os magnum and ossa parva, that the former may be considered as the principal support of the fore extremities.

Q. What is the situation of the metacarpus? — *A.* Immediately beneath the carpus.

Q. Describe the form of the metacarpi magnum. — *A.* It is a long cylindrical bone, presenting on its anterior surface a circular, smooth appearance; its posterior surface is somewhat flattened and depressed.

Q. How is the bone divided? — *A.* Into a body, and two extremities.

Q. Describe the extremities. — *A.* The superior presents a smooth articulatory surface, tapering towards its outer edges, yet more depressed on its inner and posterior part; in the anterior region is a roughened prominence, for the insertion of the extensor metacarpi, and on the lateral side of the bone are eminences which afford insertion for the lateral ligaments. The inferior extremity presents a pulley-like surface, with two unequal condyloid surfaces, separated by a semi-circular eminence, which corresponds to a counterpart found on the superior end of the suffraginis.

Q. What are the articulations of the metacarpi magnum? — *A.* It articulates superiorly with the carpus; inferiorly, with the os suffraginis; posteriorly and laterally, with the ossa sessamoidea and metacarpi parva.

OSSA METACARPI PARVA (SPLENT BONES).

Q. How many bones compose the ossa metacarpi parva? — *A.* Two: external and internal.

Q. Describe their situation. — *A.* They are attached to the lateral and posterior parts of the metacarpi magnum.

Q. How do you divide them? — *A.* Into bases, middles, and apices.

Q. Describe the base. — *A.* It is surmounted by a smooth articulatory surface, corresponding to the inferior portion of a part of the knee joint.

Q. Describe the middle. — *A.* It is trifacial: the anterior surface is roughened for the insertion of inter-articular tissue, which connects it with the cannon; the inner surface is excavated; the outer surface is rounding, and terminates, posteriorly, acuminately.

Q. Describe the apex. — *A.* It tapers, and ends in a tubercle, which curvates in an inferior and superior direction.

Q. How do the ossa metacarpi parva differ? — *A.* The *external* is generally larger than the internal, and has a broader articulatory surface.

Q. What bone does the external splent articulate with? — *A.* The unciform.

Q. What bone does the internal splent articulate with? — *A.* The trapezoid.

Q. How are the splents connected to the cannon? — *A.* By cartilago-ligamentous tissue.

Q. What changes does this cartilaginous tissue undergo, subsequent to adult life? — *A.* In a majority of cases it becomes ossified.

PASTERN BONE (OS SUFFRAGINIS).

Q. Describe the location of this bone. — *A.* It is located beneath the cannon, and takes an oblique direction from the same; it articulates superiorly with the cannon; posteriorly with the ossa sessamoidea.

Q. Describe the form of the os suffraginis. — *A.* It is a flattened cylinder, yet its superior portion is more bulky than the inferior; it is generally considered as being about one-third the length of the cannon, and is divided into a body, superior and inferior extremities.

Q. Describe the body of the os suffraginis. — *A.* The body presents two surfaces, anterior and posterior; the anterior is convex, the posterior flattened and uneven; it lessens in bulk in an inferior direction.

Q. How is the superior extremity of the pastern bone recognized from the inferior? — *A.* The superior is the largest, and presents two shallow articular cavities; between them is a groove, which receives the central eminence of the inferior extremity of the cannon bone. The inferior extremity is much smaller than the superior; it is biconvex, and consists of two articular convexities, separated by a transverse shallow depression.

SESSAMOID BONES (OSSA SESSAMOIDEA).

Q. Where are the two sessamoids situated? — *A.* At the posterior part of the articulation formed by the cannon and pastern bones.

Q. What is the form of these bones? — *A.* Trapezoid; three sides present triangular faces, whose apices unite in one point, which is directed upwards; the bases of the same form a fourth side, which is turned downwards; and are therefore divided into three sides, base, and apex.

Q. Give a general description of the faces or surfaces of these bones. — *A.* They are known as anterior, posterior, and lateral faces; the anterior are excavated, smooth, and articulatory, and along their inward borders — which are opposed to each other — are levelled off, so that the two form a groove for the reception of the central eminence of the inferior portion of the cannon. The posterior surfaces are convex and rough; the lateral surfaces are grooved and roughened; the bases are narrow and uneven.

Q. What appears to be the object in excavating the anterior surfaces of these bones? — *A.* To extend the articulatory surface of the pastern joint, and admit of extensive anterior and posterior motion.

Q. For what purposes are the posterior surfaces roughened? — *A.* For the insertion of the suspensory ligaments.

Q. What occupies the cavity which occurs in consequence of uniting the internal surfaces of these bones? — *A.* The flexor tendons.

Q. What ligaments are inserted into the bases of these bones? — *A.* The long, short, and crucial ligaments.

CORONET BONE (OS CORONÆ).

The *os coronæ* is situated beneath, or rather inferiorly, to the pastern, and may therefore be termed the inferior pastern; it occupies a location between the superior pastern and coffin bone.

Q. Describe the os coronæ, or inferior pastern. — *A.* It presents a square body; its breadth, however, somewhat exceeds its longitudinal measurement. It has four surfaces, viz., superior, inferior, anterior, and posterior; the superior surface is bi-concave, corresponding to the projections of the superior pastern; the inferior surface is biconvex, consisting of two condyloid prominences, separated by a slight transverse depression, corresponding to the articulatory surface of the coffin-bone; the anterior surface is convex, yet rough and irregular; the posterior surface is quite smooth, yet excavated.

Q. What are the connections of the os coronæ? — *A.* It connects with the pastern, coffin, and navicular bones.

BONES OF THE FOOT.

THE COFFIN BONE (OS PEDIS).

The coffin bone is considered as the base of the osseous structure of the fore extremity.

Q. What is the form of the coffin bone? — *A.* It presents a semilunar outline; anteriorly and superiorly it is convex; posteriorly and inferiorly it is concave; it is divided into wall, sole, tendinous surface, articulatory surface, and wings.

Q. Describe the wall. — *A.* It is a miniature of the form of the hoof; it exhibits a porous and furrowed surface, and has innumerable perforations, varying in size and form; its superior part is surmounted by the coronal process; the inferior edge of the wall is somewhat oval, and is notched and serrated.

Q. What are the uses of the porosities and furrows?—*A.* They serve as so many attachments for the fibrous tunic of the sensible laminæ.

Q. What name is generally applied to the largest of the perforations found in the coffin bone?—*A.* They are termed foramina.

Q. What occupies these foramina?—*A.* Blood-vessels and nerves.

Q. Describe the sole of the coffin bone?—*A.* The sole exhibits a broad, uniform, concave surface, resembling in most cases the figure of the inferior part of the hoof; it has porosities similar to those formed on the wall; it is bounded anteriorly and laterally by the circumferent edge of the wall; posteriorly, by a sharp, uneven, semi-circular edge, which divides it from the tendinous surface.

Q. What do we find on the tendinous surface?—*A.* 1st, a rough depression in its fore and middle part, marking the insertion of the tendo perforans. 2ndly, two lateral grooves, passing obliquely inwards, and terminating each in a large foramen. 3rdly, a porous space intermediate between the two former divisions, into which is fixed the inferior navicular ligament.

Q. What occupies the lateral grooves?—*A.* The trunks of the arteries and nerves which occupy the interior of the coffin bone.

Q. What are the peculiarities of the articulatory surface of the coffin bone?—*A.* It has two lateral depressions, which extend posteriorly to the alæ; a broad eminence runs transversely between them; this eminence is terminated in front by the coronal processes, having an incurvation backwards; behind it, the surface is bevelled off, to which part is opposed the navicular bone; the depressions alluded to are deepened by the prominent edge running around the anterior and lateral parts.

Q. What portion of the coffin bone does the articulatory surface occupy?—*A.* The superior part.

Q. What is the form of this surface?—*A.* It is half-moon shaped.

Q. Describe the alæ, or wings?—They consist of a protuberance on the posterior part of each side of the coffin bone; the protuberance, however, is generally bifid; the lower portion which is the largest, is irregular and asperous, and projects in a posterior direction; the upper portion is tubercular, yet smooth; between the divisions of the alæ is a notch, which, at a certain period in the life of the animal becomes a perfect foramen.

Q. What is attached to the irregular surface of the larger division of the alæ?—*A.* The cartilage of the foot.

Q. What is affixed to the tuberculated portion of the alæ?—*A.* The coffin ligaments.

Q. What vessel passes through the notch?—*A.* The lateral artery.

Q. What is there remarkable about the structure of the coffin bone?—*A.* It has a spongy, fragile texture, pervaded in every direction by minute canals for the transmission of blood-vessels and nerves; it differs very essentially from many bones of the body, which, in healthy subjects, are remarkable for compactness and solidity.

NAVICULAR BONE (OS NAVICULARE).

Q. What is the general form and division of the navicular bone?—*A.* It is semi-lunar: its lunated border, however, only forms about one third the circle of its dimensions; it is divided into two surfaces, two borders, and two extremities.

Q. Where is this bone situated?—*A.* At the posterior part of the coffin joint.

Q. Describe the superior and inferior surfaces of the navicular bone.—*A.* The superior surface bears a corresponding aspect to the articulating surface of the coffin bone, having two superficial lateral depressions, with an eminence between them. The inferior surface is also articulatory; and exhibits lateral depressions yet more superficial than the superior; it has also an eminence across the middle, narrower yet more prominent than the former.

Q. Name the tendon which articulates over the inferior surface.—*A.* Tendo perforans.

Q. Describe the borders. — *A.* The borders are limated and straight: the limated is broadest in the centre, and narrows towards the extremities; superiorly it has a smooth narrow strip of surface along the middle, which is adapted to the bevelled portion of the articulatory surface of the coffin bone; the part beneath is fluted and porous, into which is inserted a ligament which connects it with the coffin bone. The straight border is thinner than the opposite one; superiorly it is rough and porous; inferiorly it is smooth and lipped.

Q. What is the form and direction of the extremities of the navicular bone? — *A.* They are obtusely pointed, one directed outward and the other inward.

Q. What ligaments are inserted into the extremities? — *A.* The lateral ligaments.

HIND EXTREMITIES.

Q. What is the situation of the hind extremities? — *A.* They occupy the inferior and posterior parts of the pelvis, and support the posterior parts of the trunk.

Q. How are the bones of the hind extremities divided? — *A.* They are thus divided: Femur, stifle, thigh, hock, leg, pastern, coronet, and foot.

Q. Name the bones comprising these parts. — *A.*

FEMUR.
STIFLE BONE.
Patella.

THIGH BONES.
Tibia and Fibula.

HOCK BONES.
Astragalus, Os calcis.
Cuboid bone,
Three Cuneiform bones.

BONES OF THE LEG.
Metatarsi Magnum.
Metatarsi Parva, two bones (splents).

PASTERN JOINT.
Ossa Sessamoidea (two bones),
Os Suffraginis (pastern).

CORONET.
Os Coronæ.

BONES OF THE FOOT.
Os Pedis and Os Naviculare.

Q. Where is the femur situated? — *A.* Between the pelvis and thigh bones.

Q. How is it divided? — *A.* Into a body and two extremities.

Q. What are the peculiarities of the superior extremity? — *A.* It consists of two parts: a hemispherical, smooth, articulatory head, directed upwards and inwards, and joined to the body by a flattened neck, and exhibiting on its inner side a fissure, into which is fixed the teres or round ligament. The other part is a large irregular projection at the base, and posterior to the same is a deep oval cavity; at the superior part is a roughened crest; inwardly it presents a concave, smooth surface.

Q. What is the proper name of the projection? — *A.* The great external trochanter.

Q. What muscles are inserted into the same? — *A.* The gluteii.

Q. What is inserted into the concave smooth surface? — *A.* The capsular ligament.

Q. What is the form of the body of the os femoris? — *A.* It is cylindrical.

Q. How does it correspond in size and weight with other bones of the body? — *A.* It is the longest and weightiest.

Q. What is the form of inferior extremity? — *A.* It is broad and thick, and has a trochlear prominence and two condyles.

Q. Give a description of the same. — *A.* The articular or pulley-like surface anteriorly consists of a broad, semi-circular groove bounded on either side by a prominence; the condyles much resemble each other, excepting that the external is the thickest, and the internal most projecting; they exhibit prominent, convex, articulatory surfaces; on their sides are rough eminences; between them is a deep fossa; at the base of the external condyle is a pit.

Q. What articulates over the pulley-like surface? — *A.* The patella or stifle bone.

Q. What is inserted into the rough eminences? — *A.* The lateral ligaments.

Q. What occupies the fossa? — *A.* The inter-articular ligament.

Q. What is inserted into the pit? — *A.* The tendon of the extensor pedis.

Q. What is the state of this bone during colthood? — *A.* Extremities are attached to the body of the bone by means of cartilage.

Q. What changes do the extremities undergo just prior to adult life? — *A.* They become consolidated with the body of the bone.

STIFLE BONE (PATELLA).

Q. What is the situation of the patella? — *A.* It is situated on the anterior and inferior extremity of the femur.

Q. What is its general form? — *A.* Quadrangular, convex externally, irregularly concave internally.

Q. How is it divided? — *A.* Into three surfaces and four angles.

Q. Describe the surfaces. — *A.* The anterior surface is convex, yet quite prominent in the centre; it has a roughened surface, and is porous. The superior surface is angular, uneven, and roughened. The posterior surface is articulatory, and unequally divided by an eminence running across it into two shallow concavities, which are adapted to the condyles of the inferior extremity of the femur.

Q. Describe the form of the angles of the patella. — *A.* They are obtuse.

Q. Why is the anterior surface of the bone roughened? — *A.* For the insertion of tendinous and ligamentary attachments.

Q. What is implanted into the uneven and roughened part of the superior border? — *A.* The tendons of the rectus and vasti muscles.

Q. What is inserted into the inferior and lateral angle? — *A.* The ligamentum patella.

Q. What are the connections of this bone? — *A.* It is connected to the inferior portion of the femur by tendinous and capsular ligaments; to the tibia it is connected by similar ligaments.

THIGH BONES (TIBIA AND FIBULA).

In consequence of a horse having a very large femur, and that bone appearing to enter into the composition of the haunch, the tibia and fibula are termed thigh bones, although in man they are termed bones of the leg; the fibula of the horse, however, is a very small, slender bone, affixed to the superior part of the external side of the tibia.

Q. What is the situation of the thigh bone? — *A.* It is situated between the stifle and hock.

Q. What is the form of this bone? — *A.* It is long, straight, prismatic; its superior extremity is larger than the inferior.

Q. What is its direction? — *A.* Oblique in a contrary direction to the femur.

Q. How is the tibia divided? — *A.* Into a body, superior and inferior extremities.

Q. What is the general form of the body? — *A.* It is irregularly triangular, the posterior face is broadest, the anterior angle is rounded, and the sides are roughened.

Q. What is peculiar to the superior extremity of the bone? — *A.* We find two irregular ovoid articulatory surfaces, corresponding to the eminences on the inferior extremity of the femur; these are separated by an acute elevation, and two fossa, into which is inserted the lateral ligament.

Q. Describe the inferior extremity. — *A.* It is flattened, and has two deep articular grooves running in an anterior and posterior direction; its exterior margin is roughened.

Q. What are its connections? — *A.* It connects with the femur and patella superiorly; inferiorly, with the bones of the hock.

FIBULA.

Q. What is the situation of the fibula? — *A.* At the posterior part of the tibia.

Q. How is it connected to the tibia? — *A.* By cartilago-ligamentous substance.

Q. What is the form of the two ends of the bone? — *A.* The superior is bulky, flattened from side to side, and roughened. The inferior is slender and tapering, and extends about half way down the tibia.

BONES OF THE HOCK (TARSUS).

The *tarsus*, or hock, comprises a part of the osseous structure of the horse, that

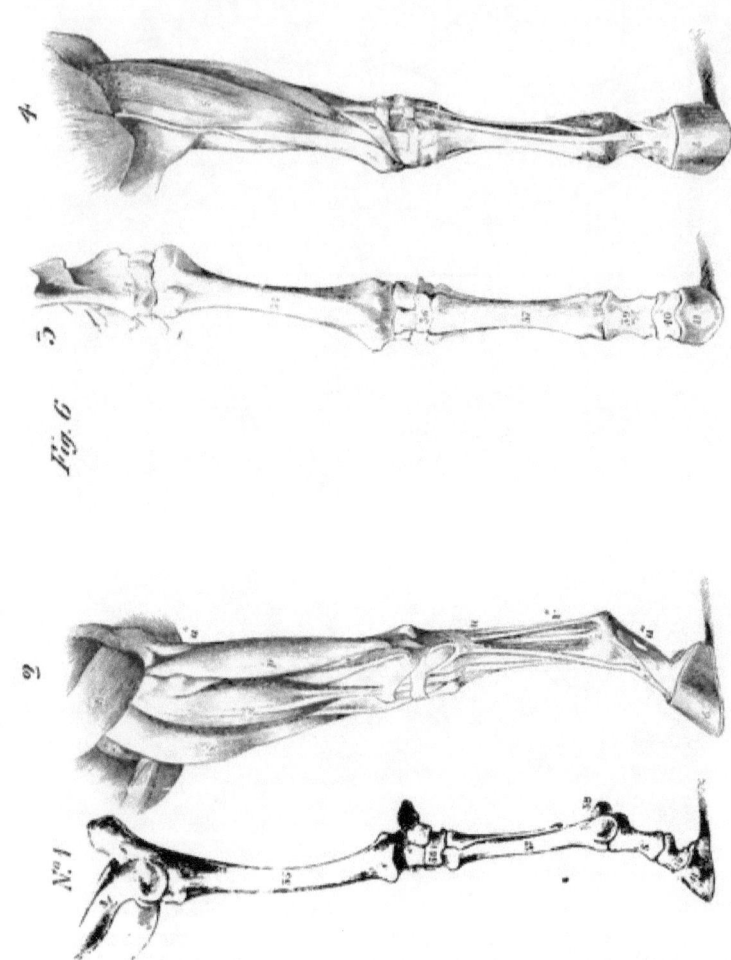

EXPLANATIONS OF FIGURE VI.

NO. 1. — FORE EXTREMITIES.

- f. The Ulna.
- 34. Humerus.
- 35. Radius.
- 36. Carpus.
- 37. Metacarpus.
- 38. Sessamoids.
- s. or 39. Os suffraginis.
- 40. Os coronæ.
- 41. Os pedis.

The above description also answers for No. 3, — the bony structure.

NO. 2. — MUSCULAR STRUCTURE.

LATERAL VIEW OF THE NEAR-FORE EXTREMITY.

- s''. Extensor metacarpi magnus.
- k''. Humero cubital.
- n''. Levator humeri.
- p''. Flexor metacarpi externus.
- x''. x''. Extensor pedis.
- u''. u''. v''. Flexors.
- v''. Flexor tendons.
- y. Extensor tendon.
- z. Suspensory ligament.
- $\&$. The hoof.

NO. 4.

ANTERIOR VIEW OF THE NEAR-FORE EXTREMITY

- s. Extensor metacarpi magnus.
- t. Extensor metacarpi obliquus.
- x''. Extensor pedis.
- y''. Extensor suffraginis.
- $\&$. The hoof.
- 8. Bifurcation of the suspensory ligament.

every veterinary student should aim to be well acquainted with; it is a part that seems to be, in this country, more liable to anchylosis and exostosis than any other region; here is the seat of spavin, and no one can possibly understand the nature of such disease unless he be conversant with the anatomical mechanism of the hock. The hock corresponds to the tarsus or instep of man, and is composed of six bones, viz., os calcis, astragalus, os cuboides, ossa cuneiformis; which comprise three small bones, viz., external, internal, and middle cuneiforme. We shall first consider the os calcis.

Q. What is the situation of the os calcis? — A. It forms the posterior projection known as the point of the hock — the superior and posterior bone of the tarsus.

Q. Give a general description of the bone. — A. Its figure is irregular; presents a body, tuberosity, posterior surface, and base; the body is most bulky at its inferior part; as a whole, it is irregularly convex; concave and expanded at its base, where it presents four surfaces for articulation with the astragalus; the tuberosity is oblong, flattened on each side, and terminates in a rough tubercle, into which is inserted the tendons of the gastrocnemii. It is situated on the superior part of the hock.

THE KNUCKLE BONE (ASTRAGALUS).

Q. What is the situation of the astragalus? — A. It is situated in the superior part of the hock, and is the principal support of the tibia.

Q. How do you distinguish it from other bones? — A. It is readily distinguished by its double pulley-like articulatory surfaces, which consist of two semicircular prominences, having between them a deep groove, well adapted to receive the projection found on the inferior extremity of the tibia.

Q. What is the appearance of the posterior surface? — A. It has four articulatory surfaces, corresponding to those of the os calcis.

Q. What is the appearance of the base or inferior extremity? — A. It has an irregularly flattened articulatory surface, which comes in contact with the large cuneiform bone.

CUBOID BONE (OS CUBOIDES).

Q. What is the situation of the cuboid bone? — A. On the outer part of the hock.

Q. How is the bone divided? — A. Into four surfaces, viz., external, internal, superior, and inferior.

Q. How do you distinguish the external from the internal surface? — A. The external surface is broad, irregular, curved, and roughened; on the other hand, the internal is excavated, and has three articulatory surfaces.

Q. How does the superior surface differ from the inferior? — A. The superior surface has two articulations, with a fossa between them; the inferior surfaces are smaller, and correspond, one to the articulatory head of the splent bone, and the other to the cannon.

LARGE CUNEIFORM BONE (OS CUNEIFORME MAGNUM).

Q. What is the situation of the cuneiform bone? — A. Directly beneath the astragalus.

Q. What is the appearance of this bone? — A. It presents a triangular form; its acute termination being in a posterior direction, it has superior and inferior surfaces, sides, and angles.

Q. How is the superior surface distinguished from the inferior? — A. The superior surface has a uniform articulatory surface, with the exception of a small, rough grove running to its centre, from the outer side, which terminates in a central pit. The inferior surface is rather convex, yet presenting a flat appearance; its posterior angle has an articulatory surface, corresponding to that of the cuboid bone.

Q. What are the articulations of this bone? — A. It articulates with the astragalus, cuboid, middle and small cuneiform bones.

MIDDLE CUNEIFORM BONE (OS CUNEIFORMI MEDIUM).

Q. What is the situation of the middle cuneiform bone? — A. It is situated beneath the large cuneiform.

Q. What is the relative size of the ossa cuneiformia? — A. The one beneath the astragalus is the largest; the middle is the medium; and that at the posterior part of the hock is the smallest.

SMALL CUNEIFORM BONE (OS CUNEIFORMI PARVUM).

Q. What is the situation of the small cuneiform bone? — A. It is situated at the posterior part of the hock.

Q. What are the articulations of this bone? — A. It articulates superiorly with the internal angle of the large cuneiform; anteriorly, with the same angle of the middle cuneiform; posteriorly, with the internal splent bone and cannon.

HIND CANNON (OS METATARSI MAGNUM).

Q. What is the popular name of the hind cannon? — A. Shank-bone.

Q. How does it compare in length with the cannon of the fore extremities? — A. It is about one-sixth part longer than the fore cannon.

Q. Is there any difference in the superior surfaces of the fore and hind cannons? — A. Yes; the superior surface of the fore cannon corresponds to the surfaces of the inferior bones of the carpus; the superior extremity of the hind cannon closely resembles the surfaces of the middle and small cuneiform bones, and also that of the cuboid.

Q. How do the hind and fore cannons differ in conformation? — A. The bone of the *hind* extremity is more circular and prominent, anteriorly, than the forward one.

HIND SPLENTS (METATARSI PARVA).

Q. What is the situation of the metatarsi parva? — A. They are situated at the posterior part of the hind cannon.

Q. How are the hind splents recognized from those of fore limbs? — A. The hind splents are longer than the fore; their bodies are more circular and prominent forward, and the superior extremities correspond to a part of the cuneiform and cuboid bones; while the superior extremities of the forward splents correspond to a portion of the inferior row of the bones of the knee.

We now come to the bones articulating beneath the inferior extremity of the hind cannon, viz., pastern, sessamoid, coronet, coffin, and navicular bones. These, according to the opinion of Mr. Percivall, "so closely resemble their fellows of the fore extremity" that we shall dispense with examinations regarding them, merely remarking that the bones of the hind feet are generally broader in a lateral and posterior direction than those of the fore; the pastern and coronet bones are somewhat longer than their fellows forward.

BONES OF THE EAR.

Q. Name the bones of the ear. — A. Malleus, incus, stapes, and orbiculare.

Q. What is the form of the malleus? — A. It appears to resemble a mallet.

Q. Name the long process or handle. — A. Manubrium.

Q. To what is the manubrium attached? — A. To the membrana tympani.

Q. Describe the form of the incus. — A. It is said to resemble a blacksmith's anvil, but, probably, approaches nearer to the figure of a molar tooth; it has a depression on its surface, which receives the head of the malleus.

Q. Describe the stapes. — A. It resembles in form a common iron stirrup, yet has a more triangular appearance.

Q. With what bone does it articulate? — A. The os orbiculare.

Q. Describe the os orbiculare. — A. It is the smallest bone of the body, not exceeding in size a grain of mustard-seed.

Q. What is its use in the mechanism of the ear? — A. It forms the medium of junction and communication between the incus and stapes, and facilitates the motions of the latter bones.

BONE OF THE TONGUE (OS HYOIDES).

Q. What is the situation of the os hyoides? — A. It is located at the root of the tongue, at the anterior part of the larynx.

Q. How is the bone divided? — A. Into a body and four horns.

Q. What is the form of the body? — A. In shape, it resembles a spur, consisting of neck and branches; the neck is inserted into the root of the tongue, and the branches are in a posterior direction, embracing the superior border of the thyroid cartilage.

Q. What is the appearance of the horns? — A. There are two long and two short horns; the short, or inferior, ascend obliquely from their articulations with the body of the bone, and terminate in oblong, smooth extremities. The long or superior horns constitute two long, flattened, thin bones, extending backward in a horizontal direction from the summits of the inferior horns.

Q. What are the connections of the os hyoides? — A. It is connected with the temporal bone, larynx, pharynx, tongue, and some of the muscles of the neck.

OF THE TEETH.

Q. How many teeth do we find in the jaws of the adult horse? — A. Forty. In the mare, however, the canine teeth are generally imperfect or undeveloped.

Q. How are the teeth divided? — A. Into three classes, viz.: incisors, or nippers; molars, or grinders; canini, or tusks.

A. Enumerate each class. — A. There are twelve incisors, twenty-four molars, and four canine.

Q. Is there anything peculiar about the development of horses' teeth? — A. Yes; the teeth with which the animal is furnished during colthood are termed temporary, and are generally shed ere the animal arrives at the age of five; the temporary teeth are twenty-four in number, twelve incisors and twelve molars; they differ from what is termed the "permanent set," in being smaller and whiter, and in having necks or contractions at the superior part of the fang, and the eminences on their face are quite sharp. The converse is the case with regard to the permanent teeth.

Q. What is the popular theory regarding the periods of cutting the teeth? — A. A foal is said, at birth, to be in the act of cutting twelve molars, three on each side of the jaw bone; at this time, there is no appearance of incisors; and when they do appear, which period will be about the second or third week from birth, sometimes sooner, the front incisors of the upper jaw are the first to show themselves, and between the fourth and fifth week, they are succeeded by the middle incisors; the side or lateral incisors make their appearance between the sixth and tenth month. The animal is then said to have a full set of temporary teeth. After the animal has attained his first year, the fourth molars make their appearance. Between the period of the first and second years, the fifth molars, in each side of the jaw, are apparent. Between the second and third years, the front permanent incisors displace the temporary, and, at the same time, the first temporary molars are shed, and replaced by the permanent. Between the third and fourth years, the middle temporary incisors are succeeded by the permanent, and about the same time the second temporary molars are shed. During the interval of the fourth and fifth years, the lateral permanent incisors appear; the sixth and last, permanent molars are up, and then the tusks also appear. At this period the horse is said to have a full mouth; a complete set of permanent teeth.*

We have now arrived at an era (or re-

---

* On this side the Atlantic we are not in possession of any reliable information as regards the periods of cutting and shedding teeth; we have to depend entirely on English authority. Their theory is, that the age of a race-horse shall be reckoned from the month of May in the year of his birth, without any inquiry whatever as to the season, month, or day of foaling; so that the produce of January are actually four months older than by reckoning, or as their ages appear on the calendar, and these are called early foals; whereas those foaled in March are denominated late. These data are more arbitrary than truthful; may suit the convenience of English turfmen, but will not pass current among our breeders, — who, generally, pay particular attention to the time of foaling, and date the birth of the colt accordingly.

markable period) in the age of the horse; have briefly considered a series of changes which the teeth of a colt undergo, up to the period of maturity, and shall now turn our attention to the changes observed in the process of wear and tear of the permanent teeth.

REMARKS ON THE CHANGES WHICH A HORSE'S TEETH UNDERGO.

The nippers or front teeth of a full-mouthed horse, just having shed all the temporary ones, present a beautiful appearance: the contrast between the lily whiteness of the teeth, and the rose-tinted color of the gums and their membranes, are never so much the subject of admiration as at this period.

Teeth, when first cut, present a sharp border externally, from which a gradual depression commences until the internal border is reached; in the course of about a year, in consequence of friction on the external, and growth of the internal, the surface presents two elliptical enamelled rims, one of which borders the face of the tooth, the other encircles the depression or pit.

Within this pit is a black incrustation, which is denominated "bean" or "mark;" at a period of about three years from the time of cutting the permanent teeth, the pit or cavity is consolidated or filled up, and the surface of the tooth is worn down so as to present a comparatively smooth one. We must not expect, however, to find the face of the teeth uniform; for cribbers, and voracious feeders, deface the surfaces very much, which gives to the teeth the appearance of age. Still, a good judge, who takes into consideration not only the appearances of surfaces, but also the form and direction of the teeth themselves, is not apt to be deceived regarding the age of a full-mouthed horse.

Pessina, from whose work Mr. Percivall quotes, concludes that—

"At the age of eight (in most horses), the disappearance of the marks is perfect: the teeth are all oval, the central enamel upon the face is triangular, and nearer to the outward than the inward border, and the cavity of the tooth appears within the outward border like a yellowish band carried from one side to the other.

"At nine years, the front teeth appear round, the middle and the lateral contract their oval faces, and the central enamel diminishes and approaches the inward border.

"At ten, the middle teeth become round, and the central enamel has approximated the inward border and is rounded.

"At eleven, the middle teeth are rounded, and the central enamel is almost worn off the posterior incisors.

"At twelve, the lateral teeth are rounded, the central enamel has quite disappeared: the yellow band has grown wider, occupies the centre of the face of the tooth, and the central enamel continues in the teeth of the upper jaw.

"At thirteen, all the incisors are rounded, the sides of the front teeth spread out, and the central enamel continues in the upper jaw, but is round and approaches the inward border.

"At fourteen, the faces of the front incisors put on a triangular appearance, the middle grow out at their sides, and the central enamel of the upper teeth diminishes, but still exists.

"At fifteen, the front teeth have become triangular, the middle enter upon that figure, and the central enamel of the upper jaw is still visible.

"At sixteen, the middle are triangular, the lateral commence that shape, and the enamel of the upper teeth has disappeared.

"At seventeen the triangular figures of the posterior jaw are completed; but their triangles are equilateral until the eighteenth year. Then their sides lengthen in succession from the front to the lateral teeth, in such a manner that—

"At nineteen, the front teeth are flattened from side to side;

"At twenty, the middle incisors have taken on the same shape; lastly—

"At twenty-one the lateral teeth are also flattened."

Professor Pessini* "systematically divides the lifetime of the horse, which he computes at thirty years, into six periods, that take their rise from and are determined by an equal number of changes the teeth naturally undergo, in regular succession.

"The first period is that during which the animal retains all or any of his milk teeth; it extends from birth to the fifth year.

"The second period includes the sixth year, and continues so long as the marks remain visible upon the faces of the posterior incisors; which is generally about three years.

"In many instances, however, and especially among horses that have been kept at pasture, the faces of the front teeth, and sometimes those of the middle, are worn off earlier.

"The third period is that during which the teeth retain the oval form. As the pits and marks degenerate, the face of the tooth slowly and gradually undergoes a deviation of figure, from that of a pretty regular ellipsis, whose long to its short axis bears the proportion of six to three, to an irregular one, in which these proportions are as five to four. This period requires, on an average, the space of six years for its completion; the front teeth enter it in the seventh and conclude it at the expiration of the twelfth; the middle pass through it one year later; and the lateral, or side teeth, one year later still.

"In the fourth period the faces of the teeth assume a circular figure, and hence have been denominated *round*. At the commencement of this period, the breadth of the face to its thickness is as 5 to 4; at the conclusion, it measures in an inverse ratio, as 4 to 5; about the middle of it, the diameters are equal. This period also endures six years; so that the front teeth, which enter it in the thirteenth year, complete it by the expiration of the eighteenth; the middle follow one year later; the lateral, one year later still.

* See Percivall's Lectures.

"During the fifth period, the face of the teeth deviates by slow degrees from the round, and passes into the triangular state. In the beginning, its thickness exceeds its breadth as 5 does 4; in the end, as 6 does 3. It is the professor's opinion, yet unconfirmed by experience, that this period, likewise, on an average, includes a space of six years; the front teeth, therefore, complete it with the twenty-fourth, the middle with the twenty-fifth, and the lateral with the twenty-sixth years.

"The sixth and last period is one, in the course of which an additional angle is projected from the anterior or inferior part of the tooth; Pessina distinguishes it by the epithet biangular; he has never met with a horse that had lost his teeth from age; but he has seen their faces elliptical contrariwise, looking outwards or forwards. This period is unlimited.

"In the anterior, or upper jaw, the *marks* disappear from the front teeth in the course of the ninth year; from the middle in the tenth; and from the lateral in the eleventh.

"What progress these upper teeth have not made in transformation during the second period, equivalent with the posterior, they gain it in the third; notwithstanding the depth of pit, their proportions are then the same. They continue three years longer in the second, and consequently are only three in the third period; so that, by the twelfth year, the third period is completed by the front upper teeth, and so on. During the fourth, fifth, and last periods, the changes are alike, and equally perceptible in either jaw.

"So far, the upper teeth are entitled to an equal share of our regard; though, in the generality of cases, they need not be inspected. In such a remarkable manner the lateral teeth of the upper jaw wear away so that they often appear as if notched or indented.

"In regard to the tusk or tush, Pessina remarks that he has found the least regularity in its changes of any tooth. The very facts that the tushes are not in all horses cut at the same age, that they have little or no attrition against each other, and that they are worn by the tongue and food, sometimes more, at others less, should lead us to draw conclusions from them with great caution; in fact, as indications of age, they can only be trusted to when they accord with the incisors. The tush or tusk makes its appearance by the fifth, and is completely evolved by the sixth year. In the seventh, the apex of the cone is worn off. In the eighth, its furrows grow shallow; in the ninth they are obliterated. Then the apex gradually wears away, in the twelfth year it becomes round; from which time, though it gradually becomes shorter, its shape varies but little. But it is not uncommon to see the tush blunted like an acorn in the ninth year, nor to find it still pointed in the sixteenth year.

"Pessina concludes his account of the changes to which the teeth are subject, by observing, that, as they are dependent on wear, which is no law of nature, but an effect of mechanical and accidental causes, they cannot, but under certain limitations, be implicitly relied on."

We are now supposed to be in possession of some of the most important facts tending to elucidate the changes which the teeth undergo; and, in view of making ourselves more conversant with this subject, we shall re-commence our examinations, for it is a matter of the highest importance that a veterinary surgeon shall understand the method of ascertaining a horse's age.

EXAMINATIONS ON THE TEETH.

Q. Does the evolution of the tush always indicate that the animal is five years of age? — *A.* No. It has been seen between the third and fourth years.

Q. Which teeth do you place the most reliance on in ascertaining the age of a horse? — *A.* The side or lateral of the lower jaw. They make their appearance last; their pits are the last to disappear; after the age of eight or nine, however, the pits in the incisors of the upper jaw are also indicative of age; they, being deeper, of course remain some time after all vestiges of the same have disappeared in the lower jaw.

Q. In adult life is there any continued accretion or after-growth of the teeth? — *A.* Yes. If it were not so, the animal would, in course of time, have to gather food, and grind the same with his gums; for, according to the law of *wear* and *tear*, destruction of the instruments — grinders of food — must more or less regularly take place.

Q. What changes take place as the horse advances in age, in the inclination of the incisors? — *A.* They acquire a horizontal direction.

Q. How is this change of direction compensated for in the grinders? — *A.* The faces of the latter are worn down by friction, and thus the nippers come in contact.

Q. Are there not times when the consumption of the faces of the teeth, by friction, is not in proportion to growth, in issue from the socket? — *A.* Yes.

Q. What is the result? — *A.* The faces of the grinders do not come in contact, and the food is, consequently, imperfectly masticated.

Q. How is this rectified? — *A.* By sawing off the nippers to their natural length.

Q. Taking it for granted that there is a time when the teeth cease to grow, how do you account for the lengthy teeth observed in aged horses? — *A.* The fang shrinks, and is carried upward in the lower and downward in the upper jaw, and the gums also shrink; thus we get length of teeth.

Q. What are the general appearances of age, unconnected with the teeth? — *A.* The muscles of the head and face condense, and give to the same a lean appearance; the cavities above the eyes are deep; the gums and palate become pale and callous; the submaxillary space is capacious, and gray hairs make their appearance in various places; the neck appears small and wiry, the withers sharp, the back curves, and the limbs appear sinewy.

MYOLOGY.

PRELIMINARY REMARKS ON THE MUSCLES.

To the naked eye, the muscles appear to be composed of fasciculi, or bundles of fibres, which are arranged side by side in the direction in which the muscle is to act, and which are united by areolar tissue. These fasciculi when separated appear like simple fibres, but when examined under a microscope are found to be themselves fasciculi, composed of minuter fibres, bound together by delicate filaments of areolar tissue. By carefully separating these, we may obtain the ultimate muscular fibre. This fibre exists under two forms, the striated and non-striated. The former is chiefly distinguished by the transversely-striated appearance which it presents. The non-striated consist of a series of filaments which do not present transverse markings. At an early stage of the development of muscular fibre, however, there is no difference in the forms of either striated or non-striated. Both are simple tubes, containing a granular matter in which no definite arrangement can be traced, yet presenting enlargements occasioned by the presence of nuclei. But, whilst the striated fibre goes on in its development, until the cells of the fibrillæ are fully produced, the non-striated fibre retains throughout life its original embryonic condition; the contents of the tube remaining granular. The non-striated muscular fibre is the kind of structure proper to the muscular coat of the alimentary canal, bladder, uterus, trachea, bronchial tubes, etc. They seem to be arranged in a parallel manner into bands or fasciculi, without any very definite points of attachment. On the other hand, striated muscular fibre has attachments to its extremities of fibrous tissue, through the medium of which it exerts its contractile power on the part it is destined to move.

At the truncated extremity of the striated muscles we find tendons. To the ordinary observer, tendons appear to unite abruptly with muscular fibre; but this is not the case, for tendinous fibres are distributed over the whole muscle, crossing it diagonally in both directions, so as to form a double-spirally extensible sheath; the tendinous fibre finally collects at the extremity of a muscle, and forms the tendon.

Each muscle is surrounded by cellular membrane, which dips into its substance, and, by means of the fat which its cells contain, lubricates the parts, and thus guards against friction.

A TABLE OF THE NAMES AND NUMBER OF MUSCLES, DIVIDED INTO REGIONS.

SUBCUTANEOUS REGION (BENEATH THE SKIN).

1. Panniculus carnosus.

AURICULAR REGION (MUSCLES OF THE EAR).

2. Attollentes maximi.
3. Attollentes anteriores.
4. Attollentes posteriores.
5. Anterior conchæ.
6. Posterior conchæ.
7. Retrahentes externi.
8. Retrahentes interni.
9. Abducens vel deprimens auris.

PALPEBRAL REGION (MUSCLES OF THE EYELIDS).

10. Levator palpebræ superioris.
11. Orbicularis palpebrarum.

OCULAR REGION (MUSCLES OF THE EYE).

12. Levator palpebræ superioris internus.
13. Levator oculi.

14. Depressor oculi internus.
15. Abductor oculi externus.
16. Adductor oculi internus.
17. Obliquus superior.
18. Obliquus inferior.
19. Retractor oculi.

ANTERIOR MAXILLARY REGION (MUSCLES OF THE NOSE AND FACE).

20. Zygomaticus.
21. Levator labii superioris alique nasi.
22. Dilator naris lateralis.
23. Nasalis longus labii superioris.
24. Caninus vel levator anguli oris.
25. Buccinator.
26. Depressor labii inferioris.
27. Levator menti.
28. Dilator narium anterior.
29. Nasalis brevis labii superioris.
30. Depressor labii superioris.
31. Orbicularis oris.

POSTERIOR MAXILLARY REGION (MUSCLES OF THE HEAD AND CHEEKS).

32. Temporalis.
33. Masseter.
34. Stylo-maxillaris.
35. Pterygoideus internus.
36. Pterygoideus externus.

HYOIDEAL REGION (MUSCLES BETWEEN THE BRANCHES OF THE LOWER JAW).

37. Digastricus.
38. Mylo-hyoideus.
39. Genio-hyoideus.
40. Hyoideus magnus.
41. Hyoideus parvus.
42. Stylo-hyoideus.

GLOSSAL REGION (MUSCLES OF THE TONGUE).

43. Hyo-glossus longus.
44. Hyo-glossus brevis.
45. Genio-hyo-glossus.
46. Lingualis.

PHARYNGIAL REGION (MUSCLES ABOUT THE PHARYNX).

47. Hyo-pharyngeus.
48. Palato-pharyngeus.
49. Stylo-pharyngeus.
49½. Constrictor pharyngis, anterior.
50. Constrictor pharyngis, medius.
51. Constrictor pharyngis, posterior.

LARYNGEAL REGION (MUSCLES ABOUT THE LARYNX).

52. Hyo-thyroideus.
53. Crico-thyroideus.
54. Crico-arytenoideus posticus.
55. Crico-arytenoideus lateralis.
56. Thyro-arytenoideus.
57. Arytenoideus.
58. Hyo-epiglottideus.

PALATINE REGION (MUSCLES OF THE PALATE).

59. Tensor palati.
60. Circumflexus palati.

MUSCLES OF THE NECK.

HUMERO-CERVICAL REGION (MUSCLES SITUATED ON THE UPPER AND LOWER PARTS OF THE NECK).

61. Rhomboideus longus.
62. Levator humeri.

LATERAL CERVICAL REGION (SIDE OF THE NECK).

63. Splenius.
64. Complexus major.
65. Trachelo-mastoideus.
66. Spinalis colli.

SUPERO-CERVICO-OCCIPITAL REGION (MUSCLES SITUATED ABOVE THE HEAD).

67. Complexus minor.
68. Rectus capitis posticus, major.
69. Rectus capitis posticus, minor.
70. Obliquus capitis, superior.
71. Obliquus capitis, inferior.

INFERIOR CERVICAL REGION (MUSCLES SITUATED IN THE ANTERIOR PART OF THE NECK).

72. Sterno maxillaris.
73. Sterno-thyro-hyoideus.
74. Sub-scapulo-hyoideus.
75. Scalenus.
76. Longus colli.

INFERIOR CERVICO-OCCIPITAL REGION (MUSCLES BENEATH THE BASE OF THE ATLAS).

77. Rectus capitis anticus, major.
78. Rectus capitis anticus, minor.
79. Obliquus capitis, anticus.

MUSCLES OF THE CHEST.

DORSO SCAPULAR REGION (MUSCLES SITUATED ABOUT THE SHOULDER BLADE).

80. Trapezius.
81. Latissimus dorsi.
82. Rhomboideus brevis.

PECTORAL REGION (MUSCLES SITUATED IN FRONT OF THE BREAST BONE).

83. Pectoralis transversus.
84. Pectoralis magnus.
85. Pectoralis parvus.

COSTAL REGION (MUSCLES SITUATED EXTERNAL AND INTERNAL TO THE RIBS).

86. Serratus magnus.
87. Intercostales externi.
88. Intercostales interni.

STERNAL REGION (MUSCLES OF THE BREAST BONE).

89. Lateralis sterni.
90. Sterno-costales externi.
91. Sterno-costales interni.

DORSO-COSTAL REGION (MUSCLES ON THE SIDES AND UPPER PART OF THE CHEST).
  92. Superficialis costarum.
  93. Transversalis costarum.
  94. Levatores costarum.

DORSAL REGION (MUSCLES OF THE BACK, ANTERIOR TO THE LUMBAR VERTEBRÆ).
  95. Longissimus dorsi.
  96. Spinalis dorsi.
  97. Semi spinalis dorsi.

DIAPHRAGMATIC REGION.
  98. Diaphragm or midriff.

MUSCLES OF THE ABDOMEN.

LUMBAR REGION (MUSCLES OF THE LOINS).
  99. Semi spinalis lumborum.
  100. Intertransversales lumborum.
  101. Sacro lumbalis.
  102. Psoas Magnus.
  103. Iliacus.
  104. Psoas parvus.

ABDOMINAL REGION (MUSCLES OF THE ABDOMEN).
  105. Obliquus externus abdominis.
  106. Obliquus internus abdominis.
  107. Transversalis abdominis.
  108. Rectus abdominis.

ANAL REGION (MUSCLES OF THE ANUS).
  109. Retractor ani.
  110. Sphincter ani.

GENITAL REGION (MUSCLES OF THE MALE ORGANS OF GENERATION).
  111. Cremaster.
  112. Erector penis.
  113. Triangularis penis.
  114. Accelerator urinæ.

The muscles in the genital regions of the female are named: Erector Clitoridis, Sphincter Vaginæ.

COCCYGEAL REGION (MUSCLES OF THE TAIL).
  115. Erector coccygis.
  116. Depressor coccygis.
  117. Curvator coccygis.
  118. Compressor coccygis.

MUSCLES OF THE FORE EXTREMITIES.

EXTERNAL SCAPULAR REGION (MUSCLES ON THE OUTSIDE OF THE SHOULDER BLADE).
  119. Antea-spinatus.
  120. Postea-spinatus.

INTERNAL SCAPULAR REGION (MUSCLE ON THE INSIDE OF SHOULDER BLADE).
  121. Subscapularis.

POSTERIOR SCAPULAR REGION (MUSCLES BEHIND THE SHOULDER BLADE).
  122. Teres major.
  123. Teres minor.

ANTERIOR HUMERAL REGION (MUSCLES IN FRONT OF THE OS HUMERI).
  124. Coraco-humeralis.
  125. Flexor brachii.
  126. Humeralis externus.

POSTERIOR HUMERAL REGION (MUSCLES BEHIND THE OS HUMERI).
  127. Caput magnum.
  128. Caput medium.
  129. Caput parvum.
  130. Anconeus.
  (Triceps Extensor Brachii.)

MUSCLES OF THE ARM AND FORE LEG.

ANTERIOR BRACHIO-CRURAL REGION (MUSCLES IN FRONT OF THE ARM).
  131. Extensor metacarpi magnus.
  132. Extensor pedis.
  133. Extensor suffraginis.
  134. Extensor metacarpi obliquus.

SUPERFICIAL POSTERIOR BRACHIO-CRURAL REGION (MUSCLES ON THE EXTERNAL SIDE OF THE ARM).
  135. Flexor metacarpi externus.
  136. Flexor metacarpi medius.
  137. Flexor metacarpi internus.
  138. Flexor accessorius sublimis.

DEEP POSTERIOR BRACHIO-CRURAL REGION. (THESE MUSCLES ARE SITUATED BENEATH THE FORMER.)
  139. Flexor pedis perforatus.
  140. Flexor pedis perforans.
  141. Flexor pedis accessorius profundus.
  142. Lumbrici, anterior.
  142½. Lumbrici, posterior.

MUSCLES OF THE HIND EXTREMITIES.

GLUTEAL REGION (MUSCLES OF THE SUPERIOR PART OF THE QUARTER).
  143. Gluteus externus.
  144. Gluteus maximus.
  145. Gluteus minimus.

PELVI-TROCHANTERIAN REGION (MUSCLES SITUATED AT THE UPPER PART OF THE THIGH BONE).
  146. Pyriformis.
  147. Obturator externus.
  148. Obturator internus.
  149, 150. Gemini.

ANTERIOR ILIO-FEMORAL REGION (MUSCLES SITUATED AT THE FORE PART OF THE HAUNCH).
  151. Tensor vaginæ.
  152. Rectus.
  153. Triceps vasti.
  154. Rectus parvus.

INTERNAL ILIO-FEMORAL REGION (MUSCLES SITUATED AT THE INNER PART OF THE HAUNCH).
  155. Sartorious.

156. Gracilis.
157. Pectineus.
158. ⎫ Adductor brevis.
159. ⎬ Adductor longus.
160. ⎭ Adductor magnus.

POSTERIOR ILIO-FEMORAL REGION (MUSCLES ON THE OUTER AND POSTERIOR PART OF THE HAUNCH).
161. Biceps abductor.
162. Abductor tibialis.

MUSCLES OF THE THIGH AND LEG.

ANTERIOR AND FEMORO-CRURAL REGION (MUSCLES IN FRONT OF THE TIBIA).
163. Extensor pedis.
164. Peroneus.
165. Flexor metatarsi.

SUPERFICIAL POSTERIOR FEMORO-CRURAL REGION (MUSCLES IN THE REGION OF THE HOCK).
166. Gastrocnemius externus.
167. Gastrocnemius internus.
168. Plantaris.

DEEP POSTERIOR FEMORO-CRURAL REGION (MUSCLES WHICH ARE FOUND BENEATH THE FORMER).
169. Popliteus.
170. Flexor pedis.
171. Flexor pedis accessorius.

The muscles of the internal ear are named:
Laxator tympani, ... 2
Membrana " 2
Tensor " 2
Stapedius, ... 2
                                    ——
Total,   8

RECAPITULATION.

We shall now recapitulate, as regards what has preceded, in reference to the number of muscles; for there exist various opinions regarding the same. It may be proper for us to bear in mind, however, that VETERINARY SCIENCE, here, is yet in its infancy; and it is well known to some practitioners, that there are several muscles which remain to be named by some future compiler of veterinary literature. But for all practical purposes we know enough of the anatomy of the horse. The industrious individual, however, who not only desires to make himself conversant with what is already known, but aims to improve in the future, will not rest satisfied with the productions of his predecessors. To such an one we bow with due deference, and encourage him to proceed in the work of progression. There is a fine field for exploration, and a discerning public are ready and willing to crown the industrious laborer with the laurel of merit.

In the preceding table, the number of muscles, including those marked 49 1-2 and 142 1-2, appears to be 173; among these are ten single ones, which are thus expressed:

Whole number, ... 173
Deduct single ones, ... 10

Pairs, ... 163
    Multiply by 2

Single muscles, ... 326
Add muscles of the internal ear, four pairs, 8
                                    ———
                                    334
Single muscles, as above added, . 10
                                    ———
                                    344

It appears, therefore, that there are in the system of the horse three hundred and forty-four muscles.

It should be borne in mind, that in the preceding classification all are considered as muscles. Among them are found tendons, which are component parts, or rather appendages, to the same. Mr. Percivall says there are, in the horse, 151 pairs, and 10 single muscles; add the four pairs of the internal ear, which he has omitted in the calculation, and we get 155 pairs. On page 72, "Hippopathology," the number of muscles is,—

                                    312
Add muscles of the ear, omitted, . 8
                                    ———
                                    320
The author's estimate, . 344
                                    ———
Difference, ... 24

Probably the above author considers the "24" as tendons.

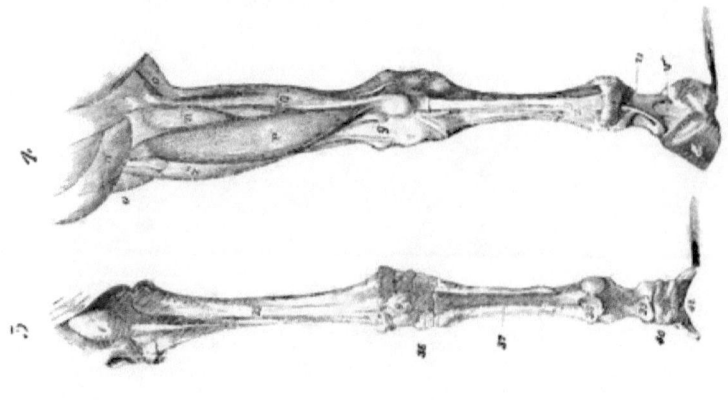

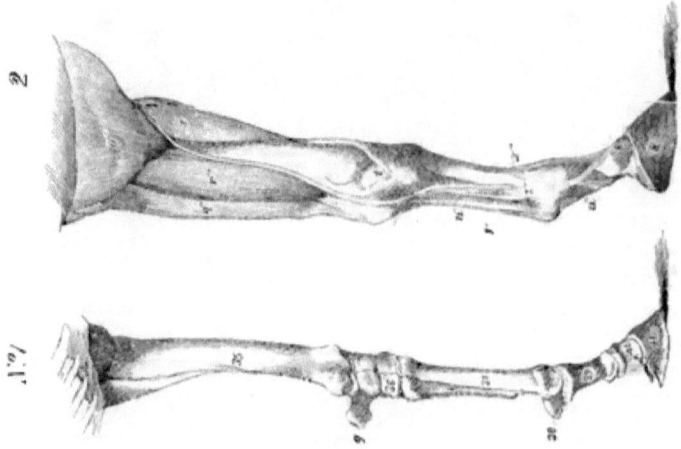

Fig. 1

EXPLANATION OF FIGURE VII.

NO. 1.—OSSEOUS STRUCTURE.

35. Radius.
g. Trapezium.
36. Lower row of the carpal bones.
37. Metacarpi magnum.
38. Sessamoids.
39. Os suffraginis.
40. Os coronæ.
41. Os pedis.

NO. 2.— MUSCULAR STRUCTURE.

INTERNAL VIEW OF THE NEAR-FORE LEG.

o''.   Pectoralis transversalis.
q''.   Flexor metacarpi medius.
r''.   "   "   internus.
s''.   Extensor metacarpi magnus.
t''.   "   "   obliquus.
u''. u''. Flexors pedis — perforatus et perforans.
v''.   Suspensory ligament.
x''.   Extensor pedis.
z. 8. Bifurcation of the suspensory ligament.
8.   The hoof.

NO. 3.

The description of No. 1 answers also for No. 3. The letter f. is intended to point out the location of the ulna, into which is inserted the triceps.

g. Region of the carpus.

NO. 4.

n'''.   Triceps extensor brachii.
o'''.   Pectoralis transversalis.
o'. l'''. Flexor metacarpi externus.
q'''.   Flexor metacarpi medius.
u'''.   Fleshy belly of the perforatus and perforans.
x'''.   Extensor pedis.
y'''.   Extensor suffraginis.
z'''. u'''. v'''. Flexor tendons.
$\&$.   Hoof.

A TABULAR SYNOPSIS OF THE NUMBER, NAME, REGION, SITUATION, INSERTION, AND ACTION OF THE MUSCLES OF THE HORSE.

NO.	NAME.	REGION.	SITUATION.	INSERTION.	ACTION.
1	Panniculus carnosus.	Subcuta-neous.	Beneath the skin.	To cellular substance, muscles, bones, and ligaments.	To corrugate the skin.
2	Attollentes, maximi.	Auricular	Subcutaneous at the ear.	To the temporal fascia and triangular cartilage.	To adduct the ear.
3	" anteriores.	"	"	To the zygoma, temporal muscles, and triangular cartilage.	To depress the ear.
4	" posteriores.	"	"	To sagittal suture, temporalis, and concha.	To elevate the ear.
5	Anterior conchæ.	"	"	To the triangular cartilage and concha.	To erect the ear.
6	Posterior.	"	"	Nearly the same.	To rotate the ear backwards.
7	Retrahentes externus.	"	Upon the side of the poll.	To the ligamentum nuchæ, obliquus capitis superior, dorsum conchæ, root of conchæ, and annular cartilage.	To retract, draw down, and rotate the cavity of the ear backwards.
8	" internus.	"	"		To adduct and depress the ear.
9	Abducens vel depromens.	"	At the root of the ear.	To the surface of parotid gland and to the outer and fore part of the conchæ.	
10	Levator palpebræ superioris.	Palpebral.	Above the orbit.	To the orbital portion of the ossa unguis et frontes; to the palpebral ligament and the skin of both lids.	To shut the eyelids.
11	Orbicularis palpebrarum.	Ditto.	Within the eyelids.	To the walls of the optic foramen and the border of the upper eyelid.	To raise the upper eyelid.
12	Levator palpebræ.	Ocular.	Between the eye-ball and the orbit.	To the optic foramen and to four opposite points, equidistant one from another of the sclerotic coat.	To elevate the eyeball.
13	Levator oculi.	"	Within the orbit, at equal distances from one another.	Ditto.	To turn the eyeball downwards.
14	Depressor oculi.	"	Ditto.	Ditto.	To turn it outwards.
15	Abductor "	"	Ditto.	Ditto.	To turn it inwards.
16	Adductor "	"	Ditto. [orbit.	Ditto.	To turn the globe obliquely.
17	Obliquus superioris.	"	Inner and upper cavity of	To the os unguis and lower and outer side of the sclerotica.	
18	Obliquus inferioris.	"	Inferior part of the orbit.		
19	Retractor oculi.	"	Posterior to the eye-ball.	To the edge of the optic foramen and posterior part of the globe of the eye.	To draw the eye within the socket.

ANATOMY AND PHYSIOLOGY OF

No.	NAME.	REGION.	SITUATION.	INSERTION.	ACTION.
20	Zygomaticus.	Anterior maxillary	Along the centre of the side of the face.	To the anterior part of the zygoma, cellular tissue, and the angle of the mouth.	Assists in retracting the angle of the mouth.
21	Levator labii superioris.	"	Just above the preceding muscle.	To the subcutaneous surfaces of the nasal and frontal bones; to the skin of the nose and false nostrils; to the side of the upper lid and angle of the mouth.	To assist in the retraction of the upper lid and angle of the mouth, and in the dilatation of both the true, and false, nostril.
22	Dilator naris lateralis.	"	Upon the side of the face.	To the zygoma, superior maxilla; spreads upon the side of the nostril and the supra-lateral parts of the upper lip.	To dilate the nostril and retract the upper lip.
23	Nasalis longus labii sup.	"	Upon the upper part of the side of the face.	To a long depression at the junction of the superior maxillary and malar bones, and to the middle of the anterior part of the upper lip.	To raise and corrugate the upper lip, and in some degree assist in the dilation of the false nostrils.
24	Caninus vel levator anguli oris.	"	Fore part of the side of the face.	To a depression on the side of the fore angle of the superior maxillary bone; to the alveolar processes of the lower jaw; to the buccal membrane; to the side of the lips and angle of the mouth.	To render the buccal membrane tense, and also to assist in elevating the angle of the mouth and side of the lip.
25	Buccinator.	"	In the space between the superior and inferior jaw, on the side of the face.	To the border of the lower jaw, near the last molar tooth; to the tuberosity of the superior maxilla; to the outer walls of the alveolar cavities of the molars; to the angle of the mouth and to the buccal membrane.	To aid in tightening the buccal membrane, and retract the angle of the mouth.
26	Depressor labii inferioris.	Posterior maxillary	Along the side of the lower jaw.	Blended with the buccinator to the tuberosity of the superior maxilla, and the superior border of the inferior maxilla, and to the infero-lateral parts of the interior of the lower lips.	To depress the lips.
27	Levator menti.	"	Under the chin.	To the alveolar processes and inferior max-	To draw forward and

THE HORSE.

28 Dilator naris anterior.	In front and between the nostrils.	To the peak and centre of the ossa nasi, and to the anterior surface of the alæ or broad cartilages of the nose.	To dilate the nostrils.
29 Nasalis brevis labii superioris.	Behind the external nares.	To the superior and anterior maxillary bones, and to their suture; to the ossa nasi, and to the nose and skin of the false nostril.	To strengthen and limit the calibre of the nasal cavity at this point.
30 Depressor labii superioris.	Side of the upper jaw.	To the alveoli of the lateral and middle incisors, extending also to the tusk; to the glandular substance of the upper lip, and to the inferior nasal cartilages.	To draw down the side of the lip, and with it the nasal cartilage, and thereby have some effect in dilating the nostril.
31 Orbicularis oris.	Within the border of the lips.	To the glandular substance and skin of the lips, and more particularly at the commissures, where the fibres coming from both lips cross one another.	To close the lips.
32 Temporalis.	Upon the parietal and temporal bones.	To the occipital, temporal, parietal, frontal bones, and to the coronoid process of the inferior maxilla.	To raise the lower jaw and close the same.
33 Masseter.	Superior & inferior maxillary Ditto. Prominent. On the side of the cheek.	To the zygomatic ridge, roughened border of the angle of the lower jaw, and to the contiguous parts of the surface of the jaws.	It co-operates with the temporalis in approximating the lower against the upper jaw.
34 Stylo-maxillaris.	Behind the lower jaw.	To the styloid process of the occipital bone and to the angle of the lower jaw.	To depress and draw the jaw backward, and aid in opening the mouth.
35 Pterygoideus internus.	On the inner side of the jaw, opposite the masseter.	To the pterygoid process; sphenoid, palate, and superior maxillary bones, and around the angle of the jaw.	To produce lateral motion in the lower jaw.
36 Pterygoideus externus.	Above and behind the former.	To the wings of the sphenoid bone, and to the inner side of the jaw, at the root of the condyle.	To assist in the elevation of the jaw and draw it forwards.

raise the prominence of the chin, and with it the under lip.

NO.	NAME.	REGION.	SITUATION.	INSERTION.	ACTION.
37	Digastricus.	Hyoideal	Along the inner side of the lower jaw.	To the styloid process of the occipital bone, and to the side of the jaw, midway between the angle and the symphysis.	To facilitate, or perhaps limit, action of the os hyoideus.
38	Mylo-hyoideus.	"	Distributed along the side of the jaw.	To the roots of the alveolar processes, and the side of the jaw, and to the os hyoideus.	To draw the os hyoideus forwards and upwards, and thereby raise the tongue within the mouth.
39	Genio-hyoideus.	"	Above the preceding muscle.	To the jaw near its symphysis, and to the spur process of the os hyoideus.	Same as preceding.
40	Hyoideus magnus.	"	In the front and upper part of the neck.	To the back and lower part or angle of the cornu of the os hyoideus, and to the middle and lower part of the same bone.	To draw the body of the os hyoideus to the side of the jaw.
41	Hyoideus parvus.	"	Above and before the preceding muscle.	To the body and appendix of the os hyoideus.	To approximate the parts.
42	Stylo-hyoideus.	"	In the lower and front part of the neck.		
43	Hyo-glossus longus.	Glossal.	Along the base and side of the tongue.	To the horns of the os hyoideus, and to the side and lower part of the tongue.	To draw the tongue within the mouth; and at the same time to depress it.
44	Hyo-glossus brevis.	"	Same as preceding.	To the side and body of the os hyoideus.	Same as preceding.
45	Genio-glossus brevis.	"	Lower part of the tongue.	To the inner part of the jaw near its symphasis; to the under part of the tongue, and to the appendix of the os hyoideus.	To project and depress the tongue.
46	Lingualis.	"	This composes the substance of the tongue.	To the root of the tongue, from the body and appendices of the os hyoideus, and receives insertions of all the other glossal muscles.	To contract the tongue lengthwise, and to draw it within the mouth.
47	Hyo-pharyngeus.	Pharyngeal.	Lower, side, and back part of the pharynx.	To the upper border of the cornu of the os hyoideus, and to the side of the pharynx.	To dilate the pharynx in the act of swallowing.
48	Palato-pharyngeus.	do.	Upon the side of the pharynx.	To the pterygoid process of the sphenoid, and to the palate bone; its fibres inter-	To assist in the dilation of the pharynx.

49	Stylo-pharyngeus.	"	In the posterior and lateral parts of the neck.	To the styloid process of the temporal bone, mixing with those of the hyo-pharyngeus and stylo-pharyngeus.	Similar to the preceding
49½	Constrictor pharyngis anterior.	"	Front and upper part of the pharyngeus.	To the cartilage of the Eustachian tube; and to the side of the pharynx.	To contract the cavity of the pharynx in the act of swallowing.
50	Constrictor pharyngis medius.	"	Behind the former.	Front and inner part of the os hyoideus and to the tendinous line uniting the cornua.	Same as preceding.
51	Constrictor pharyngis posterior.	"	In the upper and back part of the pharynx.	To the lower and side parts of the thyroid cartilage, and to the tendinous line running along the superior part of the pharynx.	Same as the two preceding.
52	Hyo-thyroideus.	Laryngial.	Upon the side of the larynx.	To the side of the cricoid cartilage, terminating in the fibres of the muscular coat of the œsophagus.	To elevate the thyroid cartilage, and with it the larynx; or to depress the os hyoides.
53	Crico-thyroideus.	Ditto.	Upon the back and side of the larynx.	To the inferior border of the semi-circular portion of the os hyoideus; and to a broad eminence upon the postero-inferior part of the side of the thyroid cartilage.	To approximate the two cartilages.
54	Crico-arytenoideus posticus.	"	Upon the upper part of the larynx.	To the borders and side of the cricoid cartilage; and to the posterior border of the thyroid, filling up the vacuity between the two cartilages.	To draw the arytenoid cartilage backwards.
55	Crico-arytenoideus lateralis.	"	At the back, side, and upper parts of the larynx, between the thyroid and cricoid cartilages.	To the upper surface of the cricoid cartilage; and to the posterior angle of the arytenoid cartilage.	To draw the arytenoid cartilages asunder, and thus dilate the glottis.
56	Thyro-arytenoideus.	"	Upon the side of the larynx between the thyroid cartilage and lining membrane.	To the front border of the cricoid cartilage; and to the side of the posterior angle of the arytenoid.	To enlarge the glottis by separating the arytenoid cartilage.
57	Arytenoideus.	"	Upon the upper part of the larynx.	To the upper and excavated parts of the two arytenoid cartilages; running across from one to the other.	To the inner side of the thyroid, and the triangular ligament; and to the side of the arytenoid cartilage. To contract the glottis by approximating these cartilages.

ANATOMY AND PHYSIOLOGY OF

NO.	NAME.	REGION.	SITUATION.	INSERTION.	ACTION.
58	Hyo-epiglottideus.	Laryngeal.	Between the epiglottis and semi-circular part of the os hyoides.	To the hollow part opposite the neck of the os hyoides; to the broadest part of the epiglottis.	To retract the epiglottis
59	Tensor palati.	Palatine.	Upon the super-lateral part of the pharynx.	To the styloid processes of the temporal and palate bones.	To dilate the palate.
60	Circumflexus palati.	"	Within the soft palate.	To the union of the palate bones and palate: its fibres also intermingle with those of the stylo-pharyngeus and tensor palati.	To contract the palate
61	Rhomboideus longus.	Humero-cervical.	In the upper part, and side of the neck.	To the entire length of the ligamentum colli, and to the superior costa and cartilage of the scapula.	To assist in raising and drawing the scapula forwards.
62	Levator humeri.	Ditto.	Lower part of the side of the neck.	To the tubercle of the occiput, mastoid process of the temporal bone, transverse process of the atlas, to the second, third, and fourth cervical vertebræ, to the ligamentum nuchæ and fascia covering the side of the neck; also to the head of the humerus, scapula fascia, to muscles about the point of the shoulder, and to a ridge on the body of the humerus, which arises from its greater tubercle.	To raise the shoulder and arm, and at the same time draw them forwards; or these parts being fixed, to turn the neck and head to one side: both acting, they depress the head.
63	Splenius.	Lateral cervical.	Superior and lateral parts of the neck.	To the ligamentum colli from the occiput to the fourth dorsal spine; to the transverse processes of all the cervical vertebræ, and to the mastoid process of the temporal bone.	Both muscles, acting simultaneously, will firmly erect the head and neck; one, acting by itself, inclines the head to one side.
64	Complexus major.	"	Deep seated beneath the former muscle.	To the spines of the first four dorsal vertebræ; to their transverse processes, and to the tubercle of the occiput.	Both to erect and curve the neck and also to extend the head.
65	Trachelo mastoideus.	"	Deep seated beneath the vertebral attachment of the splenius.		

THE HORSE.

No.	Name	Situation	Origin and Insertion	Use
66	Spinalis colli.	Lateral cervical.	Deep seated upon the side of the neck.	To the oblique and superior processes of the last five cervical vertebrae; and to the first dorsal.	To aid in the erection of the head, and flexion of the neck backward.
67	Complexus minor.	Superior cervico-occipital.	Upon the poll.	To the spinous process of the vertebra dentata, and the tendon of the complexus major.	To pull the head backwards and protrude the nose.
68	Rectus capitis posticus major.	do.	Beneath and on the outer side of the former muscle.	To the spine of the vertebra dentata, and to a scabrous depression in the occiput, below its tubercle.	Same as the preceding.
69	Rectus capitis posticus minor.	"	Beneath the preceding.	To the superior part of the atlas, and to the occiput.	To erect the head.
70	Obliquus capitis superioris.	"	Upon the side of the poll.	To the transverse process of the atlas; and to a ridge extending laterally, from the tubercle of the occiput to the mastoid process of the temporal bone.	When both muscles act, the atlas (and the head with it) will be elevated. By their alternate action, a sort of rotary motion is given to the head.
71	Obliquus capitis inferioris.	"	Upon the upper, fore part and side of the neck.	To the side of the spine of the vertebra dentata; and to the upper and back part of the body of the atlas.	
72	Sterno-maxillaris.	Inferior cervical.	Beneath the neck.	To the angle of the jaw, and to the cartilage of the sternum.	To aid in closing the mouth; and drawing the head downwards towards the breast; one muscle acting draws the head to one side.
73	Sterno-thyro-hyoideus.	Ditto.	Above the preceding.	To the spur process of the os hyoideus; to the thyroid cartilage; and to the cariniform cartilage of the sternum.	To draw the os hyoideus and larynx downwards and backwards.
74	Subscapulo-hyoideus.	"	Side and upper part of the neck.	To the inner surface of the scapula and to the centre of the body of the os hyoideus.	To draw the os hyoideus downwards and backwards.
75	Scalenus.	"	In the lower and back part of the neck.	To the transverse processes of the fifth and sixth cervical vertebrae and to the centre of the first rib.	To depress the neck and advance the first rib in a process of inspiration.
76	Longus colli.	"	Extends along the lower part of the neck and from thence to a portion of the spine.	To the bodies, transverse processes, and inferior spines of the cervical vertebrae; excepting the first; to the atlas and to the first six dorsal vertebrae.	To curve the neck.

NO.	NAME.	REGION.	SITUATION.	INSERTION.	ACTION.
77	Rectus capitis anticus major.	Inferior cervico-occipital.	Deep seated; in the front and side parts of the neck.	To the transverse processes of the second, third, fourth, fifth, and sixth cervical vertebræ, and to the cuneiform process of the occipital bone.	To bend and incline the head to one side.
78	Rectus capitis anticus minor.	Ditto.	Deep seated; at the back of the cavity of the fauces.	Same as the preceding.	Same as the preceding.
79	Obliquus capitis anticus.	"	At the posterior part of the fauces.	To the rectus minor and body of the atlas, to the postero-inferior part of the coronoid process of the occipital bone.	Somewhat similar to the two preceding muscles.
80	Trapezius.	Dorso-scapular.	Upon the side of the withers and shoulder blade.	To the spinous processes of the third, fourth, fifth, and sixth dorsal vertebræ, and to the ligament and faschia investing them; also, to a tubercle upon the spine of the scapula.	To elevate the shoulder blade and to incline the bone backward.
81	Latissimus dorsi.	Ditto.	Upon the upper part and side of the chest, behind the scapula.	To the spinous processes of the dorsal vertebræ, and bone of the lumbar spines; to the ligamentum colli, and also to the inner and upper part of the body of the humerus.	To flex the humerus and aid the motions of the trunk.
82	Rhomboideus brevis.	"	Deep seated on the side of the withers.	To four or five of the anterior dorsal vertebral spines, and to the ligament surmounting their spines, and that of the scapula; also, to the superior costa of the scapula.	To aid in supporting and elevating the scapula.
83	Pectoralis transversalis.	Pectoral.	Upon the side, front, and lower parts of the breast.	To tendinous faschia in front; to the bones of the sternum; to the front part of the body of the humerus, and to faschia covering the arm.	To abduct and limit the action of the arm, and to support weight.
84	Pectoralis magnus.	"	In the lower part and side of the thorax.	To the fourth, fifth, and sixth pieces of the sternum; to their cartilages, and also the cusiform; to the anterior white tendinous line which unites it with its fellow; to the aponeurosis of the external oblique.	To draw the head of the humerus, and along with it the lower end of the scapula, backward, throwing the latter more

THE HORSE.

	Name	Situation	Attachments	Use
85	Pectoralis parvus.	Ditto.	Below the former muscle. To the side of the front half of the sternum and to the cartilages of the first four ribs; to the fascia covering the muscles in front of the scapula and shoulder-joint; also, extending nearly as high up as the origin of the antea-spinatus.	Same as the former muscle; to the lesser tubercle of the humerus and the inner part of the lower end of the scapula. into an upright position.
86	Serratus magnus.	Costal.	Between the shoulders and lateral parts of the chest.	To the bodies and transverse processes of the fourth, fifth, sixth, and seventh cervical vertebræ, and to eight of the front ribs, and their cartilages; also, to the inner and upper part of the scapula, occupying the space between the origin of the subscapularis and the insertion of the rhomboidei. It forms the principal agent of support to the ribs and shoulder, and is more or less concerned in all the motions of the chest and shoulder.
87	Intercostales externi.	"	In the intercostal spaces.	To the external edge of the posterior border of one rib, and to the internal edge of the anterior border of the next. To dilate the thorax.
88	Intercostales interni.	"	Beneath the former, in the same spaces.	To the antero-internal edge of the border of one rib, and to the postero-internal edge of the rib immediately before it. To assist the preceding.
89	Lateralis sterni.	Sternal.	Upon the front and lower part of the outer surface of the thorax, near the sternum.	To the postero-inferior half of the first rib; to the superior parts of the cartilage of the third rib; to the cartilages of the fourth and fifth ribs, and also to the sternum. To contract the dimensions of the thoracic cavity.
90 91	{ Sterno-costales, external and internal.	"	At the lateral parts of the sternum.	To the cartilages of the true ribs (except the first) and the sternum; the fibres are also continuous with the intercostales, between the cartilages. Same as preceding.
92	Superficialis costarum.	Dorso-costal.	Over the region of the back.	To the ligamentum nuchæ, and to the posterior border of the sixth and all the remaining ribs. To elevate the ribs.
93	Transversalis costarum.	Ditto.	Upon the upper part and side of the thorax.	To the anterior border of the ribs, near the spine; to the posterior edges of their angles, and to the transverse process of the last cervical vertebra. To elevate the ribs and dilate the cavity of the thorax.

NO.	NAME.	REGION.	SITUATION.	INSERTION.	ACTION.
94	Levatores-costarum.	Dorso-costal.	Deep seated, upon the back.	To the transverse processes of the dorsal vertebræ, and to the anterior borders of the ribs; in the spaces between their tubercles and angles.	To aid in the elevation of the ribs.
95	Longissimus-dorsi.	Dorsal.	Upon the upper and side parts of the back and loins.	To the four posterior cervical vertebræ; to the transverse processes of all the dorsal: to the angles of the twelve posterior ribs; and to the spinous processes of six posterior dorsal; to the transverse processes of all the lumbar vertebræ; to the crista of the ileum, and to the side of the sacrum.	Very complicated; it will incline to a state of flexion the fore quarters upon the hind, or vice versa. In the acts of rearing, kicking, and leaping, it is markedly operative; also in erecting the head and neck. One muscle contracting, the other dilating, inclines the body in a lateral direction.
96	Spinalis-dorsi.	"	Beneath the preceding muscle; in the region of the withers.	To three or four posterior cervical vertebræ; to the spines of seven anterior dorsal; to the spinous processes of several posterior dorsal vertebræ, and to the tendinous expansion of the longissimus dorsi.	To flex and curve the back in an anterior direction, and at the same time erect the neck.
97	Semi-spinalis dorsi.	"	Between the longissimus dorsi et spinalis dorsi; in the lateral parts of the spines of the back and loins.	To the sacrum, beneath the posterior spine of the ileum; to the articular processes of the lumbar vertebræ and the transverse of the dorsal; to the lumbar and dorsal spines as far forward as the withers.	To aid in flexing the back.
98	Diaphragm.	Diaphragmatic.	It divides the cavity of the chest from that of the abdomen.	By fleshy digitations to the cartilages of the eighth pair of ribs, and to those of all the posterior ribs, excepting the two last; also, to the ensiform cartilage. The right crura is attached to the bodies of all the lumbar vertebræ; the left has several tendinous	It acts in concert with the abdominal muscles in the processes of inspiration and expiration.

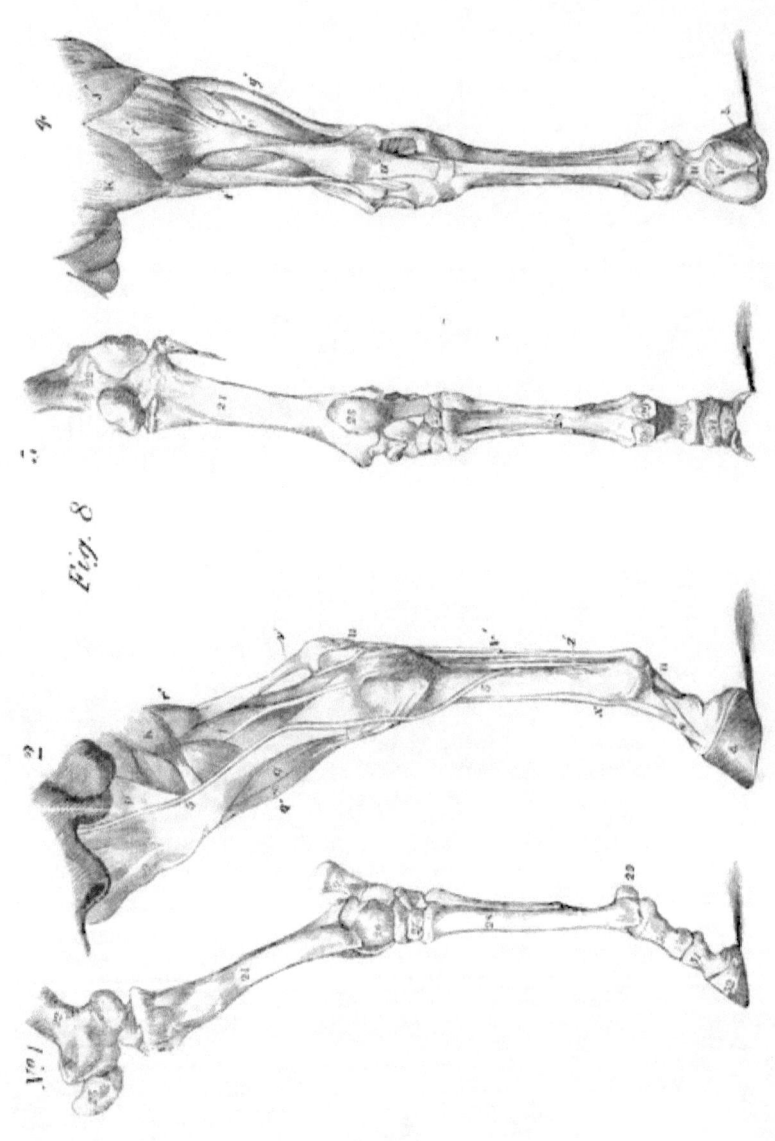

Fig. 8

EXPLANATION OF FIGURE VIII.

NO. 1.—OSSEOUS STRUCTURE.

22. Femur.
23. Patella.
24. Fibula.
25. Os calcis.
26. Astragalus.
27. Inferior row of the tarsal bones.
28. Metatarsi magnum.
29. Sessamoids.
30. Os suffraginis.
31. Os coronæ.
32. Os pedis.

The above explanations will serve to illustrate No. 3. * * are the metatarsi parva.

NO. 2.

INSIDE VIEW OF THE OFF-HIND LEG.

o. Rectus.
p'. Vastus internus.
q'. x. z. Extensor pedis.
q. Extensor metatarsi.
r'. r'. Gastrocnemius externus et internus.
t. Peroneus.
u. The insertion of the gastrocnemius.
v'. Tendon of the flexor metatarsi.
z'. Suspensory ligaments.
&. The hoof.
5. 5. The saphena vein.
K. Abductor femoris.
8. Bifurcation of the suspensory ligament.
u'. (Beneath the pastern) Perforatus et perforans.

NO. 4.

K. J. Biceps, showing the manner in which it bifurcates.
r'. Gastrocnemius internus.
t'. Flexor metatarsi internus.
v'. Flexor pedis accessorius.
5. " " externus.
u'. Insertion of the gastrocnemius.
y'. Peroneus.
w. u'. v'. Flexors of the foot.
&. The hoof.

99	Semi-spinalis lumborum.	Lumbar.	On the loins.	attachments to the first and second of these vertebræ. To the spines of the lumbar vertebræ, and to the spines of the lumbar vertebræ, and to the sacrum.	To aid in flexing the back.
100	Inter-transversalis lumborum.	"	In the lateral parts of the loins.	From the sharp border of one transverse process to that of the next.	To approximate the transverse processes.
101	Sacro lumbalis.	"	Deep-seated in the loins.	To the transverse process of the last lumbar vertebræ, body of the sacrum, transverse processes of the loins, and to the last rib.	To co-operate with the inter-transversales lumborum in approximating the transverse processes; and at the same time to draw back the last rib.
102	Psoas magnus.	"	Beneath the preceding; postero-lateral parts of the spine, and superior part of the inlet of the pelvis.	To the inner surfaces of the two last ribs, near their vertebral articulations; to the body and transverse processes of the last dorsal vertebra, and to the same of all the lumbar; also to the small, internal trochanter of the os femoris.	To bend the haunch upon the pelvis. Supposing it to act while the hind quarters remain stationary, it will produce that appearance of flexed spine known as "crouch," or "sticking up of the back."
103	Iliacus.	"	Occupying the iliac fossa, from thence to the internal part of the haunch.	To the crista of the ileum; to the anterior spinous process, venter, and inferior edge of the bone, and to the small internal trochanter, along the psoas magnus.	To flex or advance the haunch.
104	Psoas parvus.	"	On the inner side of the preceding muscle.	To the heads of the three last ribs; to the bodies of the three posterior dorsal, and all the lumbar vertebræ, and to the brim of the pelvis.	Similar to the preceding.
105	Obliquus externus abdominis.	Abdominal.	Upon the lower and side parts of the belly.	By fleshy digitations to the posterior borders of the fourteen hinder ribs, below their centres; by tendinous fibres to the two anterior thirds of the crista ilei, and to their anterior spinous processes; also, to the fascia lumborum and linea alba.	To give support to the abdominal contents; aid in the expulsion of fæces, urine and fœtus; and in forcing the diaphragm forward.

NO.	NAME.	REGION.	SITUATION.	INSERTION.	ACTION.
106	Obliquus internus abdominis.	Abdominal.	In the posterior side, and lower part of the abdomen.	To the spine of the ileum and linea semi-lunaris; external oblique muscle, and to the linea alba.	Similar to the preceding muscle.
107	Transversalis abdominis.	Ditto.	Inferior and lateral parts of the abdomen.	To the inner surface of the cartilages of all the false ribs; to the transverse process of the lumbar vertebræ; to the anterior spinous process of the ileum, and to the entire linea alba.	To support, compress, and aid in the evacuation of, the bowels.
108	Rectus abdominis.	"	At the central region of the inferior part of the abdomen.	To the cartilages of six or seven posterior true ribs; to the sternum, and to the os pubis near its symphysis.	To support the central parts of the abdomen, and to draw the sternum upwards and backwards.
109	Retractor ani.	Anal.	On the side of the rectum near its termination.	To the sacro sciatic ligament; to the os innominatum, and to the side of the rectum.	To retract the anus and draw it within the pelvis.
110	Sphincter ani.	"	It encircles the anal orifice.	To the coccyx; encircling the gut.	To close the anus.
111	Cremaster.	Genital.	See *spermatic cord*.		To suspend, and retract, the testicle.
112	Erector penis.	"	Upon the penis.	To the inner side of the infero-posterior part of the tuberosity of the ischium, and to the crest of the penis.	To promote the erection of the penis, and to retain it in a state of erection.
113	Triangularis penis.	"	Beneath the penis, anterior to the ischial arch.	To the ischial portion of the os innominatum, and to the prostrate gland.	It aids in discharging fluids from the urethra and prostrate.
114	Accelerator urinæ.	"	Around the penis at the bulb of the urethra.	It is considered as a part of the penis, and arises from the bulb of the urethra.	To aid in the discharge of urine.
115	Erector coccygis.	Coccygeal.	In the upper and side parts of the tail.	To the transverse processes of the spines of the sacrum, and to bodies and spines of the bones of the coccyx.	To erect the tail.
116	Depressor coccygis.	Ditto.	Beneath and on the lower	Within the pelvis to the sacro sciatic ligament.	To depress the tail.

THE HORSE.

	Name	Location	Origin / Insertion	Action
117	Curvator coccygis.	Coccygeal.	Upon the side of the dock between the two former muscles.	half of the side of the tail.	ment, and to the body of the sacrum, and to the inferior parts of the bodies of the coccygeal bones.	To curve the tail in a lateral direction.
118	Compressor coccygis.	Ditto.	At the side of the root of the tail.	Within the pelvis to the lateral parts of the sacrum; to the fourth and fifth lumbar vertebræ, and to the transverse processes of all the bones of the coccyx.	Same as "*depressor*."
119	Antea-spinatus.	Scapular.	Occupying the forward half of the shoulder blade; or, fossa antea spinata scapulæ.	To the surface of the fossa antea spinata; to the spine and anterior costa of the scapula; to the summit of the greater and lesser tubercles of the humerus, and to the capsular ligament of the shoulder-joint.	To extend the humerus.
120	Postea-spinatus.	"	Occupying the posterior half of the shoulder blade; or, fossa postea spinata.	To the surface of the fossa postea spinata, and spine of the bone; to the outer side of the great tubercle of the humerus; to the bony ridge extending down from it; and to the capsular ligament of the shoulder joint.	To assist in the flexion of the humerus, and turn the bone in an outward direction.
121	Subscapularis.	Internal scapular.	Beneath the scapula.	To the surface of the venter scapulæ; to the anterior and posterior costa; to the lesser tubercle of the humerus; and to the capsular ligament of the shoulder joint.	To assist in the extension of the shoulder joint, and to turn the os humeri inwards.
122	Teres major.	Posterior scapular.	Behind and below the subscapularis.	To the posterior angle of the scapula; to the posterior costa, and to the inner and upper part of the body of the humerus.	To aid in flexing the shoulder joint, and to incline the humerus inwards.
123	Teres minor.	Ditto.	Upon the outer and posterior part of the shoulder, below and behind the postea-spinatus.	To the posterior costa of the scapula; to the ridge descending from the tubercle of the humerus, and to the panniculus carnosus.	Co-operative with the postea spinatus.
124	Coraco humeralis.	Antero humeral.	Infero-internal part of the shoulder.	To the coracoid process of the scapula; to the middle third of the antero-internal part of the body of the humerus.	To extend the shoulder joint and incline the humerus inwards.

ANATOMY AND PHYSIOLOGY OF

NO.	NAME.	REGION.	SITUATION.	INSERTION.	ACTION.
125	Flexor brachii.	Anterior humeral.	Antero-inferior part of the shoulder.	To the coracoid process of the scapula; to the inner parts of the head and neck of the radius; to the capsular ligament of the elbow joint, and to the brachial fascia.	To flex the arm.
126	Humeralis externus.	Ditto.	Deep seated upon the infero-external side of the shoulder.	To the entire postero-external parts of the neck and body of the humerus; to the supero-anterior part of the body of the radius.	Co-operates with the preceding muscle.
127	Caput magnum.	Posterior humeral.	Infero-posterior part of shoulder.	To the posterior costa of the scapula; superior and posterior parts of the olecranon, and to the brachial fascia.	To extend the arm.
128	" medium.	Ditto.	Infero-external part of shoulder.	To the neck of the ridge of the outer tuberele of the humerus, and to the olecranon.	
129	" parvum.	Ditto.	Infero-internal part of shoulder.	To the inner side of the body of the humerus at its centre, and to the apex of the olecranon.	
130	Anconeus.	Ditto.	In the space between the condyles of the humerus.	To the infero-posterior parts of the body of the humerus; to the antero-external part of the olecranon, and to the capsular ligament of the elbow joint.	Similar to the preceding; also to free the ligament from the joint during action.
131	Extensor metacarpi magnus.	Antero-brachio-crural.	Fore part of the arm.	To the outer and fore parts of the external condyle of the humerus and its capsular ligament; to the antero-superior part of the os metacarpi magnum.	To extend the leg.
132	Extensor pedis.	Ditto.	Fore and outer parts of the arm.	To the external condyle of the humerus; head and superior part of the radius; capsular ligament of the elbow joint; to the inferior end of the os suffraginis, os coronæ and coronal process of the os pedis, also to the capsular ligament of the fetlock joint.	To extend the lower extremity of the limb.
133	Extensor suffraginis.	Ditto.	Back and outer parts of the arm.	To the back, outer, and upper parts of the radius, and border of the ulna; to the upper	To aid in extending the knee and fetlock.

Triceps extensor brachii.

THE HORSE. 75

No.	Name	Region	Origin	Insertion	Use
134	Extensor metacarpi obliquus.	Antero-brachio-crural.	Infero-anterior parts of the arm.	To the outer and infero-anterior parts of the body of the radius; to the supero-anterior part of the os metacarpi internus.	To aid in rotating the limb, extending the same; and to keep the extensor metacarpi tendon in its place.
135	Flexor metacarpi externus.	Superficial posterior brachio-crural.	Back, and external side of the arm.	To the postero-inferior part of the external condyle; one division is inserted into its tendon, the other to the outer small metacarpal bone.	To flex the leg
136	Flexor metacarpi medius.	Ditto.	Posterior part of the arm.	To the external condyle of the humerus; to the olecranon and trapezium.	Same as the preceding.
137	Flexor metacarpi internus.	Ditto.	Postero-internal side of the arm.	To the internal condyle of the humerus, capsular ligament of the elbow joint, and to the head of the internal metacarpal bone.	Same as above.
138	Flexor accessorius sublimus.	Ditto.	Posterior part of the arm.	To the postero-internal part of the ulna, and to the tendon of the flexor perforans.	To strengthen the perforans and aid in flexing the leg.
139	Flexor pedis perforatus.	Deep posterior brachio-crural region	Deep seated in the posterior part of the arm.	To the internal condyle; and to the upper and back part of the os coronæ.	To bend the knee and fetlock.
140	Flexor pedis perforans.	Ditto.	Same as the preceding.	Its superior attachments are the same as the preceding; inferiorly it is inserted into the posterior concavity of the os pedis.	To flex the limb and turn the sole of the foot upwards.
141	Flexor accessorius.		Ditto.	To the centre of the posterior part of the radius, its tendon joining that of the perforans.	Same as perforans.
142	Lumbrici anterior et posterior.		Deep seated, along the infero-posterior side of the arm.		

The lumbricini consist of two pairs; the anterior ones lie within the space formed by the small metacarpal bones and suspensory ligament, protected by the flexor tendons; they adhere, for some distance down the leg, to the small metacarpal bone, become tendinous, about the middle of the canon, and finally bury themselves in adipose tissue instead of the limb, connected with the extensor tendons. The lumbricini posteriores are to be found invested with adipose membrane, adhering to the inner side of the tendo perforans, about one-third of its length upward from the fetlock. About this region these muscles become broad, but narrow in descending; at the fetlock they give off slender flattened tendons, which appear to unite to form the eponastic border of the cellular and tendinous sheath at that part enclosing the tendo perforatus.

NO.	NAME.	REGION.	SITUATION.	INSERTION.	ACTION.
143	Glutæus externus.	Gluteal.	Middle and external parts of the haunch.	To the antero-superior and inferior spines of the ileum, to the spine of the sacrum; to the lumbar fasciæ, and to the trochanter minor externus.	The glutæi act as propellers of the body; they are in powerful action when the horse is kicking or rearing.
144	Glutæus maximus.	"	Anterior, middle, and external parts of the haunch.	To the spinous and transverse processes of the two or three last lumbar vertebræ; to those of two or three uppermost sacral; to the lumbar fasciæ; to the crest of the ileum, its dorsum, and posterior spine; to the sacro-sciatic ligament and to the trochanter major.	Same as preceding.
145	Glutæus internus.	"	Beneath the preceding muscle.	To the dorsum ilii and trochanter major.	Same as preceding.
146	Pyriformis.	Pelvic trochanterian.	Supero-lateral parts of the pelvic cavity.	Within the pelvis, to the transverse processes of the sacrum, and the indigo-internal part of the ileum; without the pelvis and to hollow behind the trochanter major.	To assist in the extension of the haunch.
147	Obturator externus.	Ditto.	Deep seated, in the supero-internal part of the thigh.	To the external border of the obturator foramen, and to the external surface of the obturator ligament; to the cavity behind the trochanter major, and to the upper portion of the ridge extending from the larger to the lesser trochanter.	To assist the preceding muscle, and rotate the thigh bone outwards.
148	Obturator internus.	Ditto.	Upon the lower side of the pelvic cavity.	To the internal border of the obturator foramen; to the inner surface of the obturator ligament, and to the root of the trochanter major.	To turn the thigh bone inwards.
149 150	} Gemini.	Ditto.	Without the pelvis, at its postero-inferior part.	To the supero-posterior part of the ischium and to the root of the trochanter major.	To give additional strength to the capsular ligament.
151	Tensor vaginæ.	Anterior ilio femoral.	Antero-external part of the haunch.	To the anterior spine of the ileum and to the faschia lata.	To render tense the faschia lata and extend the thigh.

Anterior ilio-femoral.	Anterior part of the haunch between the ilium and patella.	To the dorsum of the ilium, and to the upper and fore part of the patella.	To support the patella and aid in raising the limb.
Ditto.			
Ditto.	REMARKS.—The triceps vasti (including the rectus parvus) are considered by the French as one muscle, but Mr. Percivall alludes to them separately: the anterior part and sides of the haunch; the two vasti are inserted into the trochanter major, minor, externus and internus; to the femur and patella. The rectus parvus is inserted into the ilium, just above the acetabulum; to the body of the femur, and to the patella. Their action is, to extend and advance the thigh.		
Internal ilio-femoral.	Antero-internal part of the haunch.	To the bodies of the posterior lumbar vertebræ, to the brim of the pelvis; to the supero-internal part of the tibia, and to the internal condyle of the femur.	To rotate the limb inwards and aid in flexing the same.
Ditto.	Superficial on the internal part of the thigh.	To the symphysis pubis, internal condyle of the femur, and to the upper and internal part of the tibia.	Similar to the preceding muscle.
Ditto.	On the front and inner part of the haunch.	To the anterior surface of the os pubis, and trochanter internus.	To flex and adduct the femur.
Ditto.	REMARKS.—For all practical purposes the three adductors may be considered as a single muscle. Its divisions are distinguished as magnus, longus, brevis; they are all attached to one or more of the following parts; pubes, femur, condyles, sacrum, and tibia. Their action is to adduct the limb.		
Posterior ilio-femoral.	At the back and side of the haunch and thigh.	To the lateral and posterior of the spine of the sacrum, and superior parts of some of the ossa coccygis; to the sacro sciatic ligaments; to the tuberele of the ischium, and to fascialata; to the patella and external ligament; to the tibia and fascia covering the leg.	To abduct the limb.
Ditto.	At the posterior side of the haunch and thigh.	To the postero-lateral part of the spine of the sacrum; to some of the upper bones of the coccyx; to the antero-inferior side of the tuberosity of the ischium; to the upper, front, and internal part of the tibia.	Similar to the preceding muscle.
Anterior femoro-crucial.	In the anterior and superficial part of the leg.	To the external condyle of the femur, and to the coronal process of the os pedis.	To extend the pastern and foot.

No.	Name.	Region.	Situation.	Insertion.	Action.
164	Peroneus.	Anterior femoro-crucial.	On the front and outer parts of the limb.	To the head and body of the fibula and coronal process of the os pedis.	Same as the preceding.
165	Flexor metatarsi.	Ditto.	In the front and internal parts of the limb.	To the condyle of the femur, upper, and external part of the tibia; to the head of the large metatarsal bone, and to the internal head of the small one.	To flex the hock and direct its joint upwards.
166	Gastrocnemius externus.	Superficial posterior femoro-crural.	At the posterior part of the leg.	Into a depression above the external condyle of the femur; to the internal condyle, and to the os calcis.	To extend the hock.
167	Gastrocnemius internus.	Ditto.	In the same situation.	Into the upper and posterior part of the femur, passing over the os calcis; it is inserted into the os coronæ.	It assists the femur and flexes the pastern.
168	Plantaris.	Ditto.	At the back and outer part of the thigh.	To the head of the fibula, uniting with the tendo-gastrocnemii; and to the os calcis.	Same as the preceding.
169	Popliteus.	Deep posterior femoro-crural.	At the posterior part of the stifle joint.	To the back and outer side of the external of the femur; to the capsular ligament of the stifle-joint, and to the superior part of the body of the tibia.	To flex the stifle and turn the tibia inwards.
170	Flexor pedis.	Ditto.	Similar to "plantaris."	To the postero-external side of the head of the tibia and to the body of the same; to the posterior part of the fibula terminating in the os pedis.	To flex the inferior parts of the limb.
171	Flexor pedis accessorius.	Ditto.	In the back and internal parts of the leg.	To the back and outer parts of the leg.	Same as the preceding.

ON DISSECTION.

No man can ever expect to become a practical anatomist nor pathologist, unless he practise dissection. It is the only possible way by which he can familiarize himself with the healthy structural organization of the horse. Having made himself acquainted with the healthy aspect of the various parts, their uses, etc., he next is able to judge of the various grades of textural change which occur and exist, between the part that has been studied under its healthy aspect, and that which has now departed from its healthy condition. Thus, in the prosecution of the study of anatomy, the student finally becomes a pathologist; and, although he may be a *beginner*, he places himself in a position only a few removes from the old and experienced practitioner, and can venture to "measure a lance" with the *renowned* knights of the healing art.

DISSECTING INSTRUMENTS.

The dissector should supply himself with a beak-pointed scalpel (which is one of German origin), for superficial dissection, and a myology knife, strong and rounded at its point. For the dissection of blood-vessels and nerves, a more delicate and pointed scalpel is needed. The forceps should be strong, and armed at the points with teeth; two pair of scissors are needed, one pointed and the other blunt; a saw and blunt chisel, for opening the cranium. A blow-pipe, curved needles, and a few extra scalpels, are all that the student requires.

SUBJECTS SUITABLE FOR DISSECTION.

For demonstration of the muscular system, a well-proportioned and fully-developed subject should, if possible, be selected, and one that has died suddenly, or been killed in consequence of some accident, is to be preferred. For making wet and dry preparations, lean, emaciated subjects should be selected. The lymphatic system is best shown on animals of a flabby and œdematous organization. Young animals are the best subjects for dissection, in view of demonstrating the circulatory and nervous systems.

RULES IN REFERENCE TO DISSECTION OF THE MUSCLES.

As there are abundance of subjects to be had in the United States, and it being inconvenient for one individual to dissect a whole subject, he had better divide it into six parts, viz.: 1st, The head and neck. 2nd and 3rd, The anterior extremities, which include the thorax, its contents, and the diaphragm. 4th and 5th, The posterior extremities, to which belong the pelvic and abdominal viscera. 6th, Those viscera which cannot be advantageously divided, as the heart, stomach, bladder, organs of generation, &c. Should the dissector decide to commence on the *whole* subject, he first removes the skin, in order to expose the panniculus carnosus; this will require some care, as some of the fibres of this subcutaneous muscle are intimately connected with the former. There are various ways of removing the skin: the author prefers to commence on the back, and dissect off towards the feet. Supposing the subject to lie on the off-side, we commence an incision at the anterior part of the nasal region, and continue the same upward until we arrive at the occiput; we then incline the scalpel from the superior part of the neck, in order to avoid the mane, and continue the incision along the lateral part of the dorso lumbar spines until the coccyx is reached; the overlapping portion can then be dissected, and turned over to the off-side, so as to expose the tendinous insertions of the panniculus into the ligamentum nuchæ, etc., etc. The panniculus being exposed, it may be divided into three parts, viz.: 1st, *The Cervical portion*, which comprises the head, neck, shoulders, and forearms. 2nd, *The Thoracic portion*. 3rd, *The Abdominal portion*.

Having traced the attachments of the panniculus, the muscles then engage our attention; they being composed of nearly parallel fibres, the manner of displaying them is indicated. The cellular tissue should if possible be detached with the skin and panniculus; without this precaution the surface of some of the coarser muscles would have a mangled appearance. The knife should always follow the direction of the muscular fibres, and the part on which a muscle is to be dissected should be placed, if possible, in such a situation as to produce a forcible extension of that muscle; thus, in tracing the origin and insertion of a muscle, the dissector becomes acquainted with its use.

After exposing the external layer of muscles they may be detached from their insertion, or divided in their centre; if divided, we thus preserve the two points of origin and insertion. The deeper seated muscles may be demonstrated in the same way.

In the dissection of muscles the scalpel should be used in a free and prompt manner; the strokes should be long and bold, using the little finger to steady the movement of the hand. In making autopsies and in examining the viscera, the subject is generally placed on his back.

ANATOMICAL PREPARATIONS.

It is highly important that every student should be acquainted with the methods of making *wet* and *dry* preparations, and of injecting the blood-vessels; for *specimens* of this kind are the best means of familiarizing us with the structures of quadrupeds, and such, when properly prepared, possess a real and practical value.

INJECTING INSTRUMENTS.

Pole describes three kinds of instruments used in making injected preparations.

The *first* consists of a brass syringe, made of various sizes; the nozzle is adapted to pipes into which the syringe is to be inserted; a short pipe, with stop-cock, also accompanies the syringe, which is to be applied between the syringe and either of the pipes.

The *second* is a similar instrument, only much smaller; its pipe is very minute, and its piston is furnished with a ring, so that the thumb may be used to throw its contents into a vessel.

The *third* instrument is generally used for injecting the glands and lymphatics with quicksilver. It consists of a glass tube, terminating with a steel end, and having an extremely fine steel pipe, which screws on to the latter. The syringe used by the author of this work is one manufactured in England (and can be found in some of our agricultural stores), for the purpose of syringing plants; it has the most accurate bore and finely-adjusted piston of any instrument now in use, and being of medium size it can be used for either large or minute injections. Some alterations, however, have to be made in the nozzle and pipes fitted accordingly.

DIRECTIONS FOR USING THE SYRINGE.

In using the syringe, a certain amount of tact or experience is necessary, and the beginner must not feel disappointed should he fail in a first or second attempt; for some little oversight might frustrate the whole process. Everything should be in readiness, such as ligatures, forceps, scissors, sponge, hot and cold water, etc.

The pipes should be inserted into the vessels, and confined there by strong ligatures; and, before the syringe is introduced, its nozzle must be turned upwards, and the piston pressed until all the air and froth are ejected; then introduce the nozzle into the stop-pipe

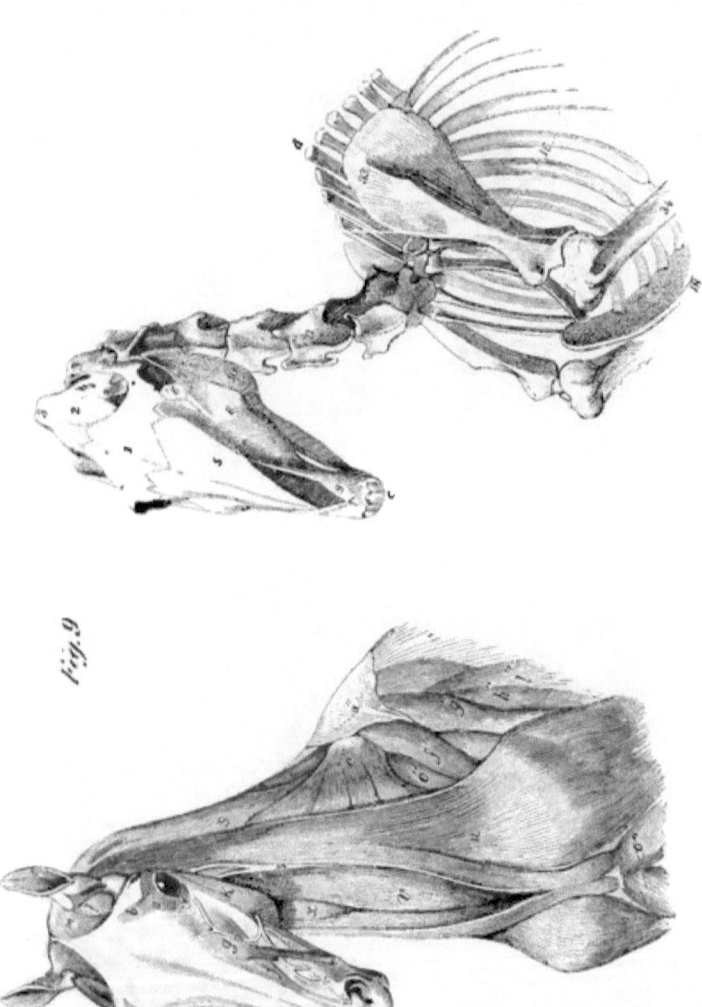

Fig. 9

EXPLANATION OF FIGURE IX.

OSSEOUS STRUCTURE.

1. Frontal bones.
2. Parietal.
3. Occipital.
4. Temporal.
5. Nasal.
6. Lachrymal.
7. Malar.
8. Superior maxillary.
9. Anterior "
10. Inferior "
11. Cervical vertebræ.
16. The true ribs.
18. The sternum.
33. The scapula.
34. The humerus.
c. The incisors.
d. Dorsal spines.

MUSCULAR STRUCTURE.

ANTERIOR VIEW.

a". Trapezius.
c". Scalenus.
e". Pectoralis transversalis.
f". Antea spinatus.
g". Postea spinatus.
h". Teres major.
i". A portion of the triceps extensor brachii.
s. The fascial covering of the splenius.
K. The masseter.
g. Levator labii superioris.
f. Nasalis longus.
s. Orbicularis oris.
c. Dilator narium lateralis.
d. Dilator narium anterior.
m. l. Attolentes et abducens aurem.
b. Levator palpebræ.
a. Orbicularis palpebrarum.
o". Pectoralis magnus.
u. Levator humeri.
V. Sterno maxillaris.
x. Subscapulo hyoideus.
2. Maxillary vein.
3. Jugular vein.

and press the piston steadily until a sensible resistance is felt. If much force be used, rupture of a vessel may take place. After a prudent force has been applied for some time, the syringe may be withdrawn, previously securing the stop-cock. A steady and uniform pressure on the piston will be more likely to secure uniform injection than force, or sudden jerks. Should the first injection fail to fill the vessels, it must be immediately followed by a second. When injecting through a very small pipe, the injector must be patient, and steadily continue the pressure on the piston.

When using *warm* injections, the syringe must be kept *warm* by immersing it in hot water, and the part to be injected must also be kept at the same temperature, by the same means.

DIFFERENT KINDS OF INJECTIONS.

There are six kinds of injections now in use, viz., the *cold*, *coarse*, and *fine* injection, the *minute*, the *mercurial*, and, finally, the plaster of Paris injection. The five first are most employed; the plaster of Paris is objectionable because it is easily fractured.

FORMULÆ FOR COARSE WARM INJECTIONS.

Red.—Beeswax, sixteen ounces; white resin, eight ounces; turpentine varnish, six ounces; vermilion, three ounces.

First liquify the wax, resin, and turpentine varnish, in an earthen pot, over a slow fire, or in a water bath; then add the vermillion, previously reducing it to a fine powder, so that the coloring ingredients may be intimately and smoothly blended, then add the same to the above ingredients, and, when they have accrued due heat, the injection is fit for immediate use.

Yellow Injection.—Take beeswax, eight ounces; resin, four ounces; turpentine varnish, three ounces; yellow ochre, one ounce and a quarter.

White Injection.—Clarified beeswax, eight ounces; resin, four ounces; turpentine varnish, three ounces; flake white, two ounces and a quarter.

Pale Blue Injection.—Take the preceding ingredients, and add to them a small portion of indigo.

Black Injection.—Beeswax, resin, and turpentine varnish in the above proportions; and add lamp-black *ad libitum*.

The same rules are to be observed in preparing all the injections.

FORMULÆ FOR FINE INJECTIONS.

Red.—Brown and white spirit varnishes, of each four ounces; turpentine varnish, one ounce; vermilion one ounce.

Yellow.—Brown and white spirit varnishes, of each four ounces; turpentine varnish, one ounce; king's yellow, one ounce and a half. To make a white injection, add to the last formula two ounces of flake-white instead of *king's yellow*.

Blue.—Brown and white spirit varnishes, of each four ounces; turpentine varnish, one ounce; Prussian blue, one ounce and a half. This may be made *black* by adding ivory black instead of Prussian blue.

FORMULÆ FOR MINUTE INJECTIONS.

The liquifying principle in minute injections is "*size*," which is made in the following manner:

Take fine transparent glue, one pound, break it into pieces; put it into an earthen pot, and pour on it three pints of cold water; let it stand twenty-four hours, stirring it occasionally with a stick; then set it over a slow fire until it is perfectly dissolved; skim off all the scum from the surface, and strain the remainder through flannel; it will then be fit for the coloring ingredients.

Minute Red Injection.—Size, one pint; vermilion, three ounces and a half.

Yellow.—Size, one pint; king's yellow, two ounces and a half.

White.—Size, half a pint; flake white, one ounce and three quarters.

Blue.—Size, half a pint; fine blue smalt, six ounces.

PLASTER INJECTION.

Before mixing the plaster of Paris, the pipes must be secured to the mouths of the vessels at which the injection is to enter. Plaster of Paris (to which some of the preceding dry coloring materials, suitable to the fancy, can be added) must be put in a mortar and rubbed with a pestle in order to pulverize it completely; water is then to be added until the mixture is of the consistence of cream; the syringe being in readiness, it is to be filled and immediately injected into the vessels. In the author's opinion, this injection is only suitable for injecting first-class vessels, for it coagulates or "sets" so quickly that it cannot be used as a *minute* injection. It is said that a small quantity of olive oil, incorporated with the liquid plaster, retards its coagulation; yet if too much were added it would spoil the preparation.

The moment the parts are injected the syringe should be washed out in cold water, and when the injection "*sets*" in the veins, the pipes must be removed and likewise cleansed.

FORMULÆ FOR COLD INJECTIONS.

Dr. Parsons recommends, for coarse cold injections, the following formula: Take coloring matter and grind it in boiled linseed oil, on a painter's marble, until it has acquired the consistence of common white lead, as sold at the stores. After being finely levigated, a little lime-water, in proportion of two table-spoonsful to a pint, is to be incorporated by stirring. At the moment of filling the syringe with the injection, there should be added to it about one-third of its measure of Venice turpentine, which should be stirred in briskly and used immediately, as it very soon hardens.

For a temporary *cold coarse* injection, white lead ground in oil answers every purpose; it requires no addition of lime-water, because the lead is generally adulterated with carbonate of lime, which hardens the mixture, and it can be colored to suit the taste, or the vessels can be colored with a pencil brush, before varnishing.

For filling the arteries, to dry and preserve, red lead is the best and cheapest material; vermilion, however, resembles more the color of arterial blood.

Whatever part we expect to inject with warm injections, must be immersed in water very hot,—not hot enough, however, to crisp the vessels. Attention to this matter is highly important, in view of successful injection.

THE COURSE OF INJECTIONS.

Injections must follow the course of the circulation; the arteries, however, having no valves, are easily injected in any direction; but the veins are furnished with numerous valves, which prevent the flow of injection *from* the heart. Sometimes it is necessary to break these valves by means of a small whalebone probang. In the region of valves are often found coagulations, which must be washed out before the injection can be introduced; and this, also, must be performed in the direction of the circulation. Small pipes are to be introduced, and warm water must be thrown in, which can be made to escape through an incision made with a lancet in a region approaching the right auricle. The incision can afterwards be closed by suture, or otherwise. Many of the veins of the horse, however, are destitute of valves, and therefore admit of injection in a direction contrary to their circulation.

To inject a *portion* of the animal,—the knee or hock, for example,—it is necessary to secure all the branches of the vessels that have been divided where it is separated from the body. The part is then to be injected in the same manner as if we were injecting the whole body.

QUICKSILVER INJECTION AND PREPARATIONS.

The fluid specific gravity and beautiful metallic lustre of quicksilver render it valuable for displaying minute vessels. Dr. Parsons remarks that the principal objection to its general use is the continuance of its fluidity, which renders dissection, after injection, almost impracticable. Yet there are some very fine specimens of quicksilver injections of glands and deep-seated lymphatics, in the Warren museum of this city, that cannot be surpassed by any other kind of injection. The same authority remarks, that the specific gravity of quicksilver, when supported in a column, is such as to exert strong pressure upon a blood-vessel or lymphatic that receives it, and therefore in some cases a syringe is unnecessary. It is to be borne in mind that the force of the injection depends upon the perpendicular height of the column, and not on its diameter, and the former may be such as to burst the vessel. The injections should always be conducted in a shallow dish or tray, so that the quicksilver may not be wasted. When injecting the lymphatics, it is necessary to be provided with small lancets, straight, curved, and delicately-pointed fine needles, which must be armed with waxed threads. For common blood-vessel preparations, glass tubes of the shape of a straight blow-pipe are needed.

INJECTING THE LYMPHATICS WITH MERCURY OR QUICKSILVER.[a]

In injecting the lymphatics, our success depends, perhaps, more on the body we choose, than on any other circumstance: bodies slightly anasarcous, if they be emaciated, are the best. From the valvular structure of the lymphatics, it is necessary to inject from the extremities towards the trunk.

It is almost impossible for one person to succeed in injecting the lymphatics without assistance; there are so many things requisite, besides merely holding the tube in the vessel, that an assistant is indispensable.

It is very necessary, before beginning, to see that the injector has within his reach sharp-pointed scissors, knives, forceps, lancets, pokers for tubes, needles, and waxed threads, so arranged that they can be used instantly, for it will often happen that it will be impossible for either the assistant or the operator to take his eye for a moment off the vessel, without losing it.

[a] Sir Charles Bell

When injecting the superficial lymphatics, we first cut off a portion of skin, so as to expose the loose cellular texture; having found a lymphatic vessel, it must be seized by the forceps and dissected from the surrounding substance. Having hold of it with the forceps, snip it half across with fine scissors, and into the incision introduce the tube containing the mercury. A *poker* or director is often necessary for the purpose of creating a vacuum; a few drops of mercury then introduced by the side of the director will open the way for more; the director being withdrawn, the mercury flows into the lymphatics.

If the vessel to be injected be a large one, it must be secured by ligature around the pipe. The quicksilver is to be pressed onward, elevating or depressing the pipe so as to regulate the force of the injection.

In injecting a gland, we must endeavor to find the vessel that has the most influence in filling it. Having found it, we secure the other vessels and fill the gland through the former channel.

The vessels or glands injected with mercury should be dried as quickly as possible, and varnished, or else preserved in spirits of turpentine.

METHOD OF INJECTING THE LACTEALS.

Take a small portion of the intestine and mesentery, and make an incision in one of the most conspicuous lacteals, as near as possible to its origin in the intestine; then introduce the point of the injecting tube, and conduct the operation agreeably to the preceding rules. When the quicksilver flows out of any of the divided vessels, they must be repaired by an assistant; when as many of the lacteals are filled as will receive the quicksilver from this orifice, introduce the pipe into another, and repeat the process as before, and so on, until as many of them are filled as can be; then inflate the intestine and suspend it in the air to dry, after which it may be preserved by varnishing both inside and out.

DIRECTIONS FOR INJECTING THE PAROTID GLAND.

This should be injected before removal, on account of the numerous vessels by which it is attached to the adjacent parts. Before commencing the operation, the skin over the region of the gland and duct must be raised, in view of searching for the duct; having found it, an opening must be made into it with the point of a lancet, sufficiently large to introduce the point of the steel injecting pipe; when introduced, confine the duct upon it by a ligature with a single knot, which shall serve when the pipe is withdrawn to secure the quicksilver in the gland. The gland having been filled, the pipe withdrawn, and the duct secured, we proceed with all possible care to dissect the gland from its situation. Any branches of vessels going off to surrounding parts must be secured by means of a small curved needle, armed with a single ligature, after which they may be divided with safety. The gland being removed, and all extraneous tissue dissected off, it should be placed in water to extract the blood, etc. This will require about thirty-six hours; the water, however, must be frequently changed; the gland can then be spread on a piece of pasteboard and exposed to dry. It makes the most beautiful preparation, when preserved in a glass vessel containing pure spirits of turpentine.

BREAKING DOWN THE VALVES.

Many of the glands, the surface of the liver included, can be injected contrary to the circulation of the lymph. When the quicksilver passes at first freely into the lymphatics, and suddenly stops, it will be necessary to force it forward by gentle pressure with the edge of a spatula, in the direction in which it seems most likely to run; by this means the valves are broken down. The valves of the superficial lymphatics of the liver are easily broken down in this way, but the valves in some of the lymphatics are much firmer, consequently not so easily broken down.

WET PREPARATIONS.

PREPARATIONS BY DISTENTION.

Hollow organs may be distended for preservation with antiseptic liquids, air, wool, hair, cotton, plaster, quicksilver, etc. *Wet preparations by distention*, with spirits of wine, oil of turpentine, etc.

The intention, in distending preparations by spirits, is to give them their natural figure, to exhibit more fully the parts of which they are composed, their vascularity, and occasionally some morbid or preternatural appearance.

METHOD OF DISTENDING AND PREPARING THE LUNGS.

The lungs taken from a sheep or calf make a very good substitute for those of the horse, which are too bulky for ordinary use.

The pulmonary arteries and veins should first be filled with red-colored injection; then immerse the lungs in oil of turpentine, contained in a vessel, large enough to admit them without compression; then inject into the trachea such a quantity of the above fluid as shall dilate them without danger of rupture. Then secure the trachea by ligature. In the same manner we proceed with other parts. If a portion only of an organ or a part of some viscera be required, we first secure the lower orifice by ligature; inject as above, and then apply a ligature to the upper opening. It can then be suspended in spirits of wine or turpentine.

ANTISEPTIC MENSTRUA FOR PRESERVING SPECIMENS.

Alcohol — Spirits of Wine. — This is one of the principal fluids now in use for the preservation of specimens. It may be used of various strengths, according to the size and thickness of the specimen to be preserved.

All those that are thick and bulky should be put into pure rectified spirits; smaller ones may require only one half the quantity of alcohol with water; and such as are thin and membranous, can be preserved in common New England rum.

Turpentine.— This also is an excellent antiseptic, and is highly recommended by Parsons and others, for cartilages, fibro-cartilages, and fibrous membranes.

The acids used are, sulphuric, nitric, muriatic, acetic and pyroligneous. Dr. Parsons states that Dr. Hayden, surgeon dentist, in Baltimore, has succeeded in preserving anatomical preparations in a superior manner, with pyroligneous acid. It should be rectified and diluted with water. Acids, however, cannot be used when the preparation contains bone.

METHOD OF PRESERVING THE BRAIN.

The following mixture is a very excellent menstruum for preserving the brain and nerves: Take alcohol, eight parts by weight; oxymuriate of mercury, one part. Rub the oxymuriate in a mortar, and gradually add the alcohol. The brain should remain in this mixture for twenty or thirty days, when it may be withdrawn from the liquid, dried, and varnished.

METHOD OF MAKING A DRY PREPARATION OF THE AIR-VESSELS OF THE LUNGS.

Throw the lungs of a horse into a barrel of water and allow them to macerate for several months, during the summer season; then, by repeated washing, cleanse the bronchia, etc., from the parenchyma, dry, and varnish them.

METHOD OF MACERATING AND CLEANING BONES.

Remove as much of the flesh, ligaments, etc., as can conveniently be done with the knife; then lay them in clean water, and change the same daily for about a week, or as long as it becomes discolored with blood. They are now to remain without changing, till putrefaction has thoroughly destroyed all the remaining flesh and ligaments, which will take from three to five months, more or less, according to the season of the year or temperature of the atmosphere. In the extremities of large cylindrical bones, holes should be bored, about the size of a quill, to give the water access to their cavities and a free exit to medullary substance. As the water evaporates from the vessel, it should be so far renewed as to keep the bones under its surface, or they will acquire a disagreeable blackness, and dust should be excluded by keeping the vessel constantly covered. When the white textures are destroyed, the bones must be scraped and again laid in water for a few days, and well washed and scrubbed with a coarse brush; then immerse them in lime-water, or a solution of pearlash, made with two ounces to the gallon of water, and after a week they are to be again washed in clear water. They are then to be bleached on the seashore, where they can be daily washed with sea-water.[*]

M. Bogros approves of the above plan of maceration, but at the conclusion of this he directs them to be boiled four hours in a strong solution of carbonate of potass, or in soap suds, adding hot water as fast as it evaporates. They are then to be washed frequently in cold water, and dried each time quickly, and then moistened (not steeped) in weak muriatic acid. The common bleaching liquor in a diluted state will whiten bones, but they should not be immersed in it any length of time.

When bleached, they may be varnished with the white of an egg.[†]

TO RENDER SOLID BONES FLEXIBLE AND TRANSPARENT.

One-half of the inferior jaw bone, or the scapula, are the most suitable bones for the above purpose. Macerate either or both until they are properly cleansed. Then immerse in a mixture consisting of water, twenty-five parts; muriatic acid, one part. If the bone is kept well covered during a period of about seven months, it will become flexible like cartilage; but as the phosphate of lime in the bone will neutralize some of the acid, a minute quantity may from time to time be added.

[*] Pole on "Cleansing Bones."
[†] Parsons on "Macerated Preparations."

When the preparation becomes flexible, immerse in warm water; then give it several washings in cold water to remove the acid; dry, and immerse in a glass vessel of oil of turpentine; it will assume a beautiful transparency, exhibiting the blood-vessels.

METHOD OF CLEANING AND SEPARATING THE BONES OF CRANIUM.

Take the head of a young colt, remove the skin and muscles, and wash out the brain, previously breaking it down with a stick or probe; macerate and cleanse it as before directed; then fill the cranial cavity with dry corn from the husk, immerse it in water, and the corn as it swells forces open the suturases, so that they can be readily separated by the hand. Wash, dry, and bleach the bones, and then cover them with colorless varnish.

A BRIEF EXPOSITION OF MR. SWAN'S NEW METHOD OF MAKING DRIED ANATOMICAL PREPARATIONS.*

The new method has been adopted by Usher Parsons, M. D., Professor of Anatomy and Physiology, from whose work the following selections are made:

DIRECTIONS FOR MAKING DRIED PREPARATIONS.

The part of a limb, chosen for injection, must be as free from fat as possible. A solution of two ounces of oxymuriate of mercury in half a pint of rectified spirits of wine, is to be injected into the arteries; the next day inject as much white spirit varnish, to which one-fifth of white spirit varnish has been added, and some vermilion; the limb is then to be put into hot water, where it is to remain until properly heated, when the coarse injection is to be thrown into the arteries and veins, if required, bearing in mind the course of the circulation; the valves of the veins can be broken down by a whalebone probe, if necessary. If the veins are to be injected, it is better to wash the blood out of them with water before the solution of oxymuriate of mercury is thrown into the arteries.

After the limb has been injected, it is to be dissected. Every time it is left, and sometimes during dissection, it is advisable to cover those parts which have been exposed, with a damp cloth. There are great advantages to be derived from previously injecting the limb in oxymuriate of mercury, for a limb thus injected undergoes very little change in many days, and, when the dissection is recommenced, the parts will be found in the same state in which they were left, and destitute of any offensive odor.

The oxymuriate of mercury is the best agent for arresting the putrefactive process.

After the dissection is finished, the limb, or part, must be immersed in a solution of oxymuriate of mercury for a fortnight or more.

The solution of oxymuriate of mercury must be contained in a wooden vessel, as metallic vessels do not answer.

The limb, or part, having been in the solution during the above period, it should be taken out, dried, varnished, and, if necessary, painted.

SOLUTION OF HARDENING THE BRAIN AND OTHER TISSUES.

Take of oxymuriate of mercury, one ounce; muriate of ammonia, thirty-five grains; pyroligneous acid, one pint. Rub the oxymuriate of mercury and muriate of ammonia together in a mortar, then add half a pint of pyroligneous acid.

OXYMURIATE OF MERCURY IS A VALUABLE ANTISEPTIC.

Dr. Parsons relates, that, when a piece of flesh had been immersed in a solution of oxymuriate of mercury until it was completely changed, and afterwards put into a large vessel containing water for some days, though the greater part of the oxymuriate of mercury was thus washed away, it did not even then appear in the least degree putrid. I procured half of the head and neck of a large horse, which I first injected

---

* Professor Chaussier claims to be the original discoverer of this method.

with the solution of oxymuriate of mercury, but as the putrefactive process was not thus sufficiently stopped, without dissecting off the skin I immersed it in the solution of oxymuriate of mercury for several days; and, as no marks of putrefaction remained (the offensive smell being entirely removed), I then put it into a vessel containing a large quantity of water for two or three days more, by which means nearly all the solution was removed from it. I was thus able to proceed with the dissection during the hot weather, without being in the least incommoded either by the smell or soreness of the hands, and without finding the instruments acted upon in any degree, that rendered the process at all objectionable. By putting a wet cloth over it when I left it, I was further enabled to make a very minute dissection of the nerves, which I could not otherwise have done, without the use of a large quantity of spirits of wine, and then not with half the convenience and pleasure I have thus experienced.

ON VARNISHES AND PAINTS.

The following are the recipes for the manufacture of paints and varnishes:

WHITE VARNISH.

Canada balsam, spirits of turpentine, of each three ounces; mastic varnish, two ounces. Put them into a bottle and shake them together until they are properly mixed.

MASTIC VARNISH.

This may be made by putting four ounces of powdered mastic into one pint of spirit of turpentine, to be kept in a stoppered bottle. It should be shaken every day until the greater part of the mastic is dissolved.

TURPENTINE VARNISH.

Turpentine varnish is made by melting Venice turpentine over a slow fire, and adding to it as much spirits of turpentine as will reduce it to the consistence of syrup.

WHITE PAINT.

Three ounces of the best white paint, and one ounce of spirit of turpentine, are to be put into a bottle and shaken together. When it is used with the varnish, a bottle of each should be mixed together.

PAINT FOR THE MUSCLES.

This is made by grinding on a slab a small quantity of "*lake*," with white varnish, to which one-fourth part of turpentine varnish has been added.

Dr. Parsons directs that varnish should be laid on with a fine camels'-hair pencil brush, as large as occasion may require. Hollow preparations should have the varnish poured into them, and, after turning them about in all directions, it is to be drained out as clear as possible.

DIGESTIVE SYSTEM.

OF THE MOUTH.

It may be observed here (as preparatory to the description of this part), that, in quadrupeds in general, the facial angle is one of very considerable obliquity, in consequence of the prolongation of that part of the head which corresponds to the face in the human subject; and this development of feature is in none more striking than in the horse and dog. Consequently, in these animals, the nose and mouth are cavities of large dimensions. And in the horse, the mouth appears to have been thus prolonged, not only to enable him to collect his food with more facility, but also that he might subject greater parcels of it at a time to the action of the grinding teeth, whereby the processes of mastication and deglutition are greatly accelerated.

"*Conformation.*— The mouth is constructed in part of bone, and in part of soft materials. The superior and anterior maxillary and the palate bones form the roof; the inferior maxilla, the lower part; the incisive teeth, the front; and the molar teeth, the sides. The lips, cheeks, soft palate, gums, and buccal membrane, constitute its soft parts. The tongue occupies its cavity, and the salivary glands are appendages to it.

"LIPS.

"*General Conformation.*— The lips, two in number, *superior* and *inferior*, are attached to the alveolar projections of the superior and inferior maxillae, by the muscles that move them; by the cellular tissue entering into their composition; and by the membrane that lines them. Their borders surround and bound the orifice of the mouth, and are united together on either side; which points of union are denominated their *commissures*, or the *angles* or *corners of the mouth*. Exteriorly, the lips are creased down the middle by perpendicular lines of division; exhibit little papillary eminences upon their surface; and present a softer and shorter coating of hair than what is found in ordinary places, out of which project several long straggling horse-hairs or *whiskers*. The inferior lip is altogether smaller, and is thinner in substance, than the superior; and is distinguished by a remarkable prominence about its centre, from which grows a tuft of long coarse hairs, vulgarly designated as the *beard*.

"*Structure.*— The lips are both muscular and glandular in their composition. Several small muscles,* arising from the maxillary bones, are inserted into them, and endow them with great self-mobility: one alone, consisting of circular fibres, is interwoven in their substance without having any other connection; this is denominated the orbicularis oris, or sphincter labiorum, from its use, which is that of closing the mouth. This muscle is an antagonist to all the others; they raise or depress the lips, or draw them to one side; but this contracts them, and occasionally projects them in such a manner that the horse can exert with them a prehensile power, which is most remarkably evinced at the time that he is picking up grain from a plain surface; indeed, the act of nibbling our hands with his lips demonstrates this faculty, and also the force with which he can employ it. The lips are lined by the same membrane that lines other parts of the cavity of the mouth.

* Percivall's Hyppopathology.

Beneath it are seated numerous mucous follicles, that elevate it everywhere into little *papillæ*, which are perforated by the mouths of these follicular glands, as may be readily seen with the naked eye by everting either the superior or the inferior lip. The skin covering the lips is extremely thin, and possesses considerable vascularity and sensibility. To the tenuity of it, and to the shortness and scantiness of their pilous covering, is to be ascribed the superior sensitive faculty of these parts.

" CHEEKS.

" The cheeks are constituted substantially of the masseter and buccinator muscles, covered by the skin upon the outside, and the *buccal membrane* upon the inside. Their internal or membranous surface is studded with scattered mucous follicles, whose excretory orifices may be seen by everting the part.

" GUMS.

" The gums consist of dense, compact, prominent, polished masses, of the nature of periosteum, adhering so closely and tenaciously to the teeth and the sides of their sockets, that it renders the one inseparable from the other, but by extraordinary mechanical force. Like other parts of the cavity of the mouth, they receive a covering from the buccal membrane.

" PALATE.

" Two distinct parts are included under this head; the *hard* and the *soft* palate. The hard palate is constituted of the palatine processes of the superior and anterior maxillary bones; and of a firm, dense, periosteum-like substance, the vaulted, inward part of which is elevated into several semicircular ridges, vulgarly called the *bars*. The fibres of this substance, which possess great tenacity, are inserted into the pores of the bone in every part, but are most numerous and dense along the palatine suture: the interstices are filled up by a dense cellular tissue, through the substance of which are dispersed the ramifications of the palatine vessels and nerves.

" The soft palate, sometimes called the *velum palati*, is attached to the superior or crescentic border of the hard palate, the border formed by the palatine bones; from which the velum extends backward and downward as far as the larynx, and there terminates over the epiglottis, in close apposition with that part, in a loose semicircular edge. In consequence of the velum palati being long enough to meet the epiglottis, the cavity of the mouth has no communication with that of the nose — these two parts forming a perfect septum between them; hence it is that a horse cannot respire and vomit by the mouth like a human being, in whom the velum is so short that there is an open space left between it and the epiglottis, through which air or aliment can pass either upward or downward. The soft palate is composed of extensions of membrane from the nose and mouth, between which is interposed a pale, thin layer of muscular fibres.

" The velum performs the office of a valve: it prevents the food, in the act of swallowing, from passing into the nose, and it conducts the air from the windpipe into that cavity, without permitting any to escape into the mouth.

" OF THE TONGUE.

" The *tongue*, the principal organ concerned in taste and deglutition, is lodged in the mouth; filling the interspace between the branches of the inferior maxilla.

" *Duplicity.* — Like the other organs of sense, it is double; being composed of two parts, whose union is marked by a longitudinal crease along its middle, the divisions having no vascular nor nervous connection nor in fact any intercommunication whatever; so that an animal has to all intents and purposes *two* tongues, and apparently for the same reason that he has two eyes, two ears, and two nostrils. Anatomy, as far as we can carry our researches, demonstrates this; perhaps we have no better proof of it, however, than what happens in hemiplegia, a disease in which only one half of the body is paralytic: under these circumstances, in the human subject, the patient can only see with one eye, use one arm, and taste with but one (and that the correspondent) side of the tongue.

"*Division.*— The tongue, in description, is commonly divided into *root*, *body*, and *apex*: by the attachments of the two former it is held in its situation; the latter is loose and unconnected.

"*Attachment.*— At its root, it is deeply and firmly inserted by several muscles which arise chiefly from the os hyoides and the inferior maxilla: it is also connected with the pharynx, and with the soft palate. From the sides of the lower jaw, separate layers of the membrane of the mouth are reflected upon its body, forming by their junction a sort of bridle, which is thence extended to the symphysis: to this part, which serves to restrain the organ in its motions, the name of *frænum linguæ* has been given.

"*Papillæ.*— The *dorsum* or anterior surface of this organ has a peculiar covering, which, though it appears to be continued from the buccal membrane, is a different structure altogether, and serves quite a different purpose. The surface of it is roughened, possessing a villous texture, everywhere studded with numerous little conical eminences, called *papillæ*, which are supposed to be formed out of the extremities of the nerves, and to be the especial seat of the sense of taste. These papillæ vary in size and figure, and are more abundant and larger upon the base and along the sides of the organ. Interspersed with them are a number of mucous follicles, whose apertures may be seen with the naked eye, through which a mucus is discharged upon the papillary surface, keeping it continually moist, and rendering its perception of taste more acute.

"*Structure.*— The tongue is said to possess a covering of common integument; and certainly its strong, compact tunic has all the appearances of skin, and presents the common tests of it: the external layer is laminated, is bloodless, is insensible; the internal or substantial part is tough, fibrous, vascular, and sensitive, in fact, is like cutis; and the intermediate or connecting material is delicate, soft, and reticular, and forms a bed for the lodgment of the papillæ. The substance of the tongue itself consists of an inter-union, or rather an incorporation, of its muscles, the fibres of which intersect one another, and take a variety of directions; but intermixed with them is a fine adipose tissue, to which is owing the flabby softness of the organ, and the peculiar aspect it exhibits when cut into.

"*Use.*— Though the tongue is emphatically denominated, from its essential character, *the organ of taste*, it is not the only part that possesses this faculty; for the palate, the pharynx, and the œsophagus, it is believed, participate in it. The tongue, in addition to possessing this faculty, disposes of the food during manducation, and, when sufficiently masticated, collects and thrusts it, portion after portion, into the pharynx; and furthermore, at the time the animal is drinking, it is not only employed as an instrument of suction, but also as a canal along which the fluid ascends into the pharynx.

"*Organization.*— Every part of this organ is plentifully supplied with blood. Its arteries are the lingual, branches of large size from the external carotids. The blood-vessels of either side are generally found free from anastomosis with one another; if either of the arterial trunks is filled with injection, it rarely happens that the opposite half of the organ receives any coloring from it. Its nerves are the ninth pair, which run to the muscles, and a considerable branch from the fifth pair, in whose extreme ramifications, which are distributed to the papillæ, the perception of taste is supposed to be inherent.

"OF THE SALIVARY GLANDS.

"*Number and Names.*— The salivary glands, properly so called, are six in number, three upon each side of the head; the

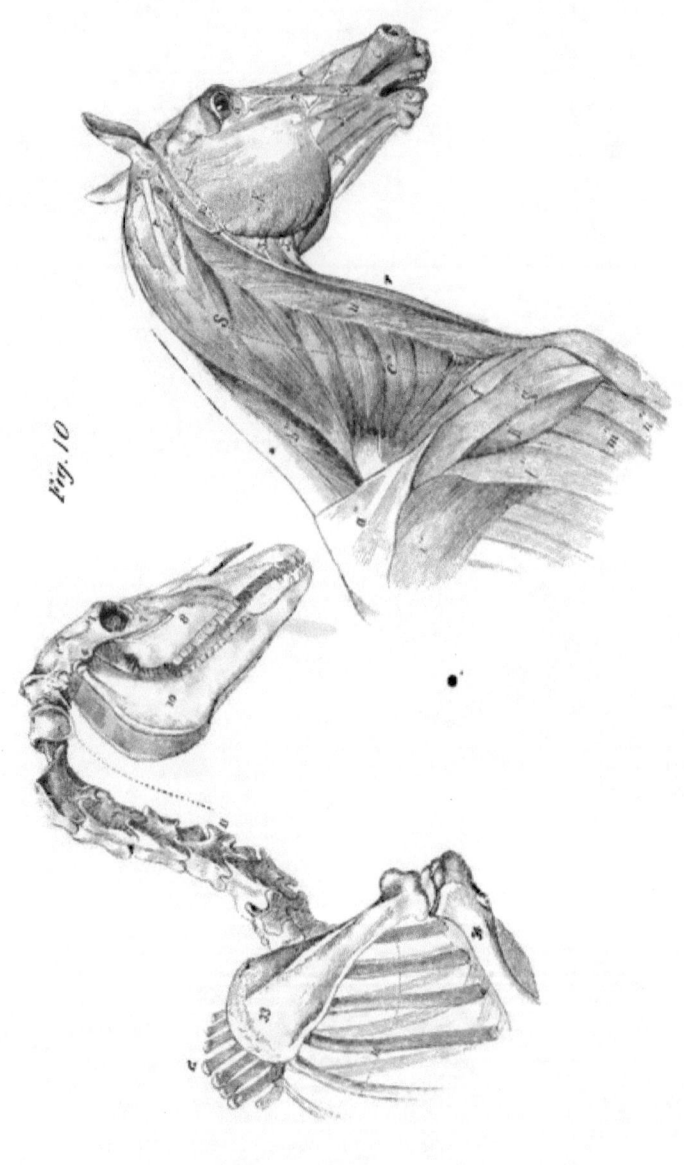

Fig. 10

EXPLANATION OF FIGURE X.

OSSEOUS STRUCTURE.
(SEE PRECEDING PLATE.)

MUSCULAR STRUCTURE.
LATERAL VIEW.

- a''. Trapezius.
- ''. Ligamentum colli.
- b''. Rhomboideus longus.
- e''. Scalenus.
- f''. Antea spinatus.
- g''. Postea spinatus.
- h''. Teres major.
- i''. Latissimus dorsi.
- l''. m''. n''. Triceps extensor bracitii.
- S. Splenius.
- K. Masseter.
- a. Orbicularis palpebrarum.
- c. Dilator naris lateralis.
- e. Orbicularis oris.
- f. Nasalis longus.
- g. Levator labii superiorus.
- h. Buccinator.
- i. Zygomaticus.
- J. Depressor labii inferiorum.
- m. Attolentes.
- n. Retrahentes aurem.
- o. Abducens vel deprimens aurem.
- q. r. Tendon of the splenius and complexus major.
- t. Obliquus capitis inferiorus.
- u. Levator humeri.
- z. Subscapulo hyoideus.

VEINS.

1. Temporal vein.
2. Facial "
3. Branch of the jugular.
10. Parotid gland.

parotid, the submaxillary, and the sublingual.

"The parotid, the largest of these glands, so called from being placed near the ear, lies within a hollow space at the upper and back part of the head, bounded by the branch of the lower jaw before, and the petrous portion of the temporal bone behind; it extends as high up as the root of the ear, and as low down as the angle of the jaw, by which latter a small portion of it is concealed. This gland, like the others of the same class, is enveloped in a case of dense cellular membrane, and is constituted, in structure, of many little lobes or *lobuli*, connected together by processes transmitted into the interior from this cellular covering. Every lobulus is composed of a distinct set of secretory vessels, from which numerous *tubuli* arise, conjoin, and at length form one main branch; these branches, which correspond in number to the lobuli, unite and re-unite until they end in one common excretory duct. The duct emerges from the inferior part of the gland, runs along the inner part of the angle of the jaw, and crosses over the posterior edge of the bone immediately above or behind the submaxillary artery and vein; in the remainder of its course it corresponds to the border of the masseter, and, about opposite to the second anterior molar tooth, pierces obliquely the buccinator, and terminates by a tubercular eminence upon the internal surface of the buccal membrane."*

"The submaxillary gland, of smaller volume than the parotid, lies in the space between the angles of the jaw, to which, and to the muscles thereabouts, it is loosely attached by cellular membrane; a portion of it is also generally found proceeding backward as far as the trachea. Its structure is similar to that of the parotid gland. The submaxillary duct issues near the centre of the gland, creeps along the under and inner border of the tongue, close to the lower edge of the sublingual gland, and terminates by a little mammiform elongation of membrane, vulgarly called the *barb* (barbillon) or pap, upon the fraenum linguae, about half an inch above its attachment to the symphysis. Among the other ridiculous and mischievous practices of farriers is that of snipping off these processes. They were seemingly designed as valves, to prevent the insinuation of alimentary matters into the ducts. The coats of this vessel are extremely thin and translucent.

"The sublingual gland is still smaller in volume than the submaxillary, though, altogether, one much resembles the other in figure. It lies along the under part of the tongue, covered by the buccal membrane, where, from the lobular unevenness it gives to the surface, its situation is well marked. Its ducts penetrate the membrane by the side of the fraenum linguae.

"*The use of the salivary glands* is to secrete a saline limpid fluid, called *saliva*; which is conveyed and poured by their ducts into the mouth during manducation: here it is mixed with the food, mollifying it, and rendering it more easy of digestion, and at the same time facilitating the passage of the alimentary bolus into the stomach.

"OF THE PHARYNX.

"The pharynx is a funnel-shaped sac, lodged in the throat for the reception of the food.

"*Situation.*—The pharynx is contiguous to the gutural pouches, superiorly; the larynx, inferiorly; and the anterior portions of the parotid glands and branches of the jaw, laterally. Posteriorly, it is continuous in substance with the esophagus; anteriorly, it presents an opening to the mouth.

"*Attachment.*—In front, to the os hyoides and palate bones; below, to the larynx; behind, it grows narrow and ends in the esophagus.

---

* To expose this duct, at or near its issue from the gland, an incision should be carried along the posterior border of the branch of the lower jaw; first, dividing the skin; secondly, the panniculus; thirdly, the cellular tissue immediately covering the duct, which is readily distinguished by its glistening pellucid aspect. By extending the incision around the angle of the jaw, directing it towards the inner edge of the bone, the duct will be found making its first turn: here, however, it is lodged in a hollow, deeply buried in cellular tissue.

"*Structure.*— The pharynx is in part muscular, and in part membranous. Of the muscles belonging to it the constrictors are those that more immediately enter into its composition. They are so disposed as to give the membrane forming the sac a complete fleshy covering, which is rendered the more uniform by their proximate fibres being indistinguishably blended; thus the muscles form the most substantial part of the pharynx. The lining membrane, which is of the mucous class, is soft and thick in substance, and palely tinged with red in color, and is papillary and in places rugose upon its surface; being perforated by the ducts of numerous follicles which discharge a mucus that preserves glibness and moisture to its interior. The membrane itself is (where it meets them) continuous both with the buccal membrane and that which lines the œsophagus.

"Although the pharynx is designed for the reception of the food, it does not open directly into the mouth; the two cavities are separated from each other by the soft palate and epiglottis. Except in the act of swallowing and coughing, they have no communication: in the former case, the velum is pressed upward by the food against the posterior openings of the nose; in the latter, the larynx is depressed by a convulsive action of the muscles in the vicinity. Into the cavity above the velum there are four openings — two of the chambers of the nose, one of the larynx, and one of the œsophagus: the eustachian tubes do not open into the pharynx; they end in two large membranous pouches at the upper part of the fauces. The opening leading into the œsophagus is constantly closed, except when alimentary matters are passing to or from the stomach; so that air received into the pharynx through the nose can pass nowhere else but into the windpipe; but if food be returned from the stomach, it will be regurgitated into the nose; at least, only that portion of it which enters the pharynx at the moment that the larynx is depressed in the act of vomiting, can be thrown into the mouth, in the same way that air is in the act of coughing.

"OF THE ŒSOPHAGUS.

"The œsophagus, or gullet, is the tube through which the food is conducted from the pharynx into the stomach.

"*Course.*— It has its beginning from the pharynx, and is there placed at the upper and back part of the larynx, taking the first part of its course above and behind the trachea, between that tube and the cervical vertebrae. Having proceeded a short way down, it inclines to the left, and soon after makes its appearance altogether on the left side of the trachea, and continues so placed during the remainder of its passage down the neck: this explains why we look for the bolus during the act of swallowing on the *left*, and not on the right side of the animal. In company with the trachea, the œsophagus enters the thorax between the first two ribs, at which place, running above that tube, it quits its companion for the superior mediastinum, which cavity it traverses below and a little to the right of the posterior aorta. Immediately beneath the decussation of the crura, the œsophagus pierces the substance of the diaphragm, and enters the stomach, at a right angle, about the centre of its upper and anterior part.

"*Structure.*— The œsophagus presents, externally, a strong, red, *muscular* coat; internally, one remarkable for its whiteness, which in its nature is *cuticular*. The muscular coat is composed of two orders of fibres — a longitudinal, forming an outward layer; and a circular, an inward layer: the former will shorten the tube, and perhaps dilate it for the reception of food; the latter, by successive contractions of the canal, will transmit the food into the stomach. The second, or internal coat, is called the cuticular, from its analogy to the cuticle of the skin. Although it is continuous with the membrane of the pharynx, it is of a totally different composition: it is thinner, but it is much more compact and stronger in its texture, and, I believe, is both insensible and inorganic. It adheres to the muscular covering by a fine cellular tissue, the extensibility of which gives full play to the latter; and admits, during the empty or collapsed state of the tube, of the former being thrown into many longitudinal *plicæ* or folds; as is demonstrated by making a transverse section of the tube: such appearances result from the contraction of the one coat, and the want of proportionate elasticity in the other. Between the two tunics, imbedded amongst the connecting cellular tissue, are numerous follicular glands, whose office is to pour forth a mucous secretion upon the internal surface of the lining membrane, to render the passage of food along it glib and free from any friction.

"NASAL FOSSÆ.*

"The nasal fossæ are the two chambers or lateral cavities, whose external openings are the nostrils. Their *walls* or *external parietes* are almost entirely osseous; and to the OSSEOUS SYSTEM (page 45) the reader must turn for a description of the manner in which the fossæ are formed, and of the bones entering into their formation. But, in addition to bone, they are *cartilaginous* in their constitution.

"The cartilages of the nose are *five* in number:—of which one (the septum nasi) is situated *internally;* the other four (entering into the composition of the nostrils) *externally.*

"The septum nasi is the vertical cartilaginous partition interposed between the nasal fossæ. It exhibits four borders. The inferior one is received into the groove of the vomer; while the superior presents a lengthened channel between two elevated edges, into which is admitted the internal crest formed by the union of the nasal bones. Its posterior border is affixed to the ethmoidal plate: its anterior serves to sustain the cartilages forming the nostrils. Both its sides are completely covered by the Schneiderian membrane.

* Hippopathology.

"*Nostrils.*—Four in number: two on each side, distinguished by the epithets *true* and *false.*

"The true nostrils are the large, ovoid, and ever-open orifices so conspicuous externally. They have for the base of their structure four pieces of fibro-cartilage, which are involved in doublings of the common integument. Each nostril is formed of two flexible *alæ* or wings: a *superior* or internal one, and an *inferior* or external. The former is supported by a broad circular cartilaginous plate; the latter is crescentic in shape, and forms a flexure outward, within which is perceptible the orifice of the lachrymal duct. They are attached to, and supported by, the nasal peak and septum nasi.

"The false nostrils are two little pouches or cavities (having the semblance of *culs-de-sacs*), situated internally, above the true nostrils, into which an external opening is found within the commissure formed by the union of the two alæ. They are formed out of duplicatures of the skin, which is here thinner, and finer, and softer in its texture; and, except at their entrance, are without hair upon their surfaces. Their use is not known.

"*Schneiderian membrane.*—The cavity of the nose is not only divided into the two nasal fossæ, but each fossa is subdivided into the three *meatus* (for a description of which, vide page —). Every part of these cavities and passages is covered by the *Schneiderian* or *pituitary membrane.* This is a membrane of the mucous class, distinguished for its thickness of substance, for its vascularity, and for its olfactory papillæ. It has two surfaces: an exposed or secreting one and an unexposed or adherent one. *The secreting surface* is smooth, and is rendered glib and shiny by the varnish it derives from the mucous secretion emitted by the numerous small rounded pores everywhere visible in the membrane, but more particularly upon the lower part of the septum, and upon the inferior turbinated bone. This surface exhibits a pale pink blush, the effect of the bloodvessels spread over it, which are here so superficial as to owe their principal defence to the mucous exudation: hence it is that the complexion of the membrane (varying with the influence of the atmosphere and other agents) is extremely fugitive and uncertain. *The adherent surface* of the membrane contracts a close and firm adherence to the parts it covers, through the insinuation of its fibres into them: indeed, to the bone it appears to supply the place of periosteum; to the cartilage, of perichondrium. The substance of the membrane exhibits a fibrous structure, interwoven with cellular tissue; and upon that—as a substratum—is spread a glandular and vascular apparatus, from which issues the mucous secretion; together with numerous *papillæ*, of small size, constituted of the terminations of those nerves from which the membrane derives ordinary sensation, as well as those that endow it with the peculiar sense of smelling. The Schneiderian membrane, inferiorly, within the nostrils, is continuous with the duplicatures of skin lining those parts; superiorly with the membrane lining the pharynx; besides which, it is continued into the several sinuses of the head, through the openings leading from them into the nose, and likewise gives them a complete covering: it is to be observed, however, that in the sinuses the membrane is thinner, and assumes a paler and more delicate aspect; its natural secretion is also found more sparing. The membrane is abundantly supplied with blood-vessels, as well as nerves; and also possesses its share of absorbent vessels. Its arteries, which ramify and anastomose so as to form a spreading network upon the secreting surface, are derived superiorly from the *lateral nasal*; inferiorly from the *facial* and *palato-maxillary*. Its nerves are furnished by the first and fifth pairs.

"*Sinuses.*— These cavities are formed in the interior of several of the bones of the cranium and face: in fact, with the exception of the membrane lining them, they are entirely osseous in their composition. This will account for their description having been already given (at page 46), to which we must again refer.

"*Ducts.*— There are two ducts belonging to, or connected with, the nose. One is the *ductus ad nasum*—a tube partly osseous and partly membranous in its composition, commencing at the inner angle or corner of the eye, within the substance of the lachrymal bone, running within a canal continued from this bone through the superior maxillary bone, and terminating at the inner and inferior part of the nasal fossa, underneath the duplicature of the inferior ala, upon the surface of the common skin, about one-fourth of an inch from its junction with the Schneiderian membrane, by an orifice large enough to admit a crow-quill. The other duct is the *ductus communis narium*, which pursues its course along underneath the vomer to the pharynx; after arising from two lateral branches springing from oblong apertures in the floor of the nostrils."

INTERNAL PARTS.

COMPREHENDING THE CAVITIES OF THE CRANIUM, ORBIT, NOSE, AND MOUTH.

I. — CAVITY OF THE CRANIUM,

CONSTRUCTED for the lodgment of the brain with its appendages, is in form ovoid, flattened inferiorly, broader anteriorly than posteriorly; its antero-posterior or long diameter measuring about seven inches; its transverse or lateral diameter about four inches; its vertical or perpendicular diameter about three and a half inches. At the same time it is to be observed, that, although the general form of the cavity is the same, its dimensions may and do vary in different heads. The eight bones composing the cranium all present internally surfaces more or less concave, which, united, form the cavity under consideration; hence it is that the interior is not regular or uniform, but presents to view different hollows, which are adapted to distinct prominences of the cerebral mass.

Division of the interior surface into roof and base of the cranium:

The roof is formed by the frontal, parietal, and occipital bones: its superficies is larger than the extent of the base, and it is without any apparently defective places, observable in the latter. It presents—1st. On the mesian line from front to back, *the sagittal groove*, for the longitudinal sinus formed by *the frontal and parietal crests*, crossed towards the front by *the coronal suture*, and bounded posteriorly by *the parietal protuberance*, to which is attached the tentorium, and behind which is *the occipital capula*, for covering the cerebellum. 2nd. On either side, along the same line, *the cerebral concavities of the frontal bone*; *the coronal suture*, the boundary line between them and *the parietal concavities*; *the transverse grooves*, for the lateral sinuses; and, sunk within them, *the lambdoidal suture*.

The base is formed by the temporal, sphenoid, ethmoid, and occipital bones. It presents—1st. On the middle line, from before backwards, *the crista galli*, and on its sides *the ethmoidal fossæ and cribriform plates*, bounded laterally by *the internal orbital plates of the frontal bones*, and there pierced by *the internal orbital foramina*; *the concave surface of the body of the ethmoid bone*; *the optic hiatus* leading to *the optic foramina*; a transverse suture between the ethmoid and sphenoid bones. Upon the sphenoid bone, *the pituitary fossa*, bounded laterally by the two *optic fossæ*; the latter leading to the foramina lacera orbitalia, over which are *the spinal foramina*; a transverse elevated line denotes the place of junction of the sphenoid with the occipital bone. Belonging to the occipital bone, are *the basilar fossæ* and the *occipital hole*. 2d. On either side, in the same direction, *the internal surface of the wing of the ethmoid bone*, rather more convex than concave, for the support of the anterior lobe of the cerebrum; *the concavity of the wing of the sphenoid bone*, for the reception of the middle lobe; *the concavity of the squamous part of the temporal bone*, for lodging the posterior lobe: and the sutures bounding these three cerebral surfaces. The *foramen lacerum basis cranii*, formed between the wing of the sphenoid anteriorly, the basilar process of the occipital bone internally, and the petrous portion of the temporal bone externally and posteriorly: it is wide and irregular before, narrow behind, and is distinguished into *the spheno-occipital* and *temporo-occipital hiatus*. The petrous portion of the temporal bone, presenting a narrow triangular surface forwards and upwards, which contributes to the posterior cerebral concavity; a broad, smooth, but uneven surface inwards, against which inclines the cerebellum, and upon which we distinguish—*a*, the orifice of *the meatus auditorius internus*; *b*, a *transverse prominence*, and several *cerebral indentations*; *c*, an irregular convexity downwards, which forms the boundary wall of the labyrinth; *d*, a *fissure* separating it from the former. Lastly, *the sutures*, uniting the petrous to the squamous portion and to the occipital bone. Of the occipital bone a part of the internal surface assisting in the formation of a concavity for the cerebellum, by the convolutions of which it is indented; the surface even and smooth, and slightly excavated below this, for the support of the medulla oblongata; still lower, *the condyloid foramina*, through which the ninth pair of nerves pass out.

II.—THE ORBITS,

Two in number, are formed for the lodgment, attachment, and protection of the eyes and their appendages.

Figure.—Symmetrical. The cavity, which is extended horizontally backward and inward, has, viewed in front, a pyramidal aspect: the base, represented by the front, has four sides, and four angles; one only of the sides, however, is sufficient in extent to reach the apex, the others being all more or less imperfect. A line drawn in a horizontal direction through the axis of this figure, inclines more outwards than forwards, more forwards than downwards, intersecting another horizontal line projected directly forward at an angle of about 70°, and one extended laterally, directly outward, (at right angles with the former), at about 20°: the inclination downward, however, will in course vary with the erect position of the head.

Structure.— The orbit is composed of unequal portions coming from four of the bones of the cranium, and from three of those of the face: viz., the frontal, ethmoid, sphenoid, and temporal bones; the malar, lachrymal, and palate bones.

Division— Into sides, angles, base, and apex.

Sides.— The superior side or roof of the cavity consists only of *the frontal arch;* which is concave and smooth internally, to make room for the lachrymal gland, and has anterior and posterior borders, sharp and slightly curved. The inferior side or floor of the orbit is formed by *the orbital surfaces of the lachrymal and malar bones,* is broader than the roof, though, like it, is deficient as a whole. It comprises *the orbital portion of the lachrymal suture:* it is terminated in front, by *a smooth, rounded, curvated border;* behind, nearly midway between the base and apex, by *a shorter and straighter border.* The internal or nasal side, the broadest and only complete one, is formed principally by *the internal orbital process of the frontal bone,* into the notch of which is received *the os planum:* the ethmoid bone further contributes, and also *the sphenoid* and *palate bones,* the three constituting that irregular termination of the cavity behind which represents the apex. The frontal orbital plate is smooth and slightly concave, and is united below by a continuation of *the transverse suture* with the lachrymal bone. Its *border in front,* though slightly curvated, is *very irregular,* having several notches and one or two small *foramina* in it; it also presents *a little tubercle,* to which the lachrymal caruncle is attached. The external or zygomatic side is formed principally by *the zygomatic process of the malar bone,* that of the temporal contributing but little: it is concave, and smooth internally, somewhat broader below than upwards; is intersected obliquely by the zygomatic suture, and has an interior border, smooth and curvated, a posterior one, sharp and straight.

Angles.— The supero-internal angles, one before, the other behind, are formed by the beginning of the frontal arch, through which, midway between them, passes the supra-orbital foramen. The infero-internal angle includes *the lachrymal fossa.* The supero-external angles, one anterior, the other posterior, are intersected by the *suture* uniting the frontal and zygomatic arches. The infero-external angles, particularly the anterior, are rounded and smooth.

Base.— Of the circumferent border, the superior and internal parts, about two-fifths of the entire circle, are formed by the os frontis; the inferior and internal parts, about one-fifth, by the lachrymal bone; and the remaining two-fifths by the malar and temporal bones, in the proportion of three parts of the former to one of the latter.

The apex or back of the orbit, formed by the ethmoid, sphenoid and palate bones, is pierced by five foramina: the two round are *the internal orbital* and *optic,* which are ranged in a row with two oval and larger in size, *the supero-posterior* and *infero-posterior orbital;* the one behind is *the spinal foramen.*

III.—CAVITIES OF THE NOSE,

Comprehending the nasal fossæ or chambers, and the sinuses. These cavities occupy about two-thirds of the internal space of the superior maxilla, the remaining third belonging to the cranium; from which they are partitioned by the cranial septum of the frontal bone, in union with the cribriform plates and crest of the ethmoid.

The nasal fossæ may be said to include about two-thirds of the entire space devoted to the olfactory cavities. They constitute the interior of the proboscis; have four boundary walls, one above, one below, and two laterally; are separated from each other by a septum; but are open both before and behind.

The superior wall presents an irregular concave formed by the internal surfaces of the nasal bones, the cells and grooves of the ethmoid, and small portions of the nasal surfaces of the palate bones.

The inferior wall is horizontal; it extends forward beyond the superior, but is considerably overreached by that wall posteriorly: it is formed by the palatine portions of the anterior and superior maxillary, and by the palate bones. The surface is transversely concave, and presents a slight eminence a little behind its middle.

Each lateral wall or side presents an irregular concavity, and is formed by the anterior and superior maxillary and the palate bones. To it are attached the superior and inferior turbinated bones, by which the fossa is divided into three separate passages or meatus. *The superior meatus*, comprised between the nasal and superior turbinated bones, extends from the angle of the lateral nasal opening, passing over the ethmoidal cells, to the cribriform plate, following superiorly the declination of the wall. *The middle meatus*, included between the turbinated bones, leads superiorly into the ethmoidal grooves and cells, and into the sinuses of the head, and ends below, beneath the termination of the superior. This passage, like the former one, is narrow; but its greatest diameter is, obliquely, in the perpendicular direction; whereas the other measures most from side to side. It receives the apertures of the ductus ad nasum, maxillary sinus, ethmoidal grooves, and turbinated cells. *The inferior meatus* is the most capacious as well as the most direct one: it extends along the inferior wall, from the anterior to the posterior opening of the nose.

The septum nasi is the partition separating one fossa from the other. It is formed, posteriorly, by the ethmoidal plate; inferiorly and posteriorly, by the vomer; superiorly and anteriorly, (and principally) by a broad perpendicular plate of cartilage.

The openings of the nose are: *the anterior*, divided by the nasal peak and septum nasi into two, and formed by the superior borders of the anterior maxillary bones: *the posterior*, divided after the same manner by the vomer and septum, and formed by the nasal surfaces and crescentic borders of the palate bones.

The sinuses of the head communicate with, and may be said to constitute part of, the nasal cavities. They are the frontal, nasal, maxillary, sphenoidal, ethmoidal, and palatine.

The frontal sinuses, formed within the frontal bones, are situated so that a straight line extended between the supero-internal angles of the orbits passes opposite to about the angular or deepest parts of their cavities. The sinus (on either side) has a triangular figure. *The superior side* or roof is flat, and (barring the septa) even upon its surface; whereas *the posterior side* is irregular, being convex inwardly, where it is formed by the cranial septum; concave outwardly, where it is opposed to the part composing the temporal fossa. The inferior side slants from behind forward, and from below upward, is irregular on its surface, and open or deficient outwardly, where the cavity communicates with the maxillary sinus. Of the *angles*, one is directed upward; another downward, terminating in the nasal sinus, with which it is conjoined, the two forming one continuous cavity; the third points backward, and is directly opposite to the imaginary transverse line above alluded to. The cavity is traversed and divided into several unequal open compartments and recesses by *septa*; the principal of which is one extended between the superior and inferior sides; it is partitioned from the opposite sinus by the nasal spine. The sinus is but small in the young compared to its proportionate dimensions in the adult subject: it continues to increase afterwards with age, and ultimately extends throughout the whole of the frontal bone.

The nasal sinuses, formed by the nasal bones above and the superior turbinated bones behind, are nothing more than the culs-de-sacs or blind terminations of the frontal sinuses.

The maxillary sinuses, the largest of these cavities, are spacious but very irregularly formed. They are situated below and in front of the frontal. Of this sinus, on either side, the posterior and external walls are formed by the malar and lachrymal bones, whose orbital processes constitute a thin partition between it and the orbit; the inferior parts consist of the excavations in the superior maxillary bone; superiorly, the sinus is open, being there continuous with the frontal: the boundary line between these cavities is marked by the suture uniting the lachrymal to the frontal and nasal bones on the outer side, and by the prominent crest formed by the junction of the superior turbinated with the ethmoid bone on the inner; underneath which part, through a curved (and in the recent subject sort of valvular) fissure, the sinus opens into the middle meatus, between the bases of the turbinated bones. The cavity is but small, and still more irregular, in the young subject, in consequence of the intrusion of the yet unent molar teeth.

The frontal sinus, then, terminates in the nasal, but both discharge themselves into the maxillary; the maxillary has also a blind termination, but empties itself into the posterior part of the middle nasal meatus.

The sphenoidal sinus is situated within the palatine portion of the body of the sphenoid bone. It has no existence in the young subject, the bone being solid throughout: but in process of growth a cavernous hollow is formed, which, from the secession and attenuation of the laminæ of the bone, continues to enlarge. It communicates, by two ovoid openings, with the ethmoidal sinuses.

The ethmoidal sinuses are two cavities, separated by the perpendicular plate, situated beneath the ethmoidal cells. They have openings in front, communicating with the lowermost and largest grooves of the same bone, and with the palatine sinuses.

The palatine sinuses are formed between the superior maxillary and palate bones; are situated below and in front of the former; are separated from each other by the vomer; and open into the maxillary sinuses: they are irregular in form and cavernous interiorly. They are not to be found in the young subject. Some might be inclined to treat them as parts of the maxillary sinuses; they are, however, as perfectly distinct from the latter as the frontal are.

IV.—THE MOUTH.

The mouth is the cavity included between the superior and inferior maxillæ, making (in the skeleton) one common vacuity with the inter-maxillary space. Its antero-posterior dimensions can be but little varied; but its supero-inferior diameter will be increased in the ratio of the distance to which the inferior maxilla recedes from the superior; the cavity during the distraction of the jaws assuming the figure of a misplaced $>$, the angle of which is turned backward.

The mouth is formed—superiorly, by the palatine and superior and anterior maxillary bones; inferiorly, by the inferior maxilla; laterally, by the molar teeth; anteriorly, by the incisive teeth. Behind, through the posterior opening of the nose, it communicates with the nasal fossæ.

PERITONEUM.*

The whole of the viscera contained within the abdomen proper, including the anterior part of the rectum, bladder, and vasa deferentia, are either entirely or partially covered by or in contact with peritoneum. This is a serous membrane reflected also over the parietes of the abdomen, so that a parietal and visceral or reflected portion require notice. Like other membranes of the same nature, it forms a closed sac, which, however, is not the case in the female, as its cavity communicates with that of the uterus, owing to the open state of the Fallopian tubes at their fimbriated edges.

It is loosely connected with the abdominal parietes by subserous cellular tissue, and the same obtains with regard to its connection with the viscera. But we find some parts more adherent than others, such as along the linea alba and cordiform portion of the diaphragm. Also on the organs it is but loosely connected with them at their attached border, where it forms generally a triangular space, occupied simply by vessels, nerves, and cellular tissue, and allowing of

* Prize Essay by Mr. Gamgee.

their distention and alteration in figure. On the other hand, it is more adherent as it extends over the free surface or margin of the various parts it is in contact with.

The peritoneum being considered as extending from the umbilicus over the abdominal parietes towards the median line of the diaphragm and spine, is found there to fold on itself, and proceed from the latter on to the intestine, forming the mesenters; and from the former on to the liver and stomach, constituting ligaments. These folds of peritoneum are also seen extending from organs to other parts of the abdominal parietes, and these also constitute ligaments. Then they may be traced from one organ to another, giving rise to the several omenta; all of which we shall more especially allude to as we speak of the peritoneal coat of each separate viscus.

STOMACH.

The stomach is the dilated portion of the alimentary canal, intermediate between the œsophagus and small intestine: through the former it receives the ingested aliment, for which it acts as a reservoir during the process of chymification, the active agent in which is the gastric secretion.

In the horse, as well as all other solipedes, this viscus is exceptional in not being the most capacious dilatation of the alimentary canal. M. Colin, in a paper published in the *Recueil de Medecine Veterinaire Pratique for June,* 1849, states that the capacity of the horse's stomach is very variable. He says, that in a very small horse he found it only nine quarts (according to his evaluation by litre, which may be considered as thirty-four fluid ounces), while in one of colossal dimensions it was as much as 33 3-4 quarts, both having died at the college (Alfort) infirmary. He gives the average as being from 13 7-20 quarts to 14 3-5. Then, considering the capacity of the stomach in relation with that of the intestines, he found it in a very small horse as one to thirteen, while in other two cases it was as one to ten. He takes the latter as the standard relative capacity between the two.

The stomach is situated transversely to the long axis of the body, in the left hypochondrium, extending into the epigastrium and during repletion into the right hypocondriac region. However, its size and situation vary under different circumstances, as to whether it be full or empty, adapting itself generally to its contents.

The stomach is fixed on its left side to the diaphragm by the œsophagus, having the spleen attached to it as well. The duodenum then, by means of the lesser or gastro-hepatic omentum, suspends the pyloric end by getting attached to the concave surface of the liver.

The shape of the stomach might be expressed as being that of a tube bent on itself, and dilated along its convex border, so as to form two cul-de-sacs: i. e., a right and a left one, whilst it has two borders or curvatures, distinguished as a lesser concave and a greater convex one. The stomach has two smooth surfaces, the anterior one being in contact with the liver and diaphragm, whilst the posterior one corresponds to the convolutions of the small intestines and gastric flexure of the colon. It has two orifices, i. e., a left œsophageal, or commonly called cardiac, and a right intestinal or pyloric one; the latter taking its name from the valve by which it is guarded.

A circular depression round the stomach, midway between the cardiac and pyloric orifices, most visible when the organ is replete, marks the external division of the stomach into a cardiac and pyloric portion, corresponding with the point where the mucous membrane varies in character internally. The sacular projection at the cardiac portion takes the name of fundus, owing to its greater magnitude as compared with a smaller cul-de-sac at the pyloric end, the analogue of which in human anatomy is characterized by the appellation of antrum pylori.

Having thus briefly described the striking peculiarities of the stomach, I proceed with more detail to the consideration of its constituent parts, such as its coats, nerves and vessels.

The coats of the horse's stomach having been generally described as four, it appears needless to alter their nomenclature, although the one which I shall allude to as third might quite as justly be described as second, or merely spoken of as connecting cellular tissue, without regarding it as a separate coat.

The external peritoneal tunic is found proceeding from the diaphragm on to the cardiac portion of the stomach, surrounding the œsophageal opening, where it is tough, and forms the gastro phrenic ligament. Thus we follow it on to the corresponding surface of the viscus, and, firstly, more especially on to the lesser curvature, where it is loosely connected with the other coats, and, the middle portion being more adherent, gives rise to two folds laterally, which seem to stretch from the cardiac to the pyloric orifices, to bind the two together, necessarily leaving a pit or cul-de-sac between them. At the pyloric end the peritoneum comes off from the concave surface of the liver on to the stomach, constituting the gastro-hepatic or lesser omentum, the anterior layer of which comes from the anterior part of the concave surface of the liver, whilst the posterior layer comes from the posterior part of the same, so that the two enclose the vessels going to and from the porta.

Having formed a covering to the corresponding surface of the stomach, the layers of peritoneum meet at the greater curvature. In following them from this point the description will be facilitated by alluding to the two separately, as they meet to form the gastro-splenic and gastro-colic omenta, as well as the omental sac. In forming the latter, they so blend as merely to constitute a fine reticulated vascular layer, inseparable into two, except near the margins of the viscera. Distinguishing the anterior or external layer as A, and the posterior or internal one as B, their arrangement admits of exposition in the following terms:—

A passes from the anterior surface of the stomach, forms the loose omentum, and gets on to the transverse colon and spleen. Reaching the latter, it is reflected over its superior surface at the posterior margin of the hilus, so as to contribute to the formation of the gastro-splenic omentum, and extends round the free posterior margin of the viscus on to the inferior surface, passing to the right on to the left kidney, and, anteriorly reaching the supero-anterior part of the spleen, is reflected from it so as to continue as the outer layer of the loose omentum. Further to the right, A is traceable on to the inferior surface of the transverse colon, and, extending round the posterior part of the latter, is found to ascend up to the spine, and then turn backward and downward to form the mesentery.

B, or the internal layer of peritoneum, passes from the posterior surface of the stomach till it reaches the infero-anterior border of the transverse colon, as well as the hilus of the spleen. After covering the anterior surface of the colon, it ascends up to the pillars of the diaphragm clothing the anterior part of the pancreas, which is thus held between A and B, or layers of the transverse meso-colon. A little to the left of this, B passes on to the anterior margin of the hilus of the spleen, forming the inner or posterior layer of the gastro-splenic omentum.

From this arrangement it results that the peritoneum, in forming the lesser or gastro-hepatic, the greater or gastro colic, and the gastro-splenic omenta, closes in a space termed the omental sac, the interior of which is inaccessible except by an opening at the posterior part of the gastro hepatic omentum, whose free margin at the right side marks the point where it may be penetrated; this passage is termed the foramen of Winslow. It is bounded anteriorly by the lesser omentum, above by the liver, and posteriorly by the transverse colon.

Thus, supposing the inner layer of the omental sac to be separable from the outer, and drawn out through the foramen of Winslow, the following parts would be deprived of peritonæum, i. e., the posterior surface of the stomach, the gastro splenic omentum of its posterior layer; so that the

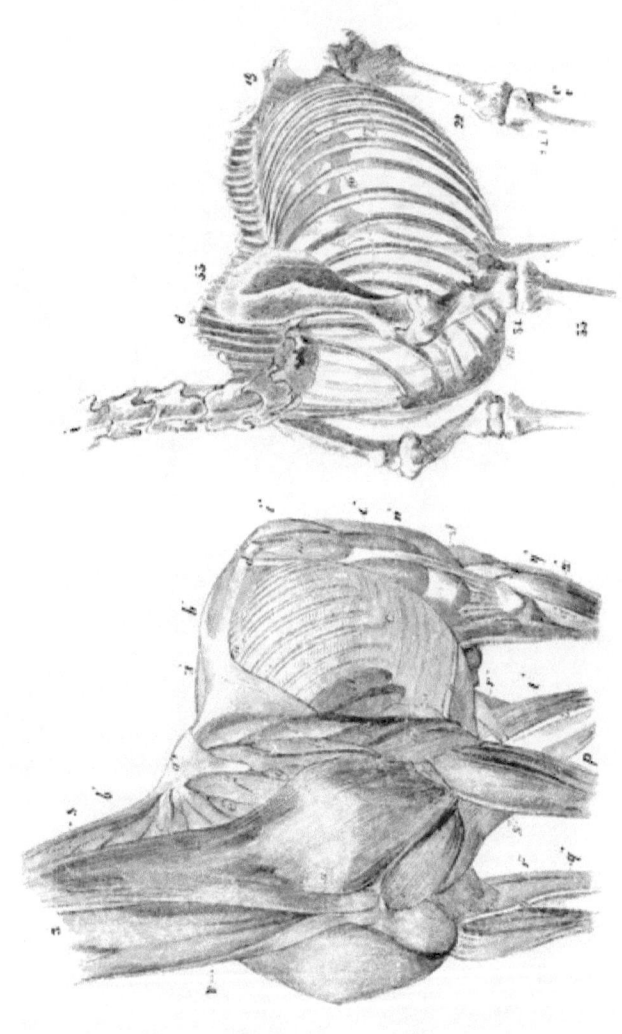

EXPLANATION OF FIGURE XI.

MUSCULAR STRUCTURE.

a''. Trapezius.
b''. Rhomboideus longus.
S. Splenius.
c''. Scalenus.
e''. Pectoralis transversalis.
f'''. Antea spinatus.
g''. Postea spinatus.
h''. Teres major.
i'''. Latissimus dorsi.
J'''. A portion of the serratus magnus.
k''. "Humero cubital."
l'', *m''*, *n''*. Triceps extensor brachii; magnum, medium, et parvum.
o'''. Pectoralis magnus.
p. Flexors.
q''. Flexor metacarpi externus.
r''.   "   "   internus.
s''. Extensor metacarpi magnus.
z''. Extensor pedis.
a'. Levatores costarum.
e'. Obliquus externus abdominis.
d'. Obliquus internus abdominis.
y'. Region of the patella.
k', *l*. Glutei muscles.
m'. Tensor vaginæ.
n'. Rectus.
o'. Vastus externus.
q'. Flexor metatarsi.
r'. Gastrocnemius externus.
t'. Flexor pedis accessorius.
n. Sterno maxillaris.
v. Internal part of the levator humeri.
y'. Peroneus.
z'. Extensor pedis.
J'. Triceps.
d. Regio.

VEINS.

3. Jugular vein.
4. Subcutaneous thoracic vein.
5. Saphena vein.
6. Radial vein.
D. Serratus magnus muscle.

OSSEOUS STRUCTURE.

d. Dorsal spines.
f. Ulnar.
c, *e*. Fibula.
16. True ribs.
17. False ribs.
18. Sternum.
19. Ileum.
2. Femur.
23. Patella.
24. Tibia.
34. Os humeri.
35. Radius.

vessels going to and from the stomach and spleen would remain uncovered, the anterior part of the transverse colon, the anterior surface of the pancreas, and inner or posterior layer of the gastro hepatic omentum.

Next to be described to the serous coat is the muscular one, which is constituted of involuntary plain fibres, whose thickness is very variable in different subjects, as well as in different parts of the same stomach. The cardiac end is more muscular than the pyloric, except at the right margin of the latter, where it is very powerful and thick, as it surrounds the pylorus. The thinnest part of the stomach is unquestionably the convex border of the lesser cul-de-sac.

The muscular coat of the stomach is intricately arranged, and authorities differ vastly from each other in the description of the several layers constituting it. The number of layers entering into its composition is three: the outer and inner ones are mostly continuations of the inner layers of the œsophagus, while the middle one is proper to the stomach.

The outer layer is composed of the longitudinal fibres of the œsophagus: as these reach the cardiac end of the stomach, they form a peculiar turn, whereby the distribution on the surfaces as a flat layer is facilitated. Some of the fibres of this layer dip down to join the deeper ones, while others continue onwards as the longitudinal fibres of the duodenum. As to the fibres which proceed on to the curvatures, they are not so intricate, as they descend directly from the portion of the œsophagus opposite the part they supply, so that the only alteration in direction is that of diverging a little from each other, and pursuing the bent course of the corresponding gastric curvature. On the lesser one they soon become scanty, and are lost in the circular fibres of the body of the stomach: very few of them are traced on to the pylorus. The fibres proceeding on to the greater curvature are mingled with other considerable bundles taking the same direction, but which are not traceable on to the œsophagus, as they seem to pass round each side of the cardia, and blend with the circular fibres on the lesser curvature.

The middle layer consists of annular fibres, which, though scanty as they encircle the extreme left end of the stomach, increase in bulk towards the middle part of the organ, and are especially developed at the lesser curvature. They again decrease over the antrum pylori, but are ultimately greatly developed for the formation of a powerful sphincter at the pylorus.

The internal or oblique fibres of the stomach have somewhat the same arrangement as the deep layer of fibres of the œsophagus, although not perfectly identical, as they are arranged like hoops placed one within the other; but while in the former the one set enters the other without intersection, in the latter there is a partial decussation by separate bundles. Thus, in reality, the oblique fibres of the stomach are constituted of two layers, the one proceeding from the left end of the stomach on to the right, which pass internally to the next layer: this one proceeds from the right of the cardia on to the fundus. Owing to the scantiness of circular fibres at the base of each cul-de-sac, the fibres are here in contact with the superficial longitudinal ones. The oblique fibres are best studied by dissecting from within, and, after removing these, the circular fibres come into view with greater ease than by attempting to expose them from without.

The third coat of the stomach consists merely of the cellular tissue existing between the muscular and mucus coats, as well as connecting the former to the outer serous tunic, in which case it is more abundant and firm nearest the curvatures. There it is situated between the muscular and mucus coats: it was named by the ancients, on account of its white aspect, the Tunica Nervosa. It is loose in some parts and firm in others; not only serving to connect parts together, but also to form a medium in which vessels ramify for the supply of the organ.

The internal or mucus coat of the stomach differs in the cardiac from the pyloric end, as in the former it is but a mere continuation of the unmodified mucus lining of the œsophagus, being characteristic for its scantiness in gland and but limited supply of blood. The most marked feature it possesses is that of being covered by a cuticular layer of extreme thickness, easily separable from the basement structure beneath after slight maceration or boiling. The cardiac portion of the gastric mucus lining is, in a healthy stomach, of a dirty white, bedewed by more or less mucus, and thrown into folds which have a radiated arrangement at the cardiac orifice, whilst at the fundus they are concentrically arranged. This portion of the membrane is also furnished with papillæ; and Sprott Boyd, in an Inaugural Essay on the structure of the Mucus Membrane of the Stomach, published in the *Edinburgh Medical and Surgical Journal* for 1836, describes a very marked peculiarity of an interposed layer between the epithelium and papillated surface of the mucus lining. This intermediate layer, he says, has a smooth equal surface, perforated by numerous foramina about the 600th of an inch in diameter, or perhaps a little smaller, the margins of which are slightly thickened. He afterwards states that he has not been able to trace in the epithelium of any other animal a structure similar to that existing in the horse. These peculiarities in the left pouch of the stomach cease abruptly midway the length of the viscus, where the cuticular lining terminates by a serrated edge.

The mucus lining of the right end of the stomach is normally of a reddish color, and presents a villous, glistening aspect, coated thickly with mucus, and also possessing a high degree of vascularity; the epithelium is here scanty, but nevertheless tabular. The villous appearance above referred to suggests itself also when the surface is examined by the naked eye and by the aid of a lens; but it is deceptive, as has been already remarked by Sprott Boyd, who correctly refers it to the raised margins of the arolæ which stud the surface. This portion of the gastric mucus membrane is also thrown into folds, which become gradually more marked towards the pylorus; whereas they are susceptible of obliteration by distention, there is one circular fold at the pylorus which is permanent, and so disposed as to fulfil the office of a valve.

The arteries of the stomach are derived from the coeliac axis, whose three divisions, i. e. gastric, hepatic, and splenic, all contribute to supply blood to the viscus; but the first is specially destined to that office. The gastric artery, being the smallest of the three divisions, takes a course downwards, forwards, and rather to the right, across the pancreas, getting between the layers of the gastro-hepatic omentum. Being then directed to the left towards the lesser curvature, it divides into an anterior left or smaller branch, and a posterior right and more capacious as well as longer one. The anterior division is destined to supply the anterior surface of the stomach, and more especially the left cul-de-sac, anastomosing with branches (sometimes called vasa breva), coming on to the stomach from the splenic. This division of the gastic also anastomoses with œsophageal twigs, which are occasionally of considerable size. The posterior or right division of the gastric artery, destined for the pyloric end of the stomach, anastomoses with some splenic branches, but more especially with the pyloric branches of the hepatic artery.

The veins returning the blood from the stomach are the gastric and splenic, which anastomose with the duodenal veins. These all have a few valves, but they may be easily injected from the porta into which they empty, owing to their very free anastomosis.

The lymphatics of the stomach are numerous, and in some parts very apparent, entering the lymphatic glands situated along the greater curvature and around the cardia, where they are numerous and large.

The stomach is supplied with nerves from both the cerebro spinal and sympathetic or ganlionic system. The pneumogastric or par vagna nerves, arising from the medulla oblongata, are the main conductors of nervous influence to and from that viscus. Their arrangement is simple, as, after they have formed various plexuses within the thorax, in which they mutually interchange fibres, they reach the diaphragm, and here are arranged as two nervous branches, i. e., a superior and an inferior one. The former is principally destined for the fundus, whilst the latter supplies the pyloric end, and sends branches off to the duodenum, with one or two to the solar plexus.

The sympathetic fibres, destined for the stomach, are derived from the solar plexus, descending on to the viscus, in company with the vessels.

INTESTINE.

This term is applied to that portion of the alimentary canal extending between the pylorus and anus, destined for the temporary retention of the chymous mass, so that its nutrient parts may be absorbed, whilst its more solid, indigestible constituents, are collected for excretion.

The intestine in all monogastria, but especially in solipeda, occupies by far the greatest part of the abdominal cavity. The bonds of attachment to the various parts of the latter are contracted by the intestine, through its peritoneal investment, more especially to the spine, constituting mesenters, which I shall especially allude to when describing with more detail each portion of this capacious tube.

Not only the attachments, but also the shape of the intestine, vary at different parts of its course, so that it has been deemed necessary to divide it, either arbitrarily or at natural demarcations. Thus we speak of the small and large intestine, the two being separated naturally by a marked change in direction, size, and confirmation.

It is also obvious that, as the situation, attachment, and shape of each portion of the intestinal canal differ, so must the relations be equally distinct, and further mention of them will therefore be reserved for fuller exposition elsewhere.

SMALL INTESTINE.

This, the smallest although longest, is also the first portion of the intestinal tube, extending from the pylorus to its sudden termination into the large intestine. In it the chymified mass is subjected to the modifying influence of important secretions, whereby its nutritive parts are fitted for absorption by the vessels, which, for this purpose, are arranged in this portion of the intestinal track.

The small intestine has been divided into three parts: this classification is, however, purely conventional. Since it does not recognize anatomical differences for its basis, it might justly be presumed that this distinction of human anatomists exhibited traces of imperfection, even when applied to the frame of man. Such being the case, it is no matter of surprise that, in referring the distinction to the intestinal canal of animals, the incongruities of the system should be still more apparent.

Extending from the pylorus, the first portion is termed the duodenum, from its being considered as twelve fingers' breadth in length: it is, however, extended round to the left side of the spine, posteriorly to the anterior mesenteric artery. The middle, or floating portion of gut, takes the name of jejunum, and the third, or cæcal portion, is distinctively designated ileum.

The duodenum forms a wide curve from the pylorus round to the right, being situated under the concave surface of the liver, passing above the transverse colon, so as to attain the posterior part of the mesentery, and, reaching the left side of the spine, comes in contact with the colon, where it is said to end in the jejunum. The duodenum is fixed by the gastro-hepatic omentum to the concave surface of the liver, the layers of which enclose the biliary and pancreatic ducts, whereby this bond of union is still further strengthened. The peritoneum coming from the right and spigelian lobes of the liver, as well as from the right kidney, forms a loose attachment for the duodenum by extending on to the hepatic flexure of the colon, after it has surrounded the first-named gut. The next portion of intestine is attached to the spine transversely to the long axis of the body; winding round the mesentery to the left of the aorta, it gets attached to the gastric flexure of the colon, and here it proceeds, under the name of jejunum, along the free borders of the mesentery.

As to the shape of the duodenum, from the pylorus to the right of the porta, we find its dimensions so very great as to have suggested to the ancients the similitude between it and the stomach, of which they regarded it in some degree as an analogue, as testified by the appellation "Ventriculus Succenturiatus," given to it by them. Further from the pylorus, we find it constricts and assumes a certain caliber, which it maintains till it loses its name for that of jejunum.

With reference to the relations of the duodenum, it may be stated that they admit of detail on account of the fixedness of that portion of the gut, an attribute with which it is endowed in contradistinction to the jejunum and ileum. In the first portion of its course, i. e., from the pylorus to the posterior part of the right lobe of the liver, the duodenum by its upper surface is in contact with the concave surface of the latter organ, crossing the vena portæ, near which it is pierced by the biliary and pancreatic ducts, which enter it at about five or six inches from the pylorus, forming an acute angle with each other. The inferior surface of the duodenum rests on the transverse colon, and its superior margin is in close contact with the anterior part of the head of the pancreas.

Round to the right, the duodenum is in contact with the hepatic flexure of the colon, right and Spigelian lobes of the liver, as well as the right kidney. To reach the spine it has to cross the direction of the right flexure of the colon, getting behind the mesentery and gastric flexure of the colon, where it is connected with the left kidney.

Alluding next to the general anatomical facts as applied to the jejunum, so called on account of its usual vacuity after death, the limit between it and the ileum is defined by imagining the small intestine, with the exception of the duodenum, divided into five equal portions, of which the first two take the name of jejunum, whilst the last three-fifths receive that of ileum.

The jejunum is suspended superiorly from the spine by an extensive fold of peritoneum, termed mesentery, which serves also as a medium for the passage of the mesenteric arteries, veins and nerves, as well as for chyliferous vessels, to take their course towards the receptaculum chyli, situated to the left of the aorta.

The width of the jejunum is far from being uniform, it being more constricted at some points than at others: its narrowest part is that which is contiguous to the ileum.

The ileum is the terminating portion of the small intestine, so called from the tortuous course it takes, emptying itself into the large intestine at the junction of the cæcum and colon, by an orifice provided with a valve.

The first portion of the ileum is simply attached by mesentery to the spine; but, in addition to this, in the last part of its course, the gut is connected with the cæcum by a fold of peritoneum, which is not large enough to prevent them deviating more than an acute angle from each other.

The ileum is, on the whole, the narrowest portion of the small intestine, but the thickest in its coats.

Having now especially to describe the structure of the small intestine, it may be taken as a whole, merely alluding to local peculiarities.

This portion of the alimentary canal has four coats, to be described in the same order by those of the stomach, i. e. peritoneal, muscular, cellular, and internal mucus.

The first, or the peritoneal, has nothing peculiar, beyond its enclosing a little triangular space all along the upper attached border of the gut. The looseness of the peritoneal folds attaching the small intestine is very marked; and Colin (Soc. cit.) notes, that the mesentery is proportionately larger in young than in adult quadrupeds, so that the gradual shortening of this explains the spontaneous reduction of exomphalus or umbilical hernia.

The second, or muscular coat, is mostly developed at the commencement of the duodenum and terminating portion of the ileum. It consists of white involuntary fibres, arranged so as to form an outer longitudinal layer, and an inner circular one, both of which completely encircle the gut.

The third, or cellular coat, is similar to that of the stomach, in being disposed in two layers, so as to connect the three coats together. It is especially condensed on the inner surface of the muscular coat, so as to take the appearance of a fibrous tunic, attached to the mucus lining by loose cellular tissue.

The fourth, or mucus, coat is thin, having a velvet appearance, due to villi, peculiarly small in the intestines of the horse, but remarkably developed in other animals, especially carnivora and fishes. The villi may be seen by a pocket lens, on a well-washed piece of intestinal mucus membrane, and between them are seen numerous foramina, which are the openings of tubular glands, known as the crypts of Lieberkuehn.

In addition to the tubular glands, by dissecting, from without, the muscular from the mucus coat, lining the commencement of the duodenum, we find clusters of vesicles, similar to the vesicular structure of the salivary and pancreatic glands. These form distinct layers, provided with ducts, which open on the free surface of the membrane; and Dr. Todd states that Brunner's glands, or, as he calls them, the *duodenal*, are more developed in the horse than in any other animal he has hitherto examined them in.

We have next to treat of the solitary glands — glandulæ solitariæ — peculiar and rather scanty bodies, visible at various parts of the small intestine. These are vesicular, and without any opening when in the perfect state, surrounded by villous processes and Lieberkuehnian follicles. Some of the villi also project from the surface of the so-called glands, which are most apparent when distended with secretion.

About the second half of the jejunum, and along the whole of the ileum, we see longitudinal patches, varying from half an inch to even three inches in length, scattered all over, but more especially situated near the superior or attached border of the small intestine, which is contrary to the faulty description of some recent authors. These patches, distinguished as Peyer's glands or patches, also as Agminated glands — Gladdulæ agminatæ seu aggregatæ — consist of an accumulation of small bodies, each resembling a glandula solitaria in miniature, being also destitute of a natural aperture. Colin (loc. cit.) states that they are first seen at a distance of about six feet and a half from the pylorus, and the least number of them he has ever counted has been 102, whilst the utmost has been 158.

The mucus membrane of the small intestine is thrown into folds, at different parts, which are transverse, and scalloped near the pylorus, whilst in other parts they are mostly longitudinal; these are all temporary folds. There is no such arrangement as the valulæ conniventes in the small intestines of the horse, though recent writers of great eminence have described them. About five inches from the pylorus, at the superior border of the duodenum, is a semicircular fold, which, if elevated, admits of the finger being thrust behind it into the wide biliary duct. The opening of the pancreatic duct is also visible beneath this fold, but it is not so capacious as the one last mentioned.

LARGE INTESTINE.

The large intestine constitutes the terminating portion of the alimentary canal, being remarkably more developed in solipedes than in any other of our domestic quadrupeds. It occupies the greater part of the abdomen, and most of it is loose, whilst its shape and other peculiarities vary considerably at different points.

It is divided into three parts — cæcum, colon, and rectum — the precise extent of each being defined by special anatomical characters.

The position of the large intestine being constant, it is necessary, for sake of precision, to speak of the whole as to the course it takes in forming the three divisions, extending thus between the small intestine and anus.

The cæcum, or blind pouch, is the first gut, which protrudes in the middle on entering through the abdominal walls at the linea alba. Its bend or blind extremity is projecting into the left hypochondriac region; its body crosses obliquely the floor of the abdomen, to reach the right iliac region, where it suddenly bends at an acute angle, being rather constricted, and forms the colon. At this part the latter receives the ileum, and extends up the right side of the abdomen to the diaphragm, where it traverses the direction of the spine, resting on the ensiform cartilage; turning round the left side, it attains the left iliac fossa posteriorly, where it forms a twist like a letter S, from which similitude it has been termed the Sigmoid Flexure of the Colon. The gut, having diminished in size, returns up the same side of the abdomen to the diaphragm, where it again crosses the spine. Being now on the right side, it continues back to a point beyond the anterior mesenteric artery, where it turns upward and forward, so as to come in front of the artery in question; then, from right to left, so as to cross the spine for the third time, constituting the transverse colon, which is more capacious than the part preceding it. The two curves which it forms, one on the right and the other on the left, are respectively called the hepatic and gastric flexures of the colon. The gut so proceeds backward along the left side of the mesentery, being diminished again in size, and constituting the single colon, till we get to the posterior mesentery artery, where, unaltered in other respects, it takes a straight course through the pelvis, out at the anus, and hence the name of Rectum.

The cæcum so called from having only one outlet, being closed at its anterior part, or cæcum caput coli, from its being the blind head of the colon, is vulgarly termed the water-bag, owing to the almost invariable fluidity of its contents.

It is situated, as I have before said, obliquely along the floor of the abdomen, extending backwards from left to right.

It is attached to the spine by a mesocæcum, which is a fold of peritoneum, coming off from the spine on to the superior part of the pouch. There is then the fold already alluded to, which stretches from the ileum on to the cæcum, and, through the medium of the mesentery, indirectly connecting the latter with the spine.

The cæcum is cone-shaped, having an apex and a broad base. The former generally protrudes the first, when a medium longitudinal incision is made into the abdominal walls, although it is situated above the left portion of the double colon, whilst the liver is directly in contact with the floor of the abdomen. Like the other divisions of the large intestine, the cæcum is sacculated. The bands producing this appearance are three in number at the apex; but between two and three inches from this, one of them bifurcates, so that four bands result, which are continuous on to the colon.

The colon arising from the cæcum, receives at first the contents of the ileum, being situated along and occupying the greater part of the floor of the abdomen.

The colon is generally distinguished as double and single. By double, is meant the flexures of the gut from its commencement to its gastric curve; whilst by the single colon, is understood the continuation of the same intestine to the part where the rectum commences.

The double colon is attached by the peritoneum coming off on to it from the cæcum, in the right iliac fossa, and continues from the outer flexure on to the inner, so as to keep the two in perfect apposition. Thus, if the abdominal parietes are cut through, the whole of the double colon may hang out, with the exception of the transverse portion. The latter is attached to the right kidney, as well as concave surface of the liver, by folds of peritoneum; to the spine by the transverse meso-colon; and still more to the left, it is loosely attached by the gastrocolic omentum to the stomach and spleen; besides which it has a peritoneal attachment to the left kidney. Then the single colon commencing, it is loosely affixed to the spine by an extensive peritoneal fold, the meso-colon, similar to the mesentery, but smaller and to its left: this fold is continuous posteriorly with the meso-rectum.

The relations of the transverse colon are important, no less than interesting, inasmuch as it is in close connection with the most important abdominal viscera. On the right, its upper surface is contiguous to the right kidney, as well as to the right and Spigelian lobes of the liver. In the middle, its superior surface is connected principally with the pancreas; and to the left, but still superiorly, it approaches the left kidney and spleen. Anteriorly, the stomach also touches it, especially during repletion.

The shape of the colon is very variable in different parts of its course. Thus, the first portion of the double colon, from the right iliac fossa till it forms the sigmoid flexure, is capacious and sacculated; the latter being due to the four bands continuous on to it from the cæcum. At the sigmoid flexure the bands are completely lost, so that the gut is smooth; but, as we extend up towards the diaphragm, the anterior band begins, and then the posterior one becomes apparent; so that the transverse and single portions of the colon are puckered by two longitudinal bands.

The Rectum, so called from its comparative straight course through the pelvic cavity, arises from the single colon, a little anteriorly to the posterior mesenteric artery, and ends at the anus, where its mucus membrane is continuous with the common tegumentary covering. It is attached in its anterior two-thirds by a meso-rectum; the posterior third is an exception to any other part of the intestinal track, in so far as it is connected to adjacent parts by special fasciæ, and at its termination by certain muscles hereafter to be dwelt upon.

The size of the rectum is much the same as the single colon. It is puckered in its anterior part by two longitudinal bands; and the sacculi, resulting therefrom, determine the shape of the fæcal matters.

The rectum is superiorly related to the spine, whilst inferiorly it comes in contact with the bladder, bulbous portions of the vasa deferentia, vesiculæ seminales, and prostate.

The structure of the large intestine does not vary essentially from that of the small, as it possesses the four coats, *i.e.* peritoneal, muscular, cellular, and internal mucus.

The peritoneal tunic forms an entire covering to the large intestine, with the exception of the superior surface of the transverse colon — which is in contact with the pancreas — and the terminating portion of the rectum. The bands by which it unites the intestine to other parts have been already described. In addition to the peritoneum forming an entire covering to the gut, at the attached margin of the flexures of the colon it constitutes folds loaded with fat, varying in width in different parts, and clustered so as to have deserved the name of appendices epiploicæ.

The muscular coat of the large intestine is differently developed in various parts. Its fibres are of the plain variety, and arranged in two orders. The outer longitudinal set is scanty in some parts, but in others forms the longitudinal bands above alluded to. These are shorter than the actual length of the gut itself, so as effectually to pucker it. The number of longitudinal bands varies from one to four in various parts of the gut, and the shape and breadth of the latter is not everywhere the same. The longitudinal fibres are abundant in the rectum, but they only form bands in the anterior two-thirds, as posteriorly to this they uniformly surround the gut. The inner layer of fibres encircles the whole of the gut, being thickest towards the apex of the cæcum, as well as in the single colon and rectum; at the end of the latter the internal sphincter-ani is formed by an accumulation of the circular fibres. The circular fibres of the colon are engaged in forming the ileo-colic valve, hereafter to be described.

The cellular coat of the large intestine resembles that of the small, only not so abundant, except at the terminating portion of the rectum, where it is much more developed.

The mucous lining of the large intestine is continuous anteriorly with that of the ileum, posteriorly with the common integument. It is thin, more or less coated with mucus, scantier in glands than the one of the small intestine; but the orifices of the Lieberkuehnian crypts are more apparent, owing to the surface here being destitute of villi. Saccular recesses, more or less capacious, exist in the membrane lining the large intestine. The difference in degree of vascularity gives rise to difference in the color of the mucus coat in various portions of the gut: thus, that lining the cæcum is generally more deeply colored than that of the colon, whilst the rectal mucus membrane is more vascular, and hence redder than the colic or cæcal one.

At the termination of the ileum is the ileo-colic or ileo-cæcal valve, which is constituted of two folds of mucus membrane, almost parallel to each other, and horizontal, leaving between them an eliptical orifice when partially drawn asunder. The folds consist of the circular fibres of the intestine, lined on the inner or ileac side by the villous membrane of the small, whilst on the cæcal and colic side they are covered by the mucus membrane proper to the large intestine. It is worthy of notice, that though muscular fibres partly enter into the construction of the valve, its efficiency is explicable on purely mechanical grounds, as proved by the fact, that it is competent in the dead body.

The anus is the outlet of the intestine, which is perfectly closed, except during the evacuation of feculent matters, and is made perceptible externally by the elevation of the tail, being situated in a space bounded superiorly by the sacrum and coccyx, laterally by the ischial tuberosities, and inferiorly by the urethra in the male and vulva in the female.

It is lined within by the mucus membrane of the rectum, which is loose and of a marked red color. Its external covering is of common integument, destitute of hairs. Lying between the skin and mucus membrane are two circular muscles, whose office is to keep the anus closed and prevent constant evacuation of fæces, whilst there are other muscular appendages situated externally to these, destined either to elevate or retract the anus, being evidently antagonistic to the sphincters.

The internal sphincter-ani is in contact with the attached surface of the intestinal mucus membrane, and separated from the integument by the external one. It is constituted of the pale circular fibres of the gut, but towards its free edge certain colored fibres are apparent on it.

The external sphincter is situated outside the internal one, and within the anal integument: it is circular, and composed of red fibres, attached superiorly under the first coccygeal bone, and inferiorly its fibres blend in the male subject in the accelerator urinæ and triangularis penis, and in the female with the constrictor vaginæ.

The levatores-ani are two pale muscles, attached on each side of the first coccygeal bones, and, spreading downward and forward on to the rectum, form an attachment for the internal sphincter, and blending with the longitudinal fibres, so as to increase the thickness of the muscular coat of the rectum. The action of these muscles must be that of elevating the anus, and shortening the rectum from before backward.

The retractors proper to the anus are one on each side attached to the inner surface of the articular extremity of the ischium. Extending from before backward, and rather upward, they blend with the external sphincter. Their action is obviously that of retracting the anal opening.

VESSELS, NERVES, AND LYMPHATICS OF THE INTESTINE.

The intestinal canal, as a whole, receives arterial blood from the anterior and posterior mesenteric arteries, hepatic branch of the cœliac axis, with branches from the internal pudic. The arteries of the small intestine are derived from the anterior mesenteric, whose divisions, varying from twenty-four to twenty-eight, proceed to the small intestine, with the exception of four, which minister to the nutrition and functions of the large intestine. The branches extending from the main trunk, at acute angles, proceed between the layers of the mesentery, to within one and a half or two inches from the gut, where they anastomose, forming vascular arches, from which the secondary branches arise, and, proceeding on to the intestine, ramify on the several coats, especially the mucus one. The anterior division of the anterior mesenteric artery, proceeding to the duodenum, anastomoses with the duodenal branch of the hepatic artery. The last iliac division inosculates with the cæcal and colic branches of the same trunk.

The cæcum and colon receive arterial blood solely from the branches derived from the anterior mesenteric, with a slight contribution from the posterior mesenteric arteries. The branches of the former originate opposite the flexure made by the cæcum and colon. The cæcal divisions, two in number, proceed downward and forward till they reach the gut. The posterior one passes round the posterior part of the border of the cæcum, to get on the under surface of the latter, extending to the apex, in somewhat a straight course, and ramifying collaterally; at its termination it forms a vascular network, by anastomosis with the superior cæcal artery. The latter one, reaching the gut, extends directly forward towards the apex, and comports itself like the former.

Thus we see the flexure, formed by the cæcum and colon, is supplied by collateral branches, from the superior and inferior cæcal mesenteric divisions, both these anastomosing on the corresponding surfaces with the colic arterial trunks.

The two branches going to the colon extend, about parallel to each other, downwards and forwards and to the left, the one gaining the cæcal end of the colon, whilst the other proceeds on to the hepatic flexure. Then these may be traced, the one backward and the other forward, relatively to the course of the gut, along its superior border, so as to reach the sigmoid flexure, where they mutually inosculate. From the mesenteric division going to the transverse colon, is a branch proceeding on to the single portion, which anastomoses posteriorly with the posterior mesenteric. This vessel divides first into two branches, i. e. an interior colic and a posterior rectal one. The anterior colic branch is directed forward and downward between the layers of the meso-colon, and divides into four or five branches, which bifurcate and form arches, like the arteries of the small intestine, for the supply of the contiguous gut. The arteries of the rectum are sometimes spoken of as hæmorrhoidals, and these are distinguished as anterior, middle, and posterior. The anterior hæmorrhoidals are formed by the hindermost branch of the posterior mesenteric artery, which, passing into the folds of the meso-rectum, supplies consecutive branches to the gut, till, posteriorly to the peritoneum, where the arteries pierce the muscular coat, and, forming a network of vessels, anastomose with the middle hæmorrhoidals, which are the ramifications of the internal pudic.* These inosculate with the posterior hæmorrhoidals derived from the same source. The anus is then supplied with blood from the last named branches, as well as from perineal twigs of the external pudic.

The veins of the intestine accompany the

---

* This artery sometimes, erroneously, goes by the name of its terminating branch — the artery of the bulb.

arteries, and are equally distributed. The posterior mesenteric vein is formed by similar divisions to those coming off from the posterior mesenteric artery, and then the main trunk extends forwards and enters the porta, near the termination of the splenic. At this spot the veins from the small intestine, as well as from the cæcum and double flexures of the colon, also contribute to form the large portal trunk.

The nerves of the intestines are derived from the solar plexus, and they are found in association with the arteries. The duodenum also receives branches from the par vagum nerves, and the rectum and anus are supplied also by divisions of the two last sacral pairs.

The lacteal and lymphatic vessels of the intestine are anatomically alike, and even physiologically they admit of being comprehended under the same term, "lymphatic," because both absorb the fluid known as lymph. But since the lymphatics of the small intestine additionally contribute to the function of chyliferous absorption, they have been distinguished as lacteals, in conformity with the color of the fluid which they take up during the digestive process.

The lacteals of the small, and lymphatics of the large intestine, enter a set of lymphatic glands, by no means numerous, and of small size, situated along the attached border of the gut. From these the lacteals ascend to about twenty-five or thirty lymphatic glands of larger size than the others, situated at a short distance from the spine, between the folds of the mesentery, from which the lymph is then conducted into the receptaculum chyli. From the large intestine the lymphatics enter, in addition to the intestinal set of glands, others situated in the lumbar region, partly between the folds of the meso-colon and meso-rectum, from which the lymph is carried into the common reservoir.

The receptaculum chyli receives the fluid from the lymphatic vessels of all the abdominal viscera, as well as from other parts. It is a membranous pouch of various calibre, lying in contact with the right crus of the diaphragm, and right psoas muscle, corresponding in situation to the second and third lumbar vertebræ. It gradually constricts anteriorly, and crossing the aorta to get on its left side, enters the thorax, and here becomes known as the thoracic duct, which empties itself into the right axillary vein.

SPLEEN.

The spleen, although, possibly, not bearing any physiological connection with the digestive process, still, from its anatomical relations, conveniently admits of description here. It is a singularly elastic organ, of a purplish grey color; smooth on its outer surface, and composed of a spongy texture, enclosed in fibrous tissue. The color of the spleen is generally darker in herbivora than in carnivorous quadrupeds, as in the latter it is more of a red color.

It is situated in the left hypochondrium, and partly in the epigastrium, being attached by its outer tunic to the stomach, left kidney, and transverse colon.

It is scythe-shaped, being small and pointed anteriorly, but broad posteriorly. It is smooth, and somewhat convex on its inferior surface, whilst its superior one is divided into two unequal halves, by a fissure termed the hilum. The anterior division is narrow, but the posterior one is broad and triangular in shape. The margin of the spleen is sharp all round. The size of the organ varies considerably in different subjects, and, according to circumstances, in the same animal.

The spleen is related, by its superior surface, to the left end of the greater curvature of the stomach, and to the diaphragm; at its broad base it is in close relation with the left kidney; its inferior surface is connected with both double and single portions of the colon.

The spleen has two coats, a parenchyma, blood vessels, nerves, and lymphatics, needing separate description.

The external coat is peritoneum, which forms not only a covering to the organ itself,

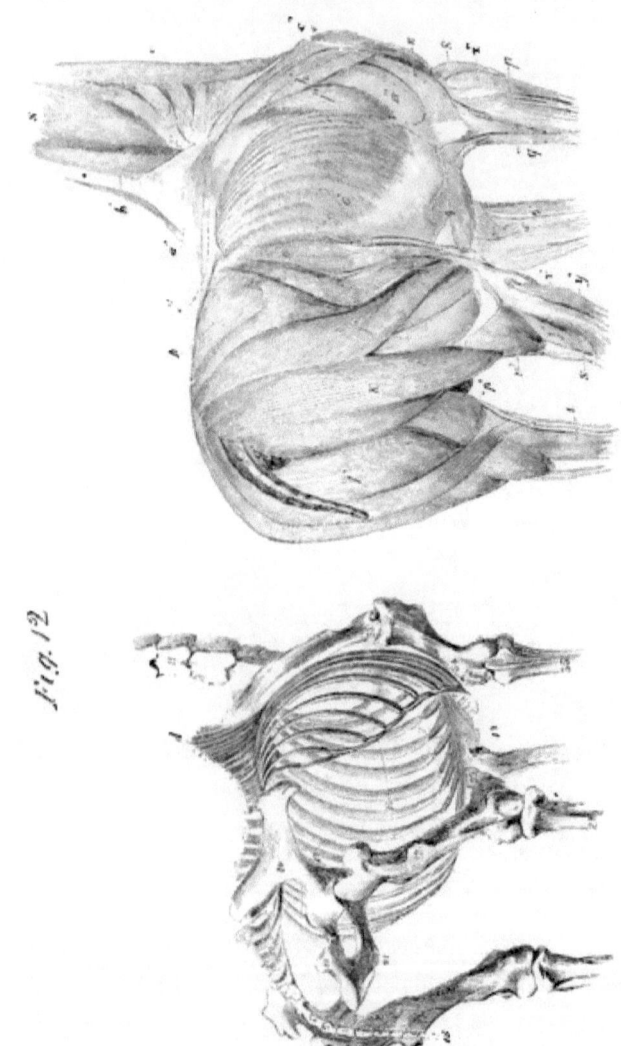

Fig. 12

EXPLANATION OF FIGURE XII.

MUSCULAR STRUCTURE.

FORWARD PARTS.

- *. Ligamentum colli.
- a''. Trapezius.
- b'. Rhomboideus longus.
- c''. Scalenus.
- g''. Postea spinatus.
- h'. Teres major.
- i''. Latissimus dorsi.
- l''. Scapulo ulnaris.
- m''. n''. Triceps extensor brachii.
- p''. Flexor metacarpi externus.
- q''. Flexor metacarpi medius.
- r. Flexor metacarpi internus.
- s''. Extensor metacarpi magnus.
- s. Splenius.
- u. Levator humeri.
- x''. Extensor pedis.
- e''. Obliquus externus abdominis.

POSTERIOR PARTS.

- h'. i'. Gluteal muscles.
- j'. Triceps.
- k. Biceps abductor tibialis, posterior.
- l'. Adductor tibialis internus.
- m. Tensor vaginæ.
- n. Region of the vastus internus.
- r. Gastrocnemius internus.
- i. s. Gastrocnemius externus and internus.
- x. Extensor metatarsi.
- y. Peroneus.
- i'. Flexor pedis accessorius.
- e'. y''. Coccygeal muscles.
- 4. Subcutaneous thoracic vein.
- 5. Saphena vein.
- 6. Radial vein.

OSSEOUS STRUCTURE.

- 10. 20. 21. The pelvis.
- 11. Cervical vertebræ.
- 15. Coccygeal bones.
- 16. The true ribs.
- 17. The false ribs.
- 18. Sternum.
- 22. Femur.
- 23. Patella.
- 33. Scapula.
- 34. Humerus.
- 35. Radius.
- d. Dorsal spines.
- f. Ulna.

but bonds of connection between it and other parts, such as the gastro-splenic omentum, and the attachment to the kidney and transverse colon heretofore described. This coat is smooth externally, rather closely attached to the fibrous coat internally, but of considerable elasticity, so as to allow the spleen sufficient freedom for distention.

The second or fibrous coat, also termed the albugineous or elastic coat, is that closely applied to the parenchyma of the organ. It consists of yellow and white fibres, and in some parts, such as in the trabeculæ, Koelliker has found plain muscular fibres, which he says do not exist in the external portion of the fibrous tunic in the horse. The covering not only envelopes the outer surface of the organ, but sends sheaths and processes into its substance. The sheaths are purposed for covering vessels, whilst the processes, termed also trabeculæ, divide the substance of the spleen into areolæ or interspaces, which contain a red matter, easily washed and pressed out, known as the splenic pulp. The trabeculæ also arise as processes from the vascular sheaths, as well as from the external tunic. When the pulp has been thoroughly washed, the outer coat, with the trabeculæ and sheaths, have the appearance of a framework or skeleton.

The splenic pulp has a medullary aspect, being composed of cells and blood vessels; and if the organ be cut clean in any direction, we see, besides the cut ends of vessels and trabeculæ, certain pearlish looking bodies, named, from their discoverer, Malpighian Corpuscles. If divided, fluid escapes from the cavity which exists in their interior. They are perfectly visible to the naked eye, being about one-thirtieth of an inch in diameter; and, with a pocket glass, they may be seen attached to the small arterial trunks, if the pulp has been previously carefully washed.

The spleen derives its arterial blood through the splenic artery, which is the main division of the cœliac axis. Winding between the folds of the gastro-splenic omentum, it not only sends numerous branches through the hilum, and on to the surface of the spleen, but also supplies the stomach, largely inosculating with the gastric artery, so that the two might mutually perform each other's office, if the main trunk of either were obstructed.

The splenic vein is similarly distributed to the artery, and it empties its blood into the vena portæ, just anteriorly to the posterior mesenteric vein.

The nerves of the spleen are derived from the solar plexus, and with the splenic artery enter the spleen.

The lymphatics of the spleen are said by Koelliker to be scanty; but Dr. Sharpey tends rather to the belief that they are abundant. They are arranged superficially and deep, both sets anastomising freely with each other, and, reaching the hilum, they enter various scattered lymphatic glands in the peritoneal folds, and then empty into the receptaculum chyli.

LIVER.

The liver is the largest gland in the body, and proportionately largest during certain periods of fœtal life. It is of a dark reddish brown color, and destined for the office of biliary secretion.

It is situated across the long axis of the body, in the right hypochondriac, epigastric, and partly in the left hypochondriac regions.

It is attached to various parts by five ligaments, four of which are peritoneal folds, and one is the remnant cord resulting from the obliteration of the umbilical vein within the abdomen. These attachments will be more fully described with the peritoneal tunic.

The external aspect of the liver is smooth, being convex superiorly and concave inferiorly, broad posteriorly, and sharp anteriorly. It has a granular appearance, and a very superficial inspection clearly shows that it is composed of lobules, about the size of a pin's head.

The hepatic substance is irregularly divided into numerous segments by fissures, which either extend through the gland from side to side, or are mere grooves of more or less depth. The different segments of the gland or lobes are three principal ones — right, middle, and left — to which smaller ones are appended.

The right lobe is the largest of the three, situated in the right hypochondrium, being thickest posteriorly and sharp anteriorly. The supero-posterior part of the right lobe is marked by a depression, for the adaptation of the anterior part of the right kidney. At the superior part of the right lobe is an excavation for the vena cava, which extends from behind forward, and marks off the division between the right and middle lobe. The vena cava is here more or less imbedded in the substance of the right lobe, but, generally speaking, it is superficial in the horse, and only an imperfect channel is formed for it.

Projecting from the inferior surface and posterior part of the right lobe, is the lobulus spigelii, which is of considerable size, being broad posteriorly, and attached by its superior and left border, so that it projects anteriorly and narrows; its apex gradually tapers, and has been capriciously designated, by the lovers of a quintuple hepatic arrangement, lobulus caudatus.

The middle lobe of the liver is the smallest of the three; it is crossed on its inferior surface by the transverse fissure or porta of the liver, at which the vessels and ducts enter into and issue from the gland. The middle lobe in the horse is divided at its anterior part into five or six portions, and Mr. Percivall, in his Anatomy of the Horse, at page 259, has termed it the lobulus scissatus. It is traversed antero-posteriorly by a channel for the remnant of the umbilical vein, which eventually joins the vena porta.

The left lobe is the thinnest of the three, but occupies an intermediate position in length and breadth. It is very thin at its left margin, and gradually thickens posteriorly. At its posterior and left side is a depression, in which the œsophagus rests. Sometimes the left lobe is divided into two at its anterior part; at others it is single.

The superior surface of the liver is convex, and in contact with the pillars and expanded portion of the diaphragm. The right as well as the Spigelian lobes, are in relation posteriorly with the right kidney and right supra-renal capsule, inferiorly with the head of the pancreas, duodenum, and transverse colon. The middle lobe is related inferiorly to the pancreas, but partially separated from it by the vena portæ. It also suspends the duodenum, and its left edge is loose and in close proximity to the flexures of the colon. The left lobe is related posteriorly to the œsophagus, and inferiorly to the left end of the stomach. The pancreas also stretches across its posterior part, partially separating it from the transverse colon.

The liver receives an incomplete covering of peritoneum. The latter, reflected from the diaphragm on to the concave surface of the middle lobe of the liver, forms a double membranous layer, known, in accordance with its shape, as the falsiform ligament, and holding in its free and concave margin the round ligament, the representative of a fœtal structure, the umbilical vein. Furthermore, the liver is provided with a coronary ligament, that surrounds the foramen dextrum of the diaphragm, through which the vena cava passes. The lateral ligaments are distinguished as right and left; they connect each lateral lobe to the diaphragm.

The only connections of the liver that remain to be mentioned are the stomach, duodenum, transverse colon, and pancreas to its inferior surface, and the right kidney to the posterior part of the right lobe.

Dissecting off the serous tunic, it is found connected with the biliary surface by cellular tissue, continuous at the porta with the so-called capsule of Glisson. The latter extends into the liver as a common sheath to blood vessels, nerves, lymphatics, and biliary ducts.

To proceed with further description of the liver would be useless, unless first examining the blood vessels and ducts in that part of their course which is external to the organ. The hepatic artery is quite subordinate in size, considering the magnitude of the organ and amount of its secretion. It is a branch of the cœliac axis, at first in contact with the pancreas, and then between the folds of the gastro-hepatic omentum, and it reaches the porta on the left side of the portal vein. After giving off pancreatic and duodenal branches, it divides into two, a right and a left one. The right, the largest and somewhat the longest, penetrates into the right lobe, giving off collateral branches, first to the middle and then to the right lobe itself. The left is the smallest division, and is distributed to the lobe corresponding to it in position, and also to the middle one.

The liver is exceptional for having, besides an artery, another afferent vessel — a vein, known as the portal vein, formed by the splenic, which also receives the gastric and mesenteries, meeting each other at the same spot near the posterior part of the pancreas. From its origin, the portal vein takes an oblique course from left to right through the pancreas, and being surrounded by nerves, it reaches the porta of the liver, and here divides into three principal branches, one for each lobe.

At the porta we also see the biliary duct coming out, formed by the union of several branches, corresponding in number to the ramification of the blood-vessels. This duct passes through the gastro-hepatic omentum, meeting the pancreatic duct at almost a right angle, and with it opening into the duodenum about five or six inches from the pylorus.

Having thus far considered the main vessels, we may examine further the internal structure of the liver. At the porta the branches of the vessels and ducts are associated together, and surrounded by cellular tissue, which sheaths grooves or canals, cut in various directions in the substance of the organ. These are the portal canals, and the cellular tissue in question is Glisson's capsule.

The vessels and ducts ramifying on the sheath acquire the name of vaginal branches, and, as they are traced between the lobules, they are termed interlobular. Here the unassisted eye ceases to take cognizance of their further relation ; but, with careful dissection, and a common pocket lens, they may be traced to the lobules, which they enter; and the blood of the hepatic artery and portal vein is emptied into a common set of vessels, the hepatic vein. The relation of these vessels in the lobules may be seen on the surface in a good injected specimen of liver, where the hepatic veins have been injected one color, and the other vessels differently. By this means the centre of the lobule is colored with the injection thrown into the hepatic veins, and the circumference with that of the portal vein.

The hepatic veins issuing from the lobules cross the structure of the liver in separate grooves, formed by the coalescence of the hepatic particles, so that their base is in contact with the veins, and hence the name of the latter is that of the sub-lobular hepatic veins. These empty into the posterior cava by several orifices, as well as by two larger ones, guarded by semi-lunar valves, situated just at the foramen dextrum of the diaphragm.

In addition to the blood-vessels and ducts of the liver, it is supplied with nerves from the solar plexus, which ramify with the vessels.

The lymphatics of the liver are abundant, and arranged, like in other organs, as a superficial and deep set, which inosculate freely in the substance of the organ, and, uniting to form several branches, they issue from the porta of the liver, passing through some lymphatic glands situated round the fissure, and from this they advance to the receptaculum chyli.

PANCREAS.

The pancreas is a compound vesicular or racemose gland, being much of the same nature as the salivary glands.

The pancreas occupies the interval between the layers of the transverse mesocolon, along the upper surface of the transverse colon.

Its attachments are merely cellular, with the exception of the pancreatic duct, which attaches it pretty closely to the duodenum.

The pancreas is spoken of as having a

body, a head, and a tail. The body of the pancreas is that part stretched across the middle lobe, while the head is longitudinally extended, being almost parallel to the vena porta, and situated below and to the right of that vessel. The head is broad anteriorly and rather narrow posteriorly, and continuous from below upward, and from right to left, then from behind forward, to gain attachment to the body, so as to form a ring for the passage of the vena porta. The part to the left of this vein is termed the tail of the pancreas.

The pancreas is related by its superior surface to the right, left, and Spigelian lobes of the liver, also to the vena cava and aorta, which separate it from the phrenic crura. The posterior part of the head of the pancreas is in relation with the right supra-renal body. The tail of the pancreas is stretched transversely to the branches of the cœliac axis, and attached to the left kidney by loose cellular tissue. The inferior surface is in contact with the transverse colon.

On examining carefully the structure of the gland, it is found to consist of clusters of cells, from which ducts arise, and these unite to form a main trunk, that is traceable back to the tail of the pancreas, increasing in size till it reaches the anterior extremity of the head, where it pierces the duodenum together with the hepatic duct. Besides these clusters of cells and ducts, the gland contains connecting cellular tissue.

The pancreas is supplied with arterial blood by branches from the three divisions of the cœliac axis, as well as from the anterior mesenteric.

The pancreatic veins empty themselves into the splenic.

The nerves are derived from the solar plexus, and the lymphatics of the pancreas, on issuing from the glandular substance, may be traced to the common reservoir of chyle and lymph.

GENITO URINARY APPARATUS.

Having already described the intra-abdominal portion of the alimentary canal, and its accessories, I proceed to the consideration of that portion of the genito-urinary apparatus as contained within the abdomen, in the widest acceptation of the latter term. By this I mean the kidneys, and with them, for anatomical convenience, I classify the supra-renal capsules, then the ureters, bladder, membranous portion of the urethra, vasa deferentia, vesiculæ seminales, prostate and Cowper's glands, with which I shall conclude.

KIDNEYS.

The kidneys are a pair of glands, whose function it is to secrete urine. They are distinguished as right and left, being both situated in the lumbar region; but, so far as concerns their topographical anatomy, notwithstanding their similarity in position, they need separate notice.

The right kidney is more anteriorly situated than the left, coming in contact with the posterior part of the right lobe of the liver, to which it is attached. It is also fixed to the abdominal parietes by peritoneum, and to the spine by blood-vessels.

Its shape is that of a bent ovoid, being more symmetrical than the left. It has two surfaces and two borders.

Though differing in these marked general characters, the kidneys resemble each other in several equally obvious points of their general anatomy. Both kidneys have a peritoneal and an albugineous coat, both have an excretory duct, vessels and nerves, with a structure also equal in the two, constituting the bulk of the organ. Externally to the peritoneal tunic is a more or less thick stratum of fat, which is more abundant in old than in young animals, when in a state of obesity.

The peritoneal covering of the kidneys is incomplete, especially that of the right one, whose inferior surface and convex border are the only parts coated by it. The left kidney is also covered on its superior surface to a considerable extent, sometimes more and sometimes less. The attachments which each organ contracts through the medium of this serous investment have already been described.

The albugineous tunic is fibrous, and partly sub-serous. It forms a distinct capsule, attached to the substance of the organ by fibrous prolongations, which are in some parts arranged in pits and depressions, so as to mark out divisions on the surface of the kidney. In addition to this, the albugineous coat surrounds the vessels and ureter at the hilus, and enters the substance of the organ.

On cutting the kidney horizontally from the convex to the concave border, there are three different parts brought into view, to be taken into consideration. Firstly, a dark content, of about half an inch or more in thickness, being generally less at the extreme ends of the kidney than at its middle, which completely encircles the central part of the gland, and is termed the cortical structure, from its being most external. This part of the kidney has somewhat a granular aspect, and, when the vessels are full of blood or injection, they appear more or less arborescent, and clustered at innumerable minute but visible spots, to form the Malpighian tufts. Next to this is a lighter colored material, rather ash-colored, but having a reddish hue, termed the medullary substance. This term is not given to it from the fact that it is medullary in consistence, but used in the metaphorical sense of being internally or centrally situated.

Approaching still nearer to the concave border of the kidney, is a funnel-shaped cavity, with its apex towards the hilus, and the base bounded by the medullary substance, which is the pelvis. The apex is tubular, and continuous with the ureter, of which the cavity is but an expansion.

The walls of the cavity are lined by a mucous membrane, which is loosely applied to the medullary substance, and thrown into folds, taking a radiated direction from the mouth of the ureter. Opposite the apex of the pelvis, the membrane is adherent to a prominent border of the medullary substance, concave from before backward, but convex from above downward, and is pierced by foramina, into which the lining membrane of the pelvis extends, so as to form the uriniferous tubes. On dissecting carefully away the mucous membranes of the pelvis, we reach to the fibrous tunic, which is not continuous on the medullary ridge, but merely attached to its sides, so as to increase the length of the boundaries of the cavity.

The ureter arising from this dilatation is continuous outward toward the spine, and then backward, being related superiorly, as it issues from the hilus, with the renal vein; and then crossing the posterior part of the kidney at its inferior surface, it gets between the peritoneum and psoas muscles, and is then traceable back to the bladder, into which it opens.

The renal arteries, one for each kidney, arise at almost right angles from the aorta, after the latter has given off the anterior mesenteric. The right one is more anteriorly situated, and is longer than the left one. After each renal artery has given off a branch or more to the supra-renal capsule of the same side, it divides, on reaching the hilus, into a variable number of branches, usually eight or ten, which pierce the kidney at different parts of the hilus, whilst a few branches proceed along the surface, supplying the capsule, and then also piercing the organ. The arterial branches entering the kidney have a definite arrangement, forming a kind of arch superiorly to the pelvis, from which secondary divisions emanate and pierce the organ in all directions, so as to reach the cortical substance, abruptly dividing into numerous branches, which eventually subdivide to form capillaries. By this it is evident that the cortical substance is more vascular than the medullary; indeed the latter is very scantily supplied with arterial blood.

From the arterial terminations the venous origins occur, and these unite to form branches, having a similar arrangement as the arteries; only as they reach the pelvis almost opposite the apex, they meet to form a wide, capacious trunk, the renal vein. This is supplied with valves, not all of which are perfect. At the opening of each renal vein into the cava is a semilunar flap, overlapping the posterior part.

The nerves of kidneys are numerous, and derived from the renal plexuses of the sympathetic; they accompany the vessels with which they penetrate their respective organs.

Lymphatics may be seen issuing from the hilus of the kidney; they enter some lymphatic glands there situate, and then convey the lymph into the receptaculum chyli.

SUPRA-RENAL CAPSULES.

These bodies, also called capsulæ suprarenales, seu atrabilariæ, are two in number, and belong to the class vascular glands, whose office is very indefinitely known.

They are situated one on each side of the spine, across the direction of the renal vessels. Their attachments are effected by vessels, as well as by the peritoneum, on their inferior surface, connecting them to the corresponding kidney and around to the spine.

The shape of the supra-renal bodies is much the same on either side, being that of a slightly bent ellipsis. They vary from three to four inches in length, and from one and a half to two inches in breadth.

Their concave border corresponds to the renal vessels, as well as to the anterior mesenteric arteries. The convex border is in contact with the inner margin of the kidney. The anterior extremity of the right one is in connection with the right hepatic lobe, whilst its inferior surface is in connection with the commencement of the colon. The left supra-renal capsule is related anteriorly to the pancreas, and inferiorly to the transverse colon.

The peritoneal coat of the supra-renal capsules is merely confined to their inferior surface. The proper substance of the organ is enclosed in a fibrous or albugineous coat, which forms a distinct covering externally, and becomes continuous as sheaths to vessels internally.

On cutting horizontally across a supra-renal capsule, it is found to consist of an outer cortical and an internal medullary substance. The cortical substance is a brownish yellow, due to fat contained in vesicles, which, according to Professor Heinrich Frey, are smaller toward the surface than more internally. The medullary substance has a greyish aspect, and vessels are apparent in it, as also a yellow tinge, due, according to the above-named author, to similar vesicles, as in the cortical substance, only much scantier in fat.

The arteries of the supra-renal capsules are offsets of the renals and anterior mesenteric, as well as of the aorta, but very variable in number and origin. They are, however, always abundant, and enter the organ principally at its concave border.

The veins are larger than the arteries, and pour their contents on the left into the renal vein, and into the vena cava on the right.

The nerves of the supra-renal capsules are very abundant, and derived from the renal plexus. Professor Frey states, that in the horse, ganglion corpuscles constitute one of the structural elements of the nervous tissue in this situation.

URETERS.

The ureters, one to each kidney, are conduits between the kidneys and the bladder, for the passage of urine. Their caliber is various, being about one-third of an inch broad, but getting narrower posteriorly.

As the ureters issue from the kidneys, they converge towards the spine; then proceed suddenly backward, till they reach the brim of the pelvis, having thus greatly diverged; here they converge again, passing downward and backward to reach the sides of the body of the bladder, which they pierce.

In their course, the ureters are attached to the kidney and psoas parvus by loose cellular tissue, and by the peritoneum, which suspends them, by being stretched across their inferior surface. After the ureters have crossed the spermatic and iliac vessels, they are received within a fold of peritoneum, constituting the false ligaments of the bladder.

They pierce the muscular coat of the bladder at a distance of about three inches from each other, if the viscus be distended. They pass between the muscular and mucous coats for about an inch, being somewhat diminished in caliber, when they suddenly open into the cavity by an elliptical orifice, so that if the bladder be distended, the sides of the orifice are stretched, and thus closed.

The ureters are externally covered by a cellulo-muscular coat, consisting of a cellular tissue, with muscular fibres arranged, partly longitudinally and partly circularly, the latter being most internally situated. The ureters are internally lined by mucous membrane, continuous anteriorly with the renal pelvis, and posteriorly with the vesical lining.

The membrane is loosely attached to the outer coat, and thrown into longitudinal effaceable folds.

BLADDER.

The bladder is a dilatable musculo-membranous viscus, destined for the temporary retention of urine. It is situated during vacuity entirely within the pelvis, but when distended, even moderately, its fundus encroaches on the proper abdominal space.

The bladder is held in situation by the peritoneum coming off from the rectum and sides of the pelvis, so as to form a serous fold, which also encloses the vasa deferentia and vesiculæ seminales. Besides this, the bladder is supplied with true ligaments, as well as bounded posteriorly through the intervention of the urethra.

The shape of the bladder is pyriform, approaching, however, to a sphere when empty or partially distended.

It presents for consideration a projecting anterior portion or fundus, a middle part or body, and a posterior one, or neck. The fundus is globular and regular, having fixed at its anterior part the two obliterated umbilical arteries, and the remains of the urachus. The body has no precise limits, but may be considered as that portion on which the bulbous portions of the vasa-deferentia rest. It is circular, but if the bladder be much distended, it bends somewhat backward and upward. The cervix vesicæ is the most constricted part of the organ, and marks the limit between the bladder and urethra.

The bladder is related by its fundus to the iliac flexures of the colon, inferiorly to the pudic and ischial bones, superiorly to the ureters, vasa deferentia, vesiculæ seminales, and middle part of the rectum.

The bladder has three coats. The peritoneal investment is merely a partial one, as it is reflected from the body on to the sides of the pelvis. It covers the superior surface almost completely, but its extent gradually declines laterally and inferiorly. The attachments contracted by the peritoneum are termed false ones. Thus we have the two umbilical arteries, one on each side, enclosed by peritoneum, forming the two lateral false ligaments. Then the vestige of the urachus is similarly enveloped by peritoneum, and constitutes the anterior false ligament. The peritoneum coming off from the rectum on to the superior surface of the bladder, gives rise to a pouch, termed the recto-vesical pouch, or cul-de-sac, and laterally to the triangular folds limiting the latter, known as the superior false ligaments. Behind the peritoneal reflection the bladder is attached to the rectum and pelvic parietes, by a continuation of the pelvic fasciæ, which, leaving the inferior surface of the pelvis at the symphisis pubis, comes on to the bladder, forming the inferior true ligaments of the latter; the fascia is then continuous on to the rectum, blending with the cellular coat. The pelvic fascia is also traced on to the prostate and sides of the bladder, from the posterior part of the obturator foramen, constituting the lateral true ligaments.

Beneath this fibro-serous coat are muscular fibres, arranged in a peculiar manner. There is an outer longitudinal set, traceable from the cervix forward toward the body, where the fibres diverge and become oblique, and some even circular; this layer is principally developed posteriorly. The inner or circular layer is not arranged in concentric rings; but its fibres, beginning at the fundus, appear to arise from various centres on the surface, and to be taking a direction more or less curved in different parts, so as to get transversely to the long axis of the viscus, and thus from the inner side have a circular appearance. These fibres are more decidedly circular at the neck, and act somewhat like a sphincter. Some of the deeper fibres at the neck of the bladder extend forward to each orifice of the ureter, marking the limit of the vesical trigon, whose office must be that of approaching the lips of the elliptical apertures.

The mucous coat of the bladder is generally more or less coated with mucus and epithelium, which guard the structure from the corroding effects of the secretion it has to come in contact with. It is thrown into numerous folds, taking various directions, but principally concentrical toward the fundus, and longitudinal at the cervix, all of which are effaceable by distention of the bladder, and are most prominent when the latter is collapsed. At the upper part of the urethral orifice of the bladder the mucous lining is smooth and free from folds, marking out a triangular space, bounded anteriorly by a line drawn between the orifices of the ureters, and laterally by two lines meeting at a spot at the superior part of the vesical orifice. This is termed the vesical trigon. At its apex is a projecting fold of mucous membrane or uvula vesicæ, which seems to moderate the flow of urine into the urethra.

The bladder is supplied with blood from the internal pudic, and its veins empty into the internal pudic vein.

The nerves of the bladder are derived from the sympathetic, and partly from the two last sacral pairs which supply the neck.

The lymphatics go to glands surrounding the origin of the iliac arteries, termed pelvic lymphatic glands, from which vessels arise, communicating anteriorly with the receptaculum chyli.

URETHRA.

This canal in the male subject is not only purposed for the passage of urine, but also transmits the products of the generative organs. It extends from the posterior part of the bladder to the glans penis; but we shall only occupy ourselves with a description of the intra-abdominal or pelvic portion, which terminates at the bulb of the penis or ischial arch.

It is continuous anteriorly with the bladder, attached to the rectum and sides of the pelvis by fascia and loose cellular tissue and muscles.

The urethra is cylindrical, of considerable length, and its coats of no mean thickness. The pelvic portion of the urethra is generally about three or four inches long, taking a direction backward and somewhat upward.

It is related superiorly to the vesiculæ seminales, middle lobe of the prostate, and posteriorly it comes in contact with the rectum, but separated from it laterally by Cowper's glands.

The first or prostatic portion of the urethra is purely membranous, strengthened by cellular tissue and a continuation of the fibres of the bladder, the circular ones in particular, which are abundant anteriorly. The posterior two-thirds of the pelvic portion of the urethra are covered by a thick red muscular layer, which completely encircles it, with the exception of that part coming in contact with Cowper's glands. This muscle is continuous behind with the muscular fibres of the penis, which constitute the accelerator urinæ. These fibres are externally mixed with longitudinal ones, a portion of which are merely the inner or inferior bundles of the retractor ani, whilst others are derived from the triangularis penis; both of these muscles tend to fix the urethra. Postero-superiorly the fibres encircling the urethra are blended with the external anal sphincter. The retractor penis, which gets attached to the sacral bone, is a white muscle also, affording fixity to the pelvic portion of the urethra.

Beneath the muscular tunic of the urethra we find a loose cellular tissue, and posteriorly also some erectile structure continuous on to the penis.

On slitting open the pelvic portion of the urethra, to examine its mucous membrane, we find that it is smooth, glistening, and thrown into longitudinal folds. It is antero-superiorly raised by the sub-mucous tissue into a permanent ridge, termed the crest of the urethra or verumontanum. This has a depression about its middle, and on each side are the elliptical orifices of the ejaculatory ducts, surrounded by the openings of the prostatic ducts. Posteriorly and laterally are little papillated projections, pierced by ducts emanating from Cowper's glands. These tubular processes are arranged in two parallel lines longitudinally to the course of the urethra.

The pelvic portion of the urethra is supplied with blood from the internal pudic, and the veins empty into the vessel of the same name. Its nerves are derived from the two last sacral and accompanying sympathetic filaments.

The lymphatics of the pelvic urethra are similarly disposed to those of the bladder.

GENERATIVE ORGANS OF THE MALE.

The last division of our subject is that of the abdominal generative organs, only a part of the generative system, and consisting in the vasa deferentia, vesiculæ seminales, prostate and Cowper's glands.

VASA DEFERENTIA.

There are two vasa deferentia, one from each testicle, for the passage of semen to seminal reservoirs.

The vas deferens arises from the posterior part of the epididymis or globus minor, passing through the inguinal canals, and reaching the abdomen; it is situated in the sub-serous tissue, taking a course upward, backward, and inward, to reach the brim of the pelvis; then, crossing the course of the ureters, it gets on to the bladder, where it is dilated, and forms the bulbous portion.

Its attachments are serous and cellular to the various parts mentioned, whilst its posterior part is connected with the urethra.

The vas deferens is related, in its course from the inguinal canal, to the bladder; after it leaves the constituents of the cord, with the parietes of the abdomen; crossing the under surface of the iliac vessels, and reaching the bladder on the inner side of the ureter; also lying internally to the seminal vesicles, and the terminating portion being covered by the prostate.

The structure of the vas deferens is similar throughout, with the exception of the greater thickness of its coats at the bulbous portion, being thinnest where it contributes to form the ejaculatory duct.

This tube, of very various length, is constituted of an outer cellular investment, not requiring peculiar notice; of an intermediate contractile and elastic tunic; and, as its name implies, is composed of muscular fibres and elastic tissue, arranged in two layers, i. e., an outer longitudinal and an inner circular one, which are easily perceived.

The internal or mucous lining is thrown into longitudinal folds, in the narrow part of the duct; but in the bulbous part it forms permanent rugæ, taking various directions, so as to enclose irregular interspaces.

The vas deferens is supplied with blood principally from the artery of the cord, although the epigastric furnishes a twig to it as well. The bulbous portion is supplied also by vessels of no small calibre from the iliacs.

Its nerves are from the sympathetic, as well as from the second and third lumbar.

VESICULÆ SEMINALES.

The seminal vesicles are one on each side of the bladder, and act as receptacles for the semen.

Each seminal vesicle extends from behind forward, upward, and outward, being external to the bulbous portion of the vas deferens. It is attached by peritoneum coming off from the sides of the pelvis and rectum on to the bladder. The posterior part is fixed by cellular tissue to the prostate and neck of the bladder.

The seminal vesicle is pyriform, being about three inches long and about an inch broad at its fundus, but more constricted at its neck. It is connected with the corresponding surface of the bladder and rectum, but partially separated from the latter by the prostate.

The seminal vesicle has an incomplete investment of peritoneum, covering only the anterior part, whilst the prostatic portion is covered by an outer cellular coat. Beneath this is an intermediate tunic, partly elastic and partly contractile. Lavocat describes this muscular coat as easily studied after maceration in dilute nitric acid, when it may be found to consist of an outer longitudinal and inner circular layer, most developed at the fundus, but very thin at the neck.

The mucous membrane is plicated, the folds enclosing similar interspaces to those seen in the bulbous portion of the vas deferens.

The vessels are supplied by the internal pudic, whilst the nerves are from the lesser splanchnic and two last sacral pairs.

EJACULATORY DUCTS.

Two in number, each being the common outlet to its corresponding vas-deferens and seminal vesicle, so that their contents may pass into the urethra by an elliptical orifice each side of the depression on the verumontanum.

The relations of these ducts are simply to the prostate and urethra. When they reach the latter, they pass between the muscular and mucous coat for some little distance, so that at first sight they appear shorter than what they really are.

The structure of the ejaculatory ducts consists in an outer cellular and inner mucous lining, both of which are very thin.

PROSTATE GLAND.

The prostate belongs to the class of secreting glands. It is situated on the commencement of the urethra and termination of the vesiculæ-seminales, being superiorly related to the rectum. Its attachments to these parts are merely cellular, although it has some connection with the sides of the pelvis, rectum and bladder, through the intervention of the pelvic fascia.

It is symmetrical in figure, and very variable in size, being quite rudimentary in aged geldings. It is of a gray color, knotty to the feel, although spongy in texture.

The prostate consists of a middle portion or body and two lateral lobes. The former is in contact with the cervix of the bladder and urethra, the latter with the ejaculatory ducts and seminal vesicles.

This gland has a posterior convex and an anterior concave margin, whilst it is flattened from above downward, although from its connection with other parts it is rendered more or less convex from side to side.

The prostate is composed of an external fibrous or cellular coat, which forms a complete covering to it. On cutting the gland in any direction, it is observed by the naked eye to have an areolar appearance, being a net-work of variously disposed fibres, the larger ones of which are found to be tubular.

The prostate opens into the urethra around the orifices of the ejaculatory ducts by numerous apertures.

It is supplied with blood from the pudic vessels, and its nerves are derived from the lesser splanchnic and two last sacral pairs.

COWPER'S GLANDS.

These also belong to the class of secreting glands, and have sometimes been called the lesser prostates. They are situated anteriorly to the bulb of the penis on each side of the membranous portion of the urethra.

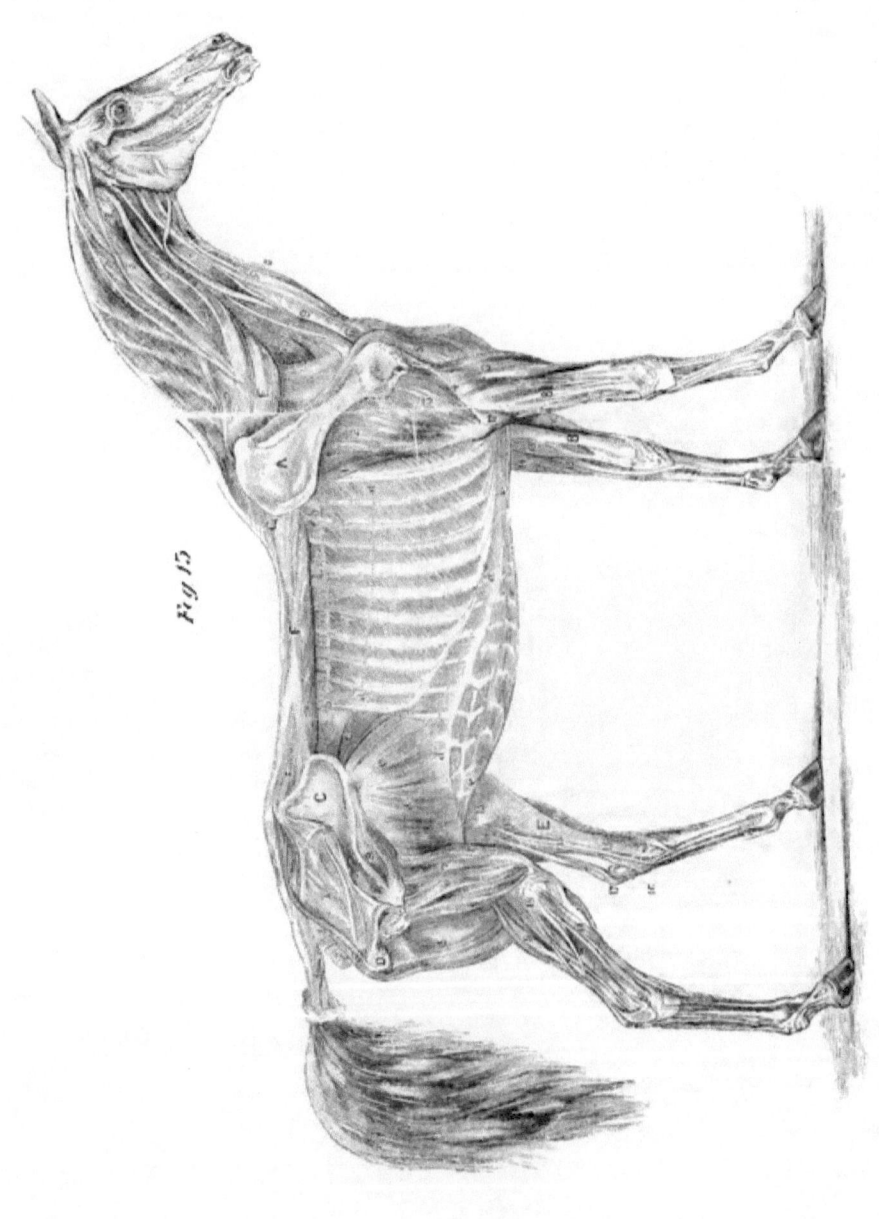

Fig 15

EXPLANATION OF FIGURE XIII.

THE SUPERFICIAL LAYER OF MUSCLES TAKEN FROM THE BODY OF THE HORSE, SO AS TO EXPOSE THOSE MUSCLES WHICH ARE MORE DEEPLY SEATED.

THE HEAD AND NECK.

- a. Buccinator.
- b. Caninus.
- c. Retractor labii inferioris.
- 1, 1. Orbicularis oris.
- 2, 2, 2. Complexus major.
- 3, 3. Trachelo mastoideus.
- 4, 4. Subscapulo hyoideus.
- 5. Sterno maxillaris.
- 6, 6. Sterno thyro-hyoideus.
- 7. Jugular vein.
- 8. Carotid artery, with the eighth pair, and sympathetic nerves.
- 9. Trachea.
- 10. Scalenus.

THE FORE LIMBS.

- 1. Scapulo ulnarius.
- 2. Caput magnum of the triceps extensor brachii.
- 3. Caput medium of the same muscle.
- 4. Anconeus.
- 5. Flexor brachii.
- 6. Extensor metacarpi.
- 7. Extensor pedis.
- 8. Extensor metacarpi obliquus.
- 10. Flexor metacarpi externus.
- 11. Perforans and perforatus.
- 12. Ulnarius accessorius.
- 13. Flexor metacarpi internus.
- 14. Flexor metacarpi medius.
- 15. Perforans and perforatus.
- 16. Extensor metacarpi.
- B, B. Radius.

THE TRUNK AND BACK.

- A. Scapula.
- F, F. Longissimus dorsi.
- G. Spinalis dorsi.
- a, a. Intercostals.
- b, b. Superficialis costarum.
- c, c. Rectus abdominis.
- d, d. Transversalis abdominis.
- e, e, e. Obliquus internus abdominis.
- f. Hollow in the longissimus dorsi, which part of the gluteus maximus once filled.

EXPLANATION OF FIGURE XIII. CONTINUED.

HAUNCH AND HIND EXTREMITY.

C. Ilium.
D. Ischium.
E. Tibia.
1. Sacro sciatic ligament.
2. Sphincter ani.
3. Depressor coccygis.
4. Muscles of the tail.
5, 5. Triceps abductor tibialis.
6. Vastus externus.
7. Rectus.
8. Gastrocnemii muscles.
9. Plantaris.
10. Extensor pedis.
11. Peroneus.
12. Flexor pedis perforans.
13. Insertion of the gracilis.
14. Gastrocnemii muscles.
15. Flexor pedis accessorius.
16. Course of the perforans tendon, inside the os calcis of the hock joint.
17. Insertion of the gastrocnemius externus into the point of the hock.
18, 18. Popliteus muscles.
19. Extensor pedis.

Cowper's glands are covered by the triangularis penis of each side: they are about the size of a filbert.

Their structure, as apparent to the naked eye, is similar to that of the prostate in every respect, only the excretory ducts are ten or twelve in number for each gland and linearly disposed on each side of the pelvic urethra.

Cowper's glands are supplied with vessels and nerves from the same source as the prostate.

ORGANS OF GENERATION.

TESTICLES AND SCROTUM.

The preparation of the seminal fluid is the office of two oval glandular bodies, called the *testes* or testicles; they are suspended in a portion of the common integument, termed the scrotum, by means of the spermatic cord and cremaster muscle.

The scrotum is composed of the common integument, sub-cellular tissue, and elastic muscle, (the fibres of the latter run in a longitudinal direction, from the cellular substance of the sheath, to the base of the penis), and lastly the tunica vaginalis, which is a prolongation of the peritoneum.

The testicle has a peritoneal covering, termed tunica vaginalis testes, and also another distinct tunic termed tunica albuginea. The substance of the testicle is extremely vascular, and the ultimate branches of its spermatic arteries are collected into small bundles of fine convoluted vessels, separated from one another by septulæ, or membranous partitions. From these the vasa seminifera, or beginnings of excretory ducts, take their origin, and gradually unite to form a smaller number of canals of larger diameter, but exceedingly tortuous in their course. The testicle is also supplied with nerves and absorbents, secretory and excretory vessels.

SPERMATIC CORD.[*]

The spermatic cord, the substance by means of which the testicle is connected with the abdomen, and by means of which it is suspended within its scrotal cavity, is composed in the following manner: 1st. It has four coverings; there is immediately underneath the skin the faschia superficialis; next, the cremaster muscle; thirdly, the tunica vaginalis; and lastly, the tunica vaginalis reflexa. Within the cavity formed by the vaginal tunic, it is that the intestine protrudes in inguinal and scrotal hernia; the hernial coverings, consequently, exclusive of the sac, will be the faschia and cremaster muscle.

2ndly. The constituent parts of the cord itself, are: *a.* The arteries, which are two in number; the artery of the cord, a small branch of the external iliac, which ramifies and expands itself upon the cord; and the spermatic artery, which, as soon as it reaches the internal ring, enters the inguinal canal, runs down the posterior part of the cord, growing tortuous as it descends, serpentines along the superior border of the testes, between it and the epididymis, winds round the anterior end of the gland, and lastly reaches the convex border, where it becomes extremely convoluted, and whereto its branches are principally distributed. In its descent it detaches small unimportant twigs to the adjacent parts; and, as it approaches the testicle, becomes surrounded by an assemblage of venous vessels. *b.* The veins accompany their corresponding arteries, and they indeed may be said to make up the principal bulk of the cord, for they are not only numerous, but large and flexuons, and, as they approach the testicle, form a sort of plexus, which has got the name of *corpus pampiniforme:* they return their blood into the posterior vena cava. *c.* The NERVES, which are derived from the hypogastric plexus, also accompany the spermatic artery; they are small, but sufficiently numerous. Though the testicle does not possess any very great sensibility in health, we may vouch for its being acutely sensitive in a state of disease. *d.* ABSORBENTS exist, both large and numerous, in the cord. They are readily found

---
[*] Percivall.

alongside of the venous trunks; and not infrequently may be filled by introducing mercury into the spermatic artery. *c*. The VAS DEFERENS, though a constituent of the cord, takes at first a solitary course, remote from the blood vessels. The duct issues from the summit of the head of the epididymis, beginning in a series of convolutions gradually unwinding as it proceeds; it takes an oblique course nearly as high as the external ring, where it joins the blood vessels, and continues to accompany them posteriorly through the inguinal canal: at the internal ring it leaves them, turns inward and ascends into the pelvis, where we find it creeping along the side of the bladder infolded in peritoneum to get to the cervix, crossing under its course first the umbilical artery and then the ureter; at length it terminates by rather a contracted orifice within the mouth of the duct of the vesicula seminalis, just behind a little eminence in the urethra — *the caput galinaginis*, about an inch posteriorly to the cervix of the bladder. Within the inguinal passage the duct is accompanied by *the artery of the vas deferens*, a long slender branch of the epigastric. Its canal, flexuous until the duct has joined the cord, but straight in its subsequent course, is not uniform throughout in caliber; the area of its tortuous part is large, but as it becomes straight it grows contracted: having entered the pelvis, it gradually enlarges again, and acquires unusual volume in running along the side of the bladder; and the canal of the enlarged portion presents a reticulated structure, which gives its exterior an irregular, tuberculated appearance; the most contracted part is that in union with the duct of the vesicula seminalis, which is a comparatively small cylindrical conduit. The parietes of the duct are so remarkably thick and firm to the feel, that we distinguish it at once by the fingers from the other parts of the cord; they consist of two tunics; the external one (in which its main thickness consists) is white, fibrous, and approaches in appearance to cartilage; the internal one is thin and fine in texture, muco-membranous in its nature, and here and there incloses a reticulated structure. The different constituent parts of the cord are connected altogether by cellular substance, destitute of any fat; and from the circumstance of the parts in general being more bulky below the ring, the cord increases in breadth and thickness as it approaches the testicle.

THE EPIDIDYMIS.

The epididymis is extended along the superior border of the testicle, upon which it rests, and to which it is connected by the tunica vaginalis reflexa. Its ends are bulky in comparison to its middle: that receiving the vasa efferentia, the smaller one, is the *caput* or *globus minor*; the other, giving rise to the vas deferens, is the *globus major*, the part farriers call the *nut*. The interior of this appendage to the testicle exhibits a structure entirely vascular. The vasa efferentia unite and re-unite until they form a single duct, of whose numberless and very remarkable convolutions the globus major is entirely constituted: these tortuosities (which, when squeezed, freely emit semen) will admit of being unwound for a considerable extent, so as to have the length of the duct calculated with very tolerable exactness from beginning to end, which has been found to amount to several yards. It is small at its formation, but grows imperceptibly larger in making its manifold windings and turnings, until at length it assumes the size of the vas deferens, in which it ends. Its various convolutions are connected together by cellular membrane, and are interspersed with a sparing supply of blood vessels.

The course of the semen is this: It is secreted by the capillary coils of the spermatic artery, from which it is received by the tubuli seminiferi: these tubes carry it into the rete, and the rete discharges it through the vasa efferentia into the epididymis, from which it is conducted by the vas deferens into the urethra.

Formation and Descent. — It is a singular fact, that the organs whose structures we have been investigating, are originally formed in a situation remote from that in which they are destined to carry on their functions; "the colt has no testicles," is the common observation of the uninformed on these matters; and we know ourselves that the purse is without them, but we know, in addition, that they exist ready-formed within the abdomen, and that they will descend at a certain period of age into the proper receptacle, the scrotum. During the fœtal state we find the testicles more or less developed, tinged with a blush of red, lodged beneath the psoas muscles, in contact with the inferior borders of the kidneys, covered and retained in their situations by peritoneum, and concealed by the intestines around them. Here they receive their arteries from the contiguous trunk — the posterior aorta; the vasa deferentia run forward to them, and the cremasters likewise turn forward instead of backward; there being at this time no such thing as a spermatic cord. Thus placed, the testicle may be regarded as one of the glands of the abdomen; indeed it has considerable similarity to the kidney — receiving its vessels from the same contiguous source, and sending a long duct backward into the cavity of the pelvis; nor does there appear any conclusive reason why it should not perform the same office in that situation that it does in the scrotum, and particularly since it is known that in birds the testicles remain within the abdomen during life. From the part where the blood vessels enter, we find growing a whitish substance, extending backward, diminishing in breadth as it recedes, passing through the ring where the hilus of the cremaster may be traced upon it, and whence it is prolonged into the scrotum, growing narrower and narrower until it vanishes; this substance, regarded by some simply as a ligament, was considered by Mr. Hunter as the gubernaculum or pilot, by means of which the testicle is directed in its passage from the abdomen into the scrotum. Quitting the spot where it has been formed and matured, the testicle gradually retrocedes, guided by the gubernaculum, until it arrives on the internal ring, which, at this time (like every other part of the parietes) is closed by peritoneum; this temporary obstruction it overcomes by drawing the membrane down along with it through the ring, and carrying the pouch made thereby down into the scrotum; the gubernaculum at the time undergoing a complete inversion. This accounts for the production of the tunica vaginalis, and explains how that membrane comes to be doubled or reflected; the testicle, receiving originally (as an abdominal viscus) one close adherent peritoneal tunic, and acquiring another which forms a loose covering as it passes through the ring, must necessarily have *two;* and since both are derived from one and the same membrane, it follows that one must be a continuation of the other. These elongations of membrane, though everywhere in contact, are prevented from adhering together by a continual exhalation of the natural serous secretion. Any interval that might subsist between them, in course, communicates with the cavity of the abdomen, through the ring, a part that remains open through life: this, however, is not the case with man — in his body the communication is cut off, after the testicles have descended, by a natural contraction and obliteration both of the ring and the inguinal passage. In many instances, one, in some few, both of the testicles, are known to have remained within the belly through life. As we are unacquainted with the immediate cause of their descent, so we are unable to give any rational explanation of this phenomenon. I have understood, that in many of these cases the glands have been found to be but *imperfectly* developed: this, however, is not without exception.

Period of Descent. — Most animals have their testicles within the scrotum at the period of birth. In the human fœtus they begin to move about the seventh month; about the eighth they reach the groins; and before birth they arrive in the scrotum. In the horse, they pass through the ring about the sixth or seventh month before birth, and are found within the scrotum at the period of parturition. In some cases, one testicle will not make its appearance for some time after the other; and as the operation for castration is seldom long delayed, this will account for the *rigs* (as horses having but one testicle are called) with which we meet every now and then. Again, instances are not wanting in which one testicle has descended to the ring and there remained through life.*

PENIS.

The penis is composed of the two corpora cavernosa; head, or glans penis; corpus musculosum urethra, and the plexus venosus. The corpore cavernosa make up the bulk of the organ, they extend from the pelvis to the glans penis; at the ischial arch they are invested with fibres of the erectors penis, and are strengthened and confined to the pubes by the suspensory ligaments. It is supplied with blood from a branch of the obturator artery by means of the internal pudic artery. Its nerves are termed pudic, so also are the veins.

The *glans* is composed of a soft spongy tissue, highly elastic and distensible, and remarkable as the seat of the plexus venosus penis: the latter structure presents itself in the form of a venous conglomeration, and in the erect state of the organ constitutes its chief bulk.

URETHRA.

The urethra is a muco-membranous canal

---

* In a communication I have been favored with from Mr. Brettargh (which I have inserted in the second volume of THE VETERINARIAN), is contained the following information on this subject: "Colts are foaled with their testicles in the scrotum, which remain there (in ordinary cases) until the fifth or sixth month, when they are taken up between the internal and external abdominal rings, and there remain until the eleventh, twelfth or thirteenth month, all depending upon the degree of keep, as in some that are well fed the testicles can at all times be found in the scrotum. Were the testicles drawn up into the abdomen, they would be too large to pass through the internal abdominal ring at the time they are wanted to prepare for secretion; which is occasionally the case, and at once accounts for our meeting with horses that are said to have but one stone. I have seen one instance where both were wanting in the scrotum at four years old."

averaging in length, in the unerected state forty-eight inches; it extends from the anterior part of the glans penis to the neck of the bladder; its use is to afford a passage for the urine and seminal fluid.

FEMALE ORGANS OF GENERATION.

The vulva or pudendum comprises the prominence and fissure, commencing immediately beneath the anus, and extending downwards some four or five inches. The fissure is longest and most conspicuous in breeding mares. The space between the anus and vulva is termed perineum. The prominences on each side of the vulva are called labia pudendia. They owe their bulk principally to muscular and fatty substance, and cellular tissue.

The commissures are the parts uniting the labia above and below. The superior or upper commissure is extended to a sharp angle, and joins the perineum; the lower portion is rounded off, and is bounded by a hollow, at the bottom of which is lodged the

Clitoris. — This is brought into view immediately after staling: it bears a close comparison to the head of the male penis, and, like the latter, is susceptible of sensual enjoyment. To the clitoris belong a pair of muscles named erectors clitordis. They take their origin from the perineum. Their office is to erect that body, and protrude it into the vagina in the act of coition.

The internal parts are the vagina, uterus, Fallopian tubes, fimbriæ, and ovaria. The vagina is a musculo-membranous canal, of large dimensions, extending from the vulva to the uterus or womb.

It is situated within the pelvis, having the bladder below and the rectum above it, to both of which it has cellular attachments, in addition to the reciprocal connection with the peritoneum. To the rectum it is closely and firmly attached by cellular membrane.

The figure of the vagina, when it is distended, is that of an oblong cylinder; but in the collapsed state, its sides are in contact, and it will vary its form according to the full or empty condition of the bladder. The largest part of the canal is the posterior; there it exceeds the dimensions of the bladder.

The length of the canal is about eighteen inches. Its course is horizontal, and rather shows an inclination to the curve of the rectum.

The vagina, at its commencement from the vulva, is much thicker in its walls than elsewhere; in composition it is partly muscular and partly membranous. The orifice of it is clothed in that strong, red, circular, fleshy band, which forms the sphincter vagina; and the adjoining part of the canal is also encircled by some considerable fleshy covering, and thickly coated with muscular fibres. Further forward than this the vagina is composed of membrane.

The Membrane of the Vagina. — The part of which it is constituted is one of the mucus class, and one that possesses considerable density, extensibility, and resistance. Its exterior surface is rough. Its interior is smooth, and has a pale pinkish cast; unless the mare be under the venereal œstrum, and then its redness is heightened, and its secretion augmented. In the ordinary state, this membrane is thrown into folds, larger in breeding mares than in others, technically called rugæ.

Considerably in advance of the clitoris is an opening leading from the lower part of the canal, large enough to admit with ease any one of the fingers: this is the orifice of the meatus urinarius, or outlet of the bladder: it is guarded by a doubling of the vaginal membrane, which hangs over it, and serves the purpose of a valve.

The large and conspicuous protuberance at the bottom of the vagina, is the mouth of the uterus.

The uterus, or womb, is a hollow musculo-membranous organ, united to the anterior part of the vagina, and is destined for the reception of the fœtus. We distinguish the uterus by the body, horns, neck, and mouth. The body is the oblong or cylindrical part, growing out of the anterior portion of the vagina, in the centre of which it is terminated internally by the os-uteri, or mouth of the womb; it gives origin, in front, to the horns. This part lies wholly within the pelvis, between the bladder and rectum, and is entirely covered by peritoneum.

The cornua, or horns, rise from the body of the uterus, and diverge towards the loins. Their length and size will be much greater in breeding mares than in others. In figure they are cylindrical; they bend upward in their course, and terminate in round extremities, to which are loosely appended the ovaries, or testicles, through the medium of the Fallopian tubes.

The cervix, or neck, of the uterus is the rugose portion, protruded backward into the cavity of the vagina, which has a flower-like appearance, and can only be seen in a virgin uterus in the undistended state; during gestation it undergoes a remarkable change.

Independently of its union with the vagina, the uterus is confined in its place by two broad portions of peritoneum, which attach it to the sides of the pelvis, named the lateral ligaments of the uterus. During the period of gestation, the uterus experiences considerable extension. The Fallopian tubes are two trumpet-shaped canals, having a remarkable serpentine course; running within the folds of the ligamenta lata, from the extremities of the horns to the ovaries.

The tube commences by an aperture in the cornu, having an elevated whitish margin, which is scarcely large enough to admit a small silver probe: from this it proceeds forward, folded in peritoneum, and extremely convoluted, until it reaches the ovary, to which it becomes attached; it then begins to enlarge in its diameter, grows less convoluted, and serpentines along the lower side of the ovary; it afterwards ends in a fringed doubling of membrane.

The internal membrane of the tubes is similar to that of the uterus.

The ovaries, or female testicles, are two egg-shaped bodies, situated further forward than the Fallopian tubes, within the cavity of the abdomen; they receive close coverings, and are loosely attached to the spine.

These bodies are about the size of walnuts. They are not regular oviform; they have deep fissures on their sides; they bear a resemblance, at first view, to the testicles and their ducts in the male.

Internally, the ovaries are composed of a whitish spongy substance, in which are, in some instances, found little vesicles, containing a yellowish glairy fluid, in others one or more dark yellow or brownish substances, named corpora lutea: the vesicles are the ova, which, from impregnation, receive further development; the corpora lutea denote the parts from which vesicles have burst, and consequently only exist in the ovaries of those mares whose organs have been engaged in the generative process. Prior to the age of sexual intercourse, these bodies are small and white; but, as soon as the season of copulation is at hand, they grow large, redden externally, and present many yellow spots or streaks through their substance.

Mammæ, though unconnected with the uterus, anatomically speaking, are in function concurring to the same important end. The mammæ, vulgarly called udder, are two flattened oval-shaped bodies, depending, between the thighs, from the posterior and inferior part of the belly. In quadrupeds, with but few exceptions, this is the situation of the mammæ.

In virgin mares the udder is so small that there hardly appears to be any. In mares who have had foals, the udder remains prominent or pendulous, and has a flabby feel

Toward the latter part of gestation, this part swells, and becomes distinctly visible. Within a few days of foaling, the udder grows turgid with milk; it does not, however, acquire its full distention until the foal has drawn it for a few days, from which time it maintains its volume, with little variation, during the period of suckling. Soon after the foal begins to forsake the teat, the secretion of milk diminishes, and is followed by a contraction of the bag, which goes on gradually, until it has resumed nearly, or quite, its former flatness.

The interior of the mammæ has a light yellowish aspect, and evidently possesses a lobulated structure, which is held together by a fine cellular tissue, interspersed with granules of fat. It is constituted of glandular masses, irregular in magnitude and form, and loosely connected one with another, each of which masses is composed of a number of lobules, closely compacted and united together. These insulated lobulous portions receive small arteries, from which the milk is secreted. The former, by repeatedly conjoining one with another, become at length several demonstrable canals, radiating from every part, and dilating to hold the milk.

When the udder becomes charged with milk, it flows into the teat and distends it. Suction is apparently an operation purely mechanical. The teat is seized and closely compressed by the lips of the foal; and the imbibing effort which follows has a tendency to produce a vacuum, or raise the valve at the upper part of the teat, and the milk passes from the reservoirs into the mouth.

PHYSIOLOGICAL CONSIDERATIONS.

ON THE REPRODUCTION OF ORGANIZED BEINGS.*

"IF the changes which living beings undergo during the period of their existence, and the termination of that existence by the separation of their elements at a period more or less remote from their first combination, be regarded as distinguishing them in a striking and evident manner from the masses of inert matter which surround them, still more is their difference manifested in the series of processes which constitute the function of *Reproduction*. A very unnecessary degree of mystery has been spread around the exercise of this function, not only by general inquirers, but by scientific physiologists. It has been regarded as a process never to be comprehended by man, of which the nature and the laws are alike inscrutable. A fair comparison of it, however, with other functions, will show that it is not in reality less comprehensible or more recondite than any one of them;—that our acquaintance with each depends upon the facility with which it may be submitted to investigation;—and that, if properly inquired into by an extensive survey of the animated world, the real character of the process, its conditions, and its mode of operation, may be understood as completely as those of any other vital phenomenon.

"It may be considered as a fundamental truth of Physiological Science, that *every living organism has had its origin in a pre-existing organism.* The doctrine of 'spontaneous generation,' or the supposed origination of organized structures *de novo* out of assemblages of inorganic particles, although at different times sustained with a considerable show of argument, based on a specious array of facts, cannot now be said to have any claim whatever to be received as even a possible hypothesis; all the facts on which it claimed to rest having either been themselves disproved, or having been found satisfactory explicable on the general principle *omne vivum ex ovo.* Thus, the appearance of Animalcules in infusions of decaying organic matter, the springing-up of Fungi in spots to which it would not have been supposed that their germs could have been conveyed, the occurrence of Entozoa in the bodies of various animals into which it seemed almost beyond possibility that their eggs could have been introduced, with other facts of a like nature, may now be accounted for, without any violation of probability, by our increased knowledge of the mode in which these organisms are propagated. Thus, it is now well ascertained that the germs of Fungi and of many kinds of Animalcules are diffused through the atmosphere, and are conveyed by its movements in every direction; and that, if to decomposing substances of a kind that would otherwise have been most abundantly peopled by these organisms, such air only be allowed to have access as has been deprived of its organic germs by filtration (so to speak) through a red-hot tube or strong sulphuric acid, no living organisms will make their appearance in them; whilst in a few hours after the exposure of the very same substances to ordinary atmospheric air, it has been found to be crowded with life.* And when it is borne in mind, in the case of the Entozoa, that the members of

---

* Carpenter's Physiology.

* See the experiments of Schulze, in the "Edinb. New Phil. Journal," 1837, p. 165.

this class are remarkable for the immense number of eggs which most of them produce, for the metamorphoses which many of them are known to undergo, and for the varieties of form under which there is reason to suspect that the same germs may develop themselves, it becomes obvious that no adequate proof has yet been afforded that they have been, in any particular case, otherwise than the products of a pre-existing living organism. This, again, is the conclusion to which all the most general doctrines of Physiology necessarily conduct us. For it is most certain that we know nothing of Vital Force, save as manifested through organized structures; whilst, on the other hand, the combination of inorganic matter into organized structures is one of the most characteristic operations of vital force; hence it is scarcely conceivable that any operation of physical forces upon inorganic matter should evolve a living organism. Nor is such a conception more feasible, if it be admitted that vital force stands in such a relation to the physical forces, that we may regard the former as a manifestation of the latter, when acting through organized structures; since no vital force can be manifested (according to this view), and no organization can take place, except through a pre-existing organism.

"It may be further considered as an established physiological truth, that, when placed under circumstances favorable to its complete evolution, every germ will develop itself into the likeness of its parent; drawing into itself, and appropriating by its own assimilative and formative operations, the nutrient materials supplied to it; and repeating the entire series of phases through which its parent may have passed, however multiform these may be.* Now the germs of all tribes of plants and animals whatever bear an extremely close relation to each other in their earliest condition; so that there is no appreciable distinction amongst them, which would enable it to be determined whether a particular molecule is the germ of a Conferva or of an Oak, of a Zoophyte or of a Man. But let each be placed in the conditions it requires; and a gradual evolution of the germ into a complex fabric will take place, the *more general* characters of the new organism preceding the *more special*, as already explained. These conditions are not different in kind from those which are essential to the process of nutrition in the adult; for they consist, on the one hand, in a due supply of aliment in the condition in which it can be appropriated; and, on the other hand, in the operation of certain external agencies, especially heat, which seems to supply the *force* requisite for the developmental process. Now, although we may not be able to discern any such ostensible differences in the germs of different orders of living beings as can enable us to discriminate them from each other, yet, seeing so marked a diversity in their operations under circumstances essentially the same, we cannot do otherwise than attribute to them distinct properties; and it will be convenient to adopt the phrase *germinal capacity* as a comprehensive expression of that peculiar endowment, in virtue of which each germ developes itself into a structure of its own specific type, when the requisite forces are brought to bear upon it, and the requisite materials are supplied to it.* Thus, then, every act of development may be considered as due to the force supplied by heat or some other physical agency, which, operating through the organic germ, exerts itself as formative power; whilst the mode in which it takes effect is dependent

---

* The apparent exceptions to this rule, which have been brought together under the collective term, "Alternation of Generations," will be presently considered, and will be shown to be only exceptional when misinterpreted.

* This term is preferred to that of "germ-power" suggested by Mr. Paget, because the latter seems to imply that the force of development exists in the germ itself. Now, if this were true, not only must the whole formative power of the adult have been possessed by its first cell-germ, but the whole formative power of all the beings simultaneously belonging to any one race, must have been concentrated in the first cell-germ of their original progenitor. This seems a *reductio ad absurdum* of any such doctrine; and we are driven back on the assumption (which all observation confirms), that the *force* of development is derived from *external* agencies.

upon the properties or endowments of the substances through which it acts, namely, the germ on the one hand, the alimentary materials on the other,—just as an electric current, transmitted through the different nerves of sense, produces the sensory impressions which are characteristic of each respectively; or, as the same current transmitted through one form of inorganic matter produces light and heat, through another, chemical change, or through another, magnetism.

"In the development of any living being, therefore, from its primordial germ, we have three sets of conditions to study—namely, *first*, the physical forces which are in operation; *second*, the properties of the germ, which these forces call into activity; and *third*, the properties of the alimentary materials which are incorporated in the organism during its development. There is evidence that each of these may have a considerable influence on the result; but in the higher organisms it would seem that the second is more dominant than it is in the lower. For among many of the lower tribes, both of plants and animals, there is reason to believe that the range of departure from the characters of its parent, which the organism may present, is considerably greater than that of the higher; and that this is chiefly due to the external conditions under which it has been developed. The forms of a number of species of the lower Fungi, for example, appear to be in a great part dependent on the nature of their aliment; so among the Entozoa, there seems strong reason to believe that those of the *Cystic* order are only *Cestoidea*, that are prevented by the circumstances under which they exist from attaining their full development; and the production of a fertile 'queen' or of an imperfect 'worker,' among the hive-bees, appears to be entirely determined by the food with which the larva is supplied. No such variations have been observed among the higher classes; in which it would seem as if the form attained by each germ is more rigidly determined by its own endowments; a modification in the other conditions, which in the lower tribes would considerably affect the result, being in them unproductive of any corresponding change. For, if such modification be considerable, the organism is unable to adapt itself to it, and consequently either perishes or is imperfectly developed; whilst, if it be less potent, it produces no obvious effect. Thus, a deficiency of food in the growing state of the higher animal will necessarily prevent the attainment of the full size; but it will not exert that influence on the relative development of different parts that it does among plants, in which it favors the production of flowers and fruit in place of leaves, or that it seems to exercise in several parallel cases among animals. So, again, a deficiency of heat may slightly retard the development of the chick; but, if the egg be allowed to remain long without the requisite warmth, the embryo dies, instead of passing into a state of inactivity, like that of reptiles or insects. The extent, indeed, to which these external conditions may affect the development of the inferior organisms, must not be in the least judged of by that to which their operation is restricted in the higher; and it is probable that we have yet much to learn on the subject. At present, it may be stated as a problem for determination, whether, from a being of superior organization, *lower* forms of living structure, capable of maintaining an independent existence, and of propagating their kind, can ever originate, by an imperfect action of its formative powers. Various morbid growths, such as cancer cells, to which the higher organisms are liable, have been looked upon in this light; these have certainly a powerful vitality of their own, which enables them to increase and multiply at the expense of the organism which they infest; and they have also an energetic reproductive power, by which they can propagate their kind, so as to transmit the disease to other organisms, or to remote parts of the same organism; but such growths are not independent; they cannot maintain their own existence, when detached from the organism in which they are developed; and they have not, therefore, the attribute of *a separate individuality*. Various phenomena hereafter to be detailed, however, respecting the 'gemmiparous' production of living beings, when taken in connection with that just cited, seem to render it by no means impossible that the individualization may be more complete in other cases, so that independent beings of a lower type *may possibly* originate in a perverted condition of the formative operations in the higher. But no satisfactory evidence has ever been afforded by experience, that such 'equivocal generation' has actually taken place; and its possibility is here alluded to only as a contingency which it is right to keep in view. That *no higher* type has ever originated through an advance in developmental power, may be safely asserted; for, although various instances have been brought forward to justify the assertion that such is possible, yet these instances entirely fail to establish the analogy that is sought to be drawn from them.^e

"The *development power* which each germ possesses, under the conditions just now detailed, is manifested, not merely in the first evolution of the germ into its complete specific type, but also in the maintenance of its perfect form, and, within certain limits, by the reproduction of parts that have been destroyed by injury or disease. This reproduction, as Mr. Paget has pointed out,* differs from the ordinary process of nutrition in this,—that 'in grave injuries and diseases, the parts that might serve as models for the new materials to be assimilated to, or as tissue-germs to develop new structures, are lost or spoiled; and yet the effects of injury and disease are recovered from, and the right specific form and composition are retained;'—and, again, 'that the reproduced parts are formed, not according to any present model, but according to the appropriate specific form, and often with a more strikingly evident design towards that form, as an end or purpose, than we can discern in the natural construction of the body.' In the reproduction of the leg of a full-grown Salamander after amputation, which was observed to take place by Spallanzani, it is clear that, whilst the process was from the first of a nature essentially similar to that by which its original development took place, it tended to produce, not the leg of a larva, but that of an adult animal. Hence it is obvious that, through the whole of life, the formative processes are so directed as to maintain the perfection of the organism, by keeping it up, so far as possible, to the model or archetype that is proper to the epoch of its life which it has attained. The amount of this regenerating power, however, varies greatly in different classes of organized beings, and at different stages of the existence of the same being; and, as Mr. Paget has pointed out,† it seems to

^e Thus, the author of the "Vestiges of the Natural History of Creation" refers to the various modifications which have taken place in our cultivated Plants and Domesticated Animals, in proof that such elevation is possible; quite overlooking the fact that these external influences merely *modify* the development, without *elevating* it, and that these races, if left to themselves, speedily revert to their common specific type. And he adduces the phenomena of metamorphosis—the transformation of the worm-like larva into an insect, and of a fish-like tadpole into a frog—as giving some analogical sanction to the same doctrine; totally overlooking the fact, that these transformations are only part of the ordinary developmental process, by which the complete form of the species is evolved, instead of being transitions from the perfected type of one class to the perfected type of one above it. So, again, he quotes the transformation of the worker-grub of the hive-bee into the fertile queen, as an example of a similar advance; without regarding the circumstance that the worker is *physically* higher (according to human ideas, at least) than the queen, whose instincts appear limited to the performance of her sexual functions; and that the utmost which the fact is capable of proving, is, that the same germ may be developed into two different forms, according to the circumstances of its early growth. It must always be borne in mind that the character of a species, to be complete, should include *all* its forms, perfect and imperfect, modified and unmodified; since in this mode alone can that "capacity for variation" be determined, which is so remarkable a feature in many cases, and is that which specially distinguishes the races of plants and animals that have been subjected to human influence.

In no instance has this variation tended to confuse the limits of well-ascertained species; it has merely increased our acquaintance with the number of diversified forms into which the same germ may develope itself.

* "Lectures on Reproduction and Repair."
† Loc. cit.

bear an inverse ratio to the degree of development which has previously taken place in each case. Thus, in the *Hydra* and other Zoophytes, it would appear (as in Plants) to be almost unlimited; for the development process in them is checked at such an early period, that both the form of the organism and the structure of its tissues retain the most simple type; and by the subdivision of one individual, no fewer than fifty were produced by Trembly. In this, as probably in all the cases in which new individuals have been obtained by artificial subdivision, there is some natural tendency to their production by the vegetative process of gemmation; but this does not always manifest itself. It is a curious fact, that the first attempt at regeneration, in some of these cases, is not always complete; but that successive efforts are made, each of which approximates more and more closely to the perfect type. This was well seen in one of Sir J. G. Dalyell's experiments; for he observed that, having cloven the stem of a *Tubularia* (a Hydroid Zoophyte), after the natural fall of its head, an imperfect head was at first produced, which soon fell off and was succeeded by another more fully formed; this in its turn was succeeded by another; and so on, until the fifth head was produced, which was as complete as the original.

"As a general statement of the amount of this regenerating power, which exists in most of the different classes of animals, has been already given, it is unnecessary here to do more than allude to some of those facts which most strongly bear out the doctrine just laid down. Next to Zoophytes, there are no animals in which the regenerative power is known to be so strong as it is in the lower Articulata (as the Cestoid Entozoa, and the inferior Annelida), and in the *Planaria*, which may perhaps be regarded as rather approximating to the Molluscous type; and here, again, we see that a low grade of general development is favorable to its exercise, and that the spontaneous multiplication which occasionally takes place in these animals by fission or gemmation, is only another form of the same process. In the higher forms of both these sub-kingdoms, as we no longer meet with multiplication by gemmation, so do we find that the reparative power is much more limited; the only manifestation of it among the fully-formed *Arrachnida* and *Crustacea* being the reproduction of limbs, and the power of effecting even this being usually deficient in perfect Insects. The inquiries of Mr. Newport, however, upon the reproductive powers of *Myriapods* and *Insects*, in different stages of their development,[*] confirm the general principle already stated; for he has ascertained that in their *larval* condition, Insects can usually reproduce limbs or antennæ; and that Myriapods, whose highest development scarcely carries them beyond the larva of perfect Insects, can regenerate limbs or antennæ, up to the time of their last moult, when, their normal development being completed, their regenerative power seems entirely expended. The *Phasmidæ* and some other insects of the order *Orthoptera* retain a similar degree of this power in their perfect state; but these are remarkable for the similarity of their larval and imago states, the latter being attained, as in Arachnida, by a direct course of development, without anything that can be called a 'metamorphosis.' Little is known of the regenerative power in the higher Mollusca; but it has been affirmed that the head of the Snail may be reproduced after being cut off, provided the cephalic ganglion be not injured, and an adequate amount of heat be supplied. In Vertebrata, again, it is observable that the greatest reparative power is found among *Batrachian Reptiles*, whose development is altogether lower, and whose life is altogether more vegetative, than that of probably any other group in this sub-kingdom. In Fishes, it has been found that portions of the fins which have been lost by disease or accident are the only parts that are reproduced. But in the Salamander, entire new legs, with perfect bones, nerves, muscles, etc., are reproduced after

[*] "Philosophical Transactions," 1844.

loss or severe injury of the original members; and in the Triton a perfect eye has been formed to replace one which had been removed. In the true *Lizards*, an imperfect reproduction of the tail takes place, when a part of it has been broken off; but the newly-developed portion contains no perfect vertebræ, its centre being occupied by a cartilaginous column, like that of the lowest Fishes. In the warm-blooded Vertebrata generally, as in Man, the power of true reproduction after loss or injury seems limited, as Mr. Paget has pointed out,* to three classes of parts, namely: (1.) 'Those which are formed entirely by nutritive repetition, like the blood and epethelia, their germs being continually generated *de novo* in the ordinary condition of the body; (2.) Those which are of lowest organization, and (which seems of more importance) of lowest chemical character, as the gelatinous tissues, the areolar and tendinous, and the bones; (3.) Those which are inserted in other tissues, not as essential to their structure, but as accessories, as connecting or incorporating them with the other structures of vegetative or animal life, such as nerve-fibres and blood-vessels. With these exceptions, injuries or losses are capable of no more than repair, in its more limited sense; *i. e.*, in the place of what is lost, some lowly organized tissue is formed, which fills up the breach, and suffices for the maintenance of a less perfect life.' Yet, restricted as this power is, its operations are frequently most remarkable; and are in no instance, perhaps, more strikingly displayed, than in the re-formation of a whole bone, when the original one has been destroyed by disease. The new bony matter is thrown out, sometimes within, and sometimes around, the dead shaft; and when the latter has been removed, the new structure gradually assumes the regular form, and all the attachments of muscles, ligaments, etc., become as complete as before. A much greater variety and complexity of actions are involved in this process, than in the reproduction of whole organs in the simpler animals; though its effects do not appear so striking. It would seem that in some individuals this regenerating power is retained to a greater degree than it is by the class at large;* and here again we find, that in the early period of development the power is more strongly exerted than in the adult condition. The most remarkable proof of its persistence even in Man, has been collected by Prof. Simpson; who has brought together numerous cases in which, after 'spontaneous amputation of the limbs of a fœtus in utero,' occurring at an early period of gestation, there has obviously been an imperfect effort at the re-formation of the amputated part from the stump.† By the knowledge of these facts and principles, we seem justified in the surmise, that the occurrence of supernumerary or multiple parts is not always due (as usually supposed) to the 'fusion' of two germs, but that it may result from the subdivision of one;

---

* "Lectures on Reproduction and Repair."

* One of the most curious and well-authenticated instances of this kind is related by Mr. White, in his work on the "Regeneration of Animal and Vegetable Substances," 1785, p. 16. "Some years ago, I delivered a lady of rank of a fine boy, who had two thumbs upon one hand, or rather, a thumb double from the first joint, the other one less than the other, each part having a perfect nail. When he was about three years old, I was desired to take off the lesser one, which I did; but to my great astonishment it grew again, and along with it the nail. The family afterwards went to reside in London, where his father showed it to that excellent operator, William Bromfield, Esq., surgeon to the Queen's household; who said, he supposed Mr. White, being afraid of damaging the joint, had not taken it wholly out, but he would dissect it out entirely, and then it would not return. He accordingly executed the plan he had described, with great dexterity, and turned the ball fairly out of the socket; notwithstanding this, it grew again, and a fresh nail was formed, and the thumb remained in this state." The Author has been himself assured by a most intelligent Surgeon, that he was cognizant of a case in which the whole of one ramus of the lower jaw had been lost by disease in a young girl, yet the jaw had been completely regenerated, and teeth were developed and occupied their normal situations in it.

† These cases were brought by Prof. Simpson before the Physiological Section of the British Association, at its meeting in Edinburgh, August, 1850. The Author, having had the opportunity of examining Prof. Simpson's preparations, as well as two living examples, is perfectly satisfied as to the fact.

for, if it be supposed that this subdivision has taken place when the developmental process has advanced no further than in a Hydra or a Planaria, it seems by no means impossible that each part might, as in those creatures, advance in its development up to the attainment of its complete form.

"There are many tribes, both of Plants and Animals, in which multiplication is effected not only *artificially* but *spontaneously*, by the separation of parts, which, though developed from the same germ in perfect continuity with each other, are capable of maintaining an independent existence, and which, when thus separated, take rank as distinct individuals. This process, which is obviously to be regarded, no less than the preceding, as a peculiar manifestation of the ordinary operations of Nutrition, may take place in either of four different modes — 1. In the lowest Cellular Plants, and the simplest Protozoa, every component cell of the aggregate mass that springs from a single germ, being capable of existing independently of the rest, may be regarded as a distinct individual; and thus every act of growth which consists in the multiplication of cells, makes a corresponding augmentation in the number of individuals. 2. In many organisms of a somewhat higher type, in which the fabric of each complete individual is made up of several component parts, we find the new growths to be complete repetitions of that from which they are put forth; and thus the composite organism presents the semblance of a collection of individuals united together, so that nothing is needed but the severance of the connection, to resolve it into a number of separate individuals, each perfect in itself. The most characteristic example of this is presented by the *Hydra*, which is continually multiplying itself after this fashion; for the buds or 'gemmæ' which it throws off are not merely structurally but functionally complete (being capable of seizing and digesting their own prey), previously to their detachment from the parent. 3. In by far the larger proportion of cases, on the other hand, the 'gemma' does not possess the complete structure of the parent, at the time of its detachment, but is endowed with the capacity for developing whatever may be deficient. Thus, the bud of a Phanerogamic Plant possesses no roots, and its capacity for independent existence depends upon its power of evolving those organs. On the other hand, the 'zoospore' of an Ulva or a Conferva is nothing else than a young cell, from which the entire organism is to be evolved after it has been set free; and, even in the 'bulbels' of the Marchantia, the advance is very little greater. The 'bulbels' of certain Phanerogamic plants, however, bear more resemblance to ordinary buds. 4. In the preceding cases, the organism which is developed by this process resembles that from which it has been put forth; but there are many cases in which the offset differs in a marked degree from the *stock*, and evolves itself into such a different form that the two would not be supposed to have any mutual relation, if their affinity were not proved by a knowledge of their history. Sometimes we find that the new individual thus budded off is in every respect as complete as that from which it proceeded, though developed upon a different type; but in other instances it is made up of little else than a generative apparatus, provided with locomotive instruments to carry it to a distance, its nutritive apparatus being very imperfect. Of the first, we have an example in the development of Medusæ from the Hydroid Polypes; and of the second in the peculiar subdivision of certain Annelida, hereafter to be described. Now it is obvious that, in this process, no agency is brought into play that differs in any essential mode from that which is concerned in the ordinary nutritive operation. The multiplication of individuals is performed exactly after the same fashion as the extension of the parent organism; and the very same parts may be regarded as organs belonging to it, or as new individuals, according to their stage of development, and the relation of dependence which they still hold to it. The essence of this operation is *the multiplication of cells by continual subdivision.*

"We have now, on the other hand, to inquire into the nature of the true *Generative* process, by which the original germ is endowed with its developmental capacity; and this we shall find to be of a character precisely the opposite of the preceding. For, under whatever circumstances the generative process is performed, it appears essentially to consist in *the re-union of the contents of two cells*,* of which the germ, which is the real commencement of a 'new generation,' is the result. This process is performed under the three following conditions: 1. All the cells of the entire aggregate, produced by the previous subdivision, may be capable of thus uniting with each other indiscriminately; there being no indication of any sexual distinction. This is what we see in the simplest Cellular plants. 2. All the component cells of each organism may, in like manner, pair with other cells, to produce fertile germs; but there are differences in the shares which they respectively take in the process, which indicate that their endowments are not precisely similar, and that a sexual distinction exists between them, notwithstanding that this is not indicated by any obvious structural character. This condition is seen in the Zygnema and its allies. 3. The generative power is restricted to certain cells, which are set apart from the rest of the fabric, and destined to this purpose alone; and the endowments of the two sets are so far different, that the one furnishes the germ, whilst the other supplies the fertilizing influence; whence the one set have been appropriately designated 'germ-cells' and the other 'sperm-cells.' Such is the case in all the higher Plants among which a true generative apparatus has been discovered; and also throughout the Animal kingdom.

"Thus, then, in the entire process in which a new being originates, possessing like structure and endowments with its parent, two distinct classes of actions participate, — namely, the act of *Generation*, by which the Germ is produced; and the act of *Development*, by which that germ is evolved into the complete organism. The former is an operation altogether *sui generis*: the latter is only a peculiar modification of the Nutritive function; yet it may give origin, as we have seen, to new individuals, by the separation (natural or artificial) of the parts which are capable of existing as such. Now, between these two operations there would seem to be a kind of antagonism. Whilst every act of Development tends to *diminish* the 'germinal capacity,' the act of Generation *renews* it; and thus the tree, which has continued to extend itself by budding until its vital energy is well-nigh spent, may develop flowers and mature seeds from which a vigorous progeny shall spring up. But the *multiplication of individuals* does not directly depend upon the act of generation alone; it may be accomplished by the detachment of *gemmæ*, whose production is a simple act of development; and the individuals thus produced are sometimes similar, sometimes dissimilar, to the beings from which they sprang. When they are dissimilar, however, the original type is always reproduced by an intervening act of generation; and *the immediate products of the true generative act always resemble one another.* Hence the phrase, 'alternation of generations,' can only be legitimately employed when the term generation is used to designate a succession of individuals, by whatever process they have originated; an application of it which cannot but lead to a complete obliteration of the essential distinction which the attempt has been here made to draw between the generative act and the act of gemmation. For when it is said that 'generation A produces generation B, which is dissimilar to itself, whilst generation B produces generation C, which is dissimilar to itself, but which returns to

---

* In very rare instances, it is the re-union of the two parts of the contents of the same cell, which had previously tended to separate from each other, as if in the process of subdivision.

the form of generation A, it is entirely left out of consideration that generation A produces (the so-called) generation B by a process of *gemmation*; whilst the process by which generation B produces generation C is one of *true generation*. So generation C developes D by gemmation, which resembles B; and D, by a true generative act, produces E, which resembles A and C. This distinction, although it may at first sight appear merely verbal, will yet be found of fundamental importance in the appreciation of the true relations of these processes, and of their resulting products. So, in the Author's opinion, the application of the term 'generation' to the *entire product* of the development of any germ originating in a generative act, whether that product consist of a single individual, or of a succession, will be found much more appropriate, and more conducive to the end in view, than the indiscriminate application of it to each succession, whether produced by gemmation or by sexual re-union. It is of great importance to the due comprehension of certain phenomena of Reproduction, which will come under consideration in the Animal kingdom, that the relations of the products of these two processes should be rightly appreciated; and this appreciation of them will, it is believed, be best gained by a careful inquiry into the phenomena of Reproduction in the Vegetable kingdom."

EXAMINATIONS RESUMED.

GLANDULAR APPARATUS.

Q. Describe the structure of a glandular body. — *A.* It consists of a collection of tubes, more or less convoluted, united by cellular substance into masses of a rounded form, constituting a lobule; each lobule has a separate investment of membrane; and the whole aggregate of lobules is furnished with a general membranous envelope or capsule. Each gland presents a complex arrangement of numerous arteries, veins, nerves, and lymphatics, and most of them are provided with an excretory duct, which conducts the secretion prepared in the gland.

Q. What glands are supposed to be destitute of a secretory duct? — *A.* The pineal gland, thyroid, thymus, and renal capsules.

Q. What function do most of the glands perform? — *A.* Their function is two-fold, namely, the separation of some material from the circulating fluid, which would otherwise prove injurious to the system, and the elaboration of a product destined to renovate the tissues.

OF THE ABDOMEN.

Q. How is the cavity of the abdomen bounded? — *A.* Anteriorly, by the diaphragm; posteriorly, by the pelvis; superiorly, by a portion of the vertebra; inferiorly and laterally, by abdominal muscles.

Q. Into how many regions is the abdomen divided? — *A.* Into nine, as follows: right and left hypochondriac; right and left lumbar; right and left iliac; epigastric, umbilical, and hipogastric.

PERITONEUM.

Q. Why is the peritoneum called "serous membrane?" — *A.* In consequence of the serous or watery fluid with which its surface is constantly moistened.

Q. What is the structure of serous membranes? — *A.* The same as that of the areolar tissue, having a very smooth and glistening inner surface, which is covered with a layer of cells; constituting a distinct tissue, termed *epithelium*. This is in contact with the primary membrane, thus isolating it from the tissues beneath. Sub-adjacent to this is a layer of condensed areolar tissue, which constitutes the chief thickness of the serous membrane, and confers upon it its strength and elasticity; this gradually passes into that base variety, by which the membrane is attached to the part it lines, and which is commonly known as the sub-serous tissue. A fibrous tissue enters into the composition of the membrane itself, and its filaments interlace in a beautiful network, which confers upon it equal elasticity in every direction.

Q. What is the purpose of this membrane? — *A.* To facilitate the movements of the contained organs, by forming smooth surfaces which shall freely glide over each other.

STOMACH.

Q. What effect does the gastric fluid have upon the food? — *A.* It is supposed to have the property of dissolving the albuminous and gelatinous constituents of the food.

Q. What is the real solvent of the gastric fluid? — *A.* Either hydrochloric, acetic, or lactic acid.

Q. Is not the solvent action of the gastric fluid aided by some mechanical means? — *A.* Yes. By the movements of the walls of the stomach, which are produced by the successive contractions and relaxations of their muscular fibres, the contents of the stomach are thus kept in a state of constant agitation, which is considered favorable to their chemical solution.

Q. Does absorption of nutritious matter take place in the stomach? — A. Yes. A portion of the nutritious matter dissolved by the gastric fluid is at once absorbed into the blood-vessels of the stomach, and never passes into the intestinal tube, nor into the special lacteal system of vessels.

Q. What term is applied to the food after its reduction, in the stomach, to a pulpy mass? — A. Chyme.

Q. Gas is frequently evolved in the stomach and intestines during digestion; how do you account for this? — A. It is owing to a disturbed or morbid condition of that process, and by no means a necessary attendant upon healthy digestion.

Q. Does violent exercise immediately after a feed tend to retard the formation of chyme? — A. It does. The circumstances most favorable to perfect digestion are, a short period of rest, followed by gentle exercise.

Q. Does any portion of the food ever pass unchanged through the pylorus along with the chyme? — A. Yes. Whole oats are frequently found in the horse's excrement.

INTESTINES.

Q. The aliment now being converted into chyme, and having passed the pylorus, what becomes of it? — A. It enters the duodenum.

Q. Having entered the duodenum, with what does the chyme mingle? — A. The biliary and pancreatic secretions.

Q. What effect do they have on the gastric secretion and the chyme? — A. The biliary and pancreatic secretions are supposed to contain an excess of alkali; this neutralizes the acid of the gastric juice, so that there is no further solution of albuminous compounds, but the conversion of starch into sugar, which was interrupted in the stomach, now recommences.

Q. What are the uses of the bile? — A. The chief uses of the bile appear to be those of a chemical agent promoting the decomposition of the chyme, and also stimulating the secretion of mucus, and the peristaltic action of the intestines.

Q. What effect has the pancreatic juice on chyme or the elements of digestion? — A. It forms an emulsion with oil and fat.

Q. The chyme, having been acted on by the preceding secretions, what name is then given to it? — A. Chyle.

Q. Describe the properties of chyle? — A. If chyle be taken from the thoracic duct of an animal a few hours after it has taken food, it has very much the appearance of cream, being a thick fluid of an opaque white color, without smell, and having a slightly acid taste, accompanied by a perceptible sweetness. It restores the blue color of litmus, previously reddened by acetic acid, and appears, therefore, to contain a preponderance of alkali. When subjected to microscopic examination, chyle is found to contain a multitude of globules, of smaller diameter than those of the blood, and corresponding in size and appearance to those of milk. In about ten minutes after it is removed from the thoracic duct, it coagulates into a stiff jelly, which in the course of twenty-four hours separates into two parts, providing a firm and contracted coagulum, surrounded by a transparent colorless fluid.

Q. What are the principal ingredients of chyle? — A. A large proportion of albumen, a smaller one of fibrin; a fatty substance or emulsion, which gives to chyle the appearance of milk; and several salts, such as carbonate of potassa, muriate of potassa, and prophosphate of iron.

Q. What change does the chyle undergo in its passage along the various vessels? — A. Its resemblance to blood increases in each of the successive stages of its progress towards the heart and lungs.

Q. How are the chemical changes, and the contents of the intestines propelled through the tract of the alimentary canal? — A. By the peristaltic action of the muscular coat of the same.

Q. What becomes of the chyle after it has been prepared in the duodenum and first intestines? — A. It is received by absorption into the lacteals, and by them conveyed to the thoracic duct, which transmits it to large veins in the vicinity of the heart. (See distribution of lymphatics.)

Q. What do you understand by the "absorbent system?" — A. The absorbent system of vessels consists of two principal divisions, which may be compared to two sets of roots proceeding from a common trunk; one of these commences upon the walls of the intestines, and is termed the "lacteal" system; whilst the other takes its origin in various parts of the substance of the organism at large, especially in the skin and subcutaneous textures, and is known as the "lymphatic" system.

Q. Where do the lacteals most numerously abound? — A. In the small intestines, below the point at which the liver and pancreas discharge their secretions.

Q. Where do the lacteals commence? — A. Near the free extremities of the villi of the intestines.

Q. In what way do they commence? — A. It was formerly supposed that they commenced by orifices upon the internal surface of the intestine; but Carpenter, and other physiologists, contend that the lacteal vessels form loops by anastomosis with each other, so that they have no free extremity.

Q. What are the functions of the large intestines? — A. They are engaged in the conveyance and expulsion of feculent matter, and there are certain changes which take place in their contents, in aid of the object of nutrition, the exact nature of which has never been clearly determined. According to the best authority, it appears that some important changes are effected in that enlarged portion of the canal, termed cæcum, and which has, by some, been regarded as a kind of supplementary stomach, in which fresh chyme is formed, and fresh nutriment extracted from the materials that have passed through the small intestines. The large intestines also extract nutriment from their contents, which is proved by the fact that nutritious matter injected into them has been known to support life for a certain time.

SPLEEN.

Q. What is the function of the spleen? — *A.* It serves as a kind of diverticulum, to relieve the vessels of the digestive viscera when they are compressed by undue accumulation of the contents of their cavities, or when they are congested by obstruction to the flow of blood, through the liver or heart. It may also be considered as a lymphatic gland, for, in some instances in which animals have been allowed to survive longest after removal of the spleen, the lymphatic glands of the vicinity have been found greatly enlarged and clustered together, so as nearly to equal the original spleen in volume; hence, in such case we infer that its function must be similar to that of the enlarged lymphatic glands.

LIVER.

Q. What comprises the principal bulk of the liver? — *A.* It is made up of a vast number of minute lobules of irregular form, but about the average size of a millet seed; and each of them contains the elements of which the entire organ is composed, viz., a plexus of biliary ducts connected with their main trunks, and a mass of biliary cells; each of which are connected in like manner with the three blood-vessels which unite to the circulation of this organ.

Q. What are the vessels of the liver? — *A.* The hepatic artery, vena portæ, and hepatic veins, to which may be added the excretory ducts and absorbents.

Q. Of what use is the hepatic artery? — *A.* It is the nutrient artery of the liver.

Q. Of what use is the vena portæ? — *A.* It acts both as a vein and artery: as a vein, it receives the blood from most of abdominal viscera; as an artery, it ramifies through the liver for the secretion of bile.

Q. What is the use of the hepatic veins? — *A.* They return blood to the vena cava.

Q. What is the function of the liver? — *A.* It is an organ of excretion, designed to remove from the circulating fluid that portion of the products of disintegration, of which the principal component of the biliary, is the largest.

Q. Into what substance is the greater part of the excrementitious matter converted? — *A.* Biline.

KIDNEYS.

Q. What is the embryotic condition of the kidneys? *A.* The kidneys are preceded in the embryo by a substance first noticed by Wolff, and called after him the Wolffian bodies, or false kidneys, which originally extend along the spine from the heart to the end of the intestines; but they afterwards become shorter, and after a time diminish by absorption, and wholly disappear.

Q. What is the function of the kidneys? — *A.* Their principal function is to separate from the blood certain matters which would be injurious to it if retained.

Q. What does the secretory surface of the kidneys consist of? — *A.* It is composed of epithelial cells which line the tubuli urinifera, which draw the peculiar elements of the urinary excretion from the vascular plexus which surrounds the exterior of the tubes, carrying off the same to their terminations in the ureter.

Q. What other arrangement is provided within the kidneys for the elimination of the superfluous fluid of the blood? — *A.* A process of transudation takes place by the function of malpighian bodies, whose thin-walled capillaries allow the transudation of water to take place, under a certain pressure, into the tubuli urinifera.

SUPRA-RENAL CAPSULES.

Q. What is the function of the *supra-renal capsules*? — *A.* Their function has hitherto been involved in obscurity, and was supposed to be identical with other glands destitute of ducts or outlets; but, lately, *M. Brown Sequard* has demonstrated that they play a very important part in the nervous system of the horse.

VASA DEFERENTIA.

Q. What is the function of the vas deferens? — *A.* It is the excretory duct of the testicle, and conveys the semen to the vesiculæ seminales.

The author, instead of introducing examinations on the reproductive organs, has thought it best to substitute the opinions of that eminent physiologist, Dr. Carpenter; and therefore the reader's attention is now directed to " *Physiological considerations on the reproduction of organized beings.*"—page 128.

REMARKS AND EXAMINATIONS ON THE EYE.

The parts which compose the eye are divided into external and internal. The external parts are: First, the eyelashes, or cilia, which, in the horse, can scarcely be reckoned more than one, there being very few hairs in the under eyelid. Secondly, the eyelids, or palpebræ, upper and under; where they join outwardly, it is termed the external canthus, and inwardly toward the nose, the internal canthus; they cover and defend the eyes. The cartilaginous margin or rim of the eyelid, from which the eyelashes proceed, is named tarsus. In the tarsus and internal surface of the eyelid there are small glands, which secrete a fluid, to prevent friction of the eye and its lids, and facilitate motion. Thirdly, the lachrymal gland, which is placed on the upper part of the eyelid toward the external canthus; from this gland the tears are secreted, and conveyed to the inner surface of the upper eyelid by several minute ducts, or canals, named lachrymal ducts. There is another small body, having a glandular appearance, in the inner corner of the eye; on each side of which there are small orifices which are called puncta lachrymalia: these are the mouths or openings of two small canals, which, joining together, form a membranous tube; and this, passing through a small opening in the bone, extends to the lower part of the nostril, where its termination may be distinctly seen in the horse. As the lachrymal gland is constantly forming tears, it must be obvious that some contrivance is necessary to convey them off, and prevent them flowing over the cheek: this purpose is answered by the canal just described.

When any irritating matter is applied to the eye, the tears are formed too abundantly to be carried off in this way; they then flow over the cheek. In the human eye, the puncta lachrymalia terminate in a small sac, from which the lachrymal duct proceeds: this is not the case in the horse. In the inner corner of the horse's eye is placed a body commonly termed the haw, no resemblance to which is to be found in the human eye. The horse has the power, by means of the muscles of the eye, to bring the haw completely over its surface; it serves, therefore, as a second eyelid, and effectually wipes off any dust, hay, or seeds, or other matter which may have fallen upon the eye. The conjunctivial membrane or tunica conjunctiva, lines the inner surface of the eyelids, and covers the white part of the globe of the eye. This membrane has numerous blood-vessels, which are conspicuous when it is inflamed. The bulb or globe of the eye is composed of several coats and humors. The transparent cornea, which, in the horse, forms the front part of the eye, comprehends a larger part of the globe than in the human subject; on removing this cornea, a fluid, which is named the aqueous humor, escapes, and the iris appears. The iris is a muscular curtain, having a hole in the centre, which is termed the pupil. This divides the fore part of the eye into two parts, named chambers, which are occupied by the aqueous humor. The pupil is of a dark bluish cast; is of an oval, or rather of an oblong, form. The iris regulates the quantity of light that is required to pass through the pupil. For this purpose, it is composed of two sets of muscular fibres: by means of one the pupil is enlarged, and by the other it is diminished. Thus, if the pupil is first examined in the stable, where there is a moderate light, and immediately after in the sunshine, it will be found quite altered; being so small, in a strong light, as to be nearly closed. On removing the iris, the second humor, or crystalline lens, appears: this is retained in its situation by a transparent membrane, named its capsule, between which and the lens is a minute quantity of fluid. The third humor of the eye is the vitreous. This humor is not contained in one general sac, but in numerous minute and perfectly transparent cells, and resembles pure water: this humor serves to produce a small degree of refraction in the rays of light, and occupies and distends all the posterior part of the globe of the eye. The next coat to the conjunctival is the sclerotica, or white of the eye, a strong, thick membrane, which extends from the transparent cornea to the optic nerve. The next coat to the sclerotic is the choroid. This is a delicate and very vascular membrane. In the human eye it appears of a black color, and it is this which causes the pupil of the human eye to appear black; but the choroid coat of the horse's eye is variegated in color; in some parts black, in others blue, and in others green. The next coat is the retina: this is a delicate expansion of the optic nerve over the choroid coat, which it accompanies to the margin of the crystalline lens, and there terminates. The use of the retina is to receive certain impressions made by the light reflected from objects, so as to produce in the mind an idea of their figure and color; the optic nerve being the medium of communication between the retina and brain. From the above explanation of the mechanism of the eye, it will readily appear that many circumstances may occur to render vision imperfect, or to destroy it altogether. If the transparent cornea, for example, became white, light could not pass through it, and the animal would be blind, however perfect the other parts of the eye might be. The cornea may be either too convex or too flat; in the former case, causing the animal to be near-sighted; in the latter, producing an indistinctness of vision with respect to objects that are near. The iris may, in consequence of disease, become fixed, or lose its power of motion; in which case, the pupil would be always of the same size, and the animal would not have the power of adapting it to the various distances or objects; or, as sometimes happens, the pupil may become quite closed, by which light would be perfectly excluded from the retina. Supposing the cornea and iris to be healthy, the crystalline lens, or its capsule, may become opaque, and thereby cause total blindness. But in this part, as in the cornea, we meet with different degrees of opacity: sometimes it is very slight, the pupil appearing of a lighter color, and unusually large: in this state, the pupil is said to look dull or muddy, which causes the horse to start; but when the opacity is complete, it constitutes the disease termed cataract. There is another disease, to which the reader's attention is called; it is named gutta serena, or amaurosis. This disease is known by the pupil being unusually large or open, and by its continuing so when the eye is exposed to a strong light.

EXAMINATIONS OF THE ORGANS OF SIGHT AND THEIR APPENDAGES.

Q. Where are the eyes located?—*A.* Within the orbits.

Q. By what foramina is each orbit perforated?—*A.* By the optic foramen.

Q. From whence is the lining membrane of the orbit derived?—*A.* From the dura mater and periosteum.

Q. Enumerate the appendages of the eye.—*A.* The eyelids, eyelashes, muscles of the eyelids, tarsal cartilages, meibomian glands, tunica conjunctiva, membrana nictitans, lachrymal gland, puncta lachrymalia, lachrymal sac, ductus ad nasum, and the muscles of the eyeball.

EYELIDS.

Q. What parts do the eyelids occupy?—*A.* The circumference of the orbits and front of the eyeball.

Q. What are the eyelids composed of?—*A.* In composition they are cuticular, muscular, cartilaginous, and membranous; also glandular, vascular, and nervous.

Q. What muscle enters into the composition of the eye?—*A.* The orbicularis palpebrarum.

Q. How are the lids separated?—*A.* By a transverse fissure, bounded by the angles or canthi of the eye.

Q. What is attached to the superior or temporal angle?—*A.* The tarsal ligament.

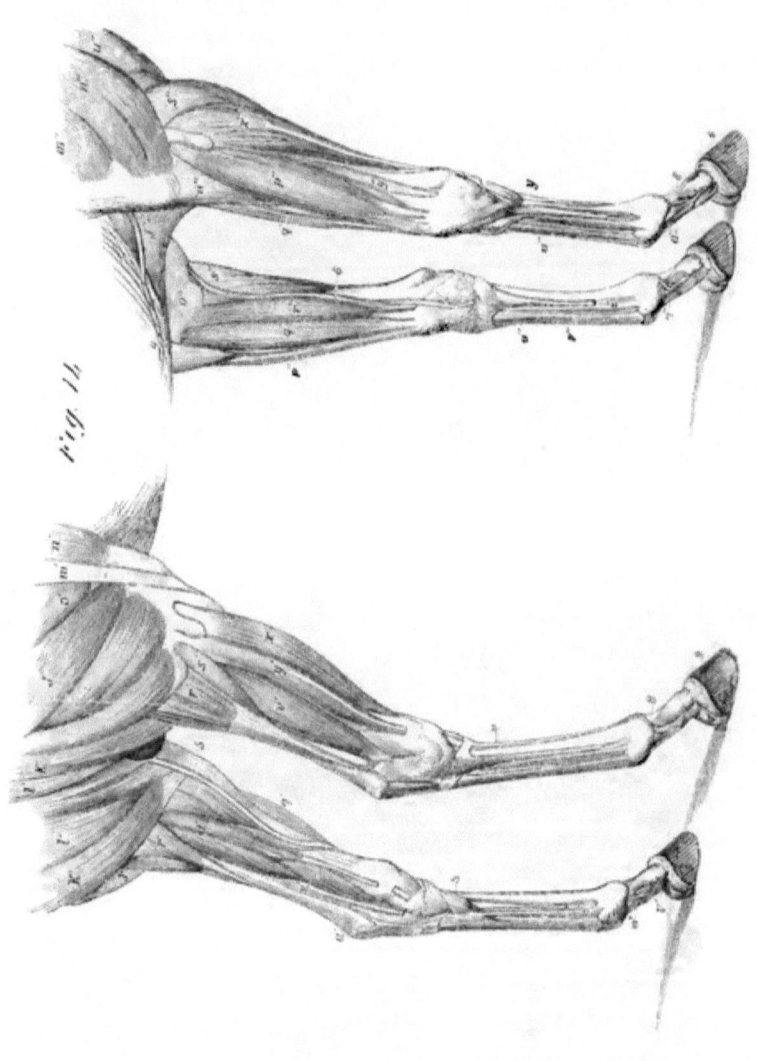

Fig. 14.

EXPLANATION OF FIGURE XIV.

HIND EXTREMITIES.

- *m'*. Tensor vaginæ.
- *n'*. Rectus.
- *o'*. Vastus externus.
- *q'*. Flexor metatarsi.
- *r'*, *r*. Gastrocnemius internus.
- *s'*.   "   externus.
- *t'*. Flexor pedis accessorius.
- *u'*. Insertion of the gastrocnemius.
- *r'*. Flexor metatarsi.
- *x'*. Extensor pedis.
- *y*, *y'*. Extensors.
- *u*, *v*. Tendo perforans and perforatus.
- *K'*, *K'*. Abductors tibialis.
- *J'*, *J'*. Triceps.
- *l'*, *l'*. Adductors.
- *&c.* Hoof.
- 5, 5. Saphena vein.
- 8. Bifurcation of the suspensory ligament.
- *v'*. (Off-hind leg.) Plantaris.
- *z*. Suspensory ligament.

FORE EXTREMITIES.

- *J"*. Pectoralis magnus.
- *m"*, *n"*. Triceps extensor brachii.
- *o"*. Pectoralis transversalis.
- *p"*, *p"*. Flexor metacarpi externus.
- *q"*.   "   "   medius.
- *r"*.   "   "   internus.
- *s"*. Extensor metacarpi magnus.
- *u"*. (At the upper part of the figure.) Levator humeri.
- *u"*, *u"*, *u"*, *u"*, *v*. (Beneath the olecranon and carpus.) Flexors perforans and perforatus.
- *x"*. Extensor pedis.
- *y"*, *y*. Extensor suffraginis.
- *z"*. Suspensory ligament.
- *&c.* The hoof.
- 4. Subcutaneous thoracic vein.
- 6. Radial vein.
- 8. Bifurcation of the suspensory ligament.

Q. What is fixed to the inferior angle?—*A.* The tendon of the orbicularis.

Q. From whence is the loose portion of skin, entering into the composition of the upper lid, derived?—*A.* It is a prolongation of the skin covering the forehead.

Q. From whence is that of the lower lid derived?—*A.* From the integuments of the face.

Q. How are the internal surfaces of the lids shaped?—*A.* Into concavities which adapt them to the convexity of the globe of the eye.

Q. By what membrane are the lids lined?—*A.* By the conjunctival.

TARSAL CARTILAGES.

Q. What are the tarsal cartilages?—*A.* They enter into the substance of the borders of the lids, imparting to them both firmness and elasticity.

Q. Describe the tarsal cartilages?—*A.* The superior cartilage is broader and more convex than the inferior; they correspond in shape and size to their respective lids; they are convex outwardly and concave inwardly, and are inserted into the rims of the orbits.

Q. What is the texture of the tarsus?—*A.* Their texture is fibro-cartilaginous.

MEIBOMIAN GLANDS.

Q. Describe the meibomian glands?—*A.* They have the appearance of white follicular bodies, vertically ranged in parallel lines; they vary both in calibre and length, and are in the upper rather than the lower lid.

Q. What is the function of the meibomian glands?—*A.* To secrete a fluid which guards against friction between the eye and its appendages.

TUNICA CONJUNCTIVA.

Q. What is the situation of the tunica conjunctiva?—*A.* It is the lining membrane of the eyelids, membrana nictitans, caruncula lachrymalis, puncta lachrymalia, and is reflected to the globe of the eye.

Q. Describe the conjunctival surface?—*A.* The adherent one is rough, lax, and flocculent; the outer surface is smooth, glossy, and humid with secretion.

Q. What are the peculiarities in the organization of the conjunctiva?—*A.* It is a continuous membrane, yet varies in texture, as follows: 1st. That portion which gives a covering to the conjunctiva palpebralis is highly organized with blood-vessels, and is often tinged a deep red color. 2d. The conjunctiva sclerotica is not so highly organized, yet has a few straggling vessels of larger calibre than those of the former, and its texture is more dense. 3d. The conjunctiva corneæ is thin and transparent, more of a horny texture, and has no appearance of vascularity.

MEMBRANA NICTITANS.

Q. What is the common name for the above membrane?—*A.* The haw.

Q. What is its structure?—*A.* Cartilaginous.

Q. What is its situation?—*A.* It is located behind the inferior canthus, between the eyeball and side of the orbit.

Q. What is its figure?—*A.* It approaches that of an extended triangle, of which the short side is turned forwards, and the lengthened angle backwards.

Q. Describe the anterior part?—*A.* It is thin and elastic, and bounded by a crescentic edge, terminating in two salient angles; it increases in substance, but grows narrow posteriorly, and there ends in an obtuse conical point, which appears in the adipose tissue at the bottom of the orbit.

Q. What is the form of its surfaces?—*A.* Inwardly concave; outwardly convex.

Q. What is the body of the nictitating membrane clothed with?—*A.* By a portion of conjunctival membrane.

Q. What is the function of the membrana nictitans?—*P.* To protect the eyeball, in the removal of foreign bodies from its surface.

LACHRYMAL APPARATUS.

Q. What parts compose the lachrymal apparatus?—*A.* The lachrymal gland, caruncula lachrymalis, lachrymal puncta and conduits, lachrymal sac, and ductus ad nasum.

LACHRYMAL GLAND.

Q. Where is the lachrymal gland situated?—*A.* In a depression, beneath the process of the orbital arch.

Q. What are its coverings, and with what is it in contact?—*A.* It is covered by the common aponeurotic lining of the orbit; it is in contact with the levator palpebræ, and is enveloped in fat and cellular membrane.

Q. What is its form?—*A.* It is irregular, slightly convex superiorly; inclining to the concave inferiorly. It is a conglomerate gland, constituted of many lobules.

Q. Have the lobules any further organization?—*A.* Yes, they are composed of minute granules.

Q. What vessels do the granules receive and what springs from them?—*A.* They receive the terminating ramifications of the supplying arteries, and from them spring the radicles of the excretory ducts.

Q. What do the radicles terminate in, and where is their outlet?—*A.* The radicles unite with one another into a set of tubes, which open upon the conjunctival lining of the upper lid in the source of seven visible orifices near its superior angle; this is their outlet.

Q. What is the function of the lachrymal gland?—*A.* To secrete the tears.

Q. What becomes of the superfluous tears?—*A.* They either fall over the lower lids, or pass into the lachrymal sac; from thence, by the ductus, to their outlets within the nostrils, at their inferior parts.

CARUNCULA LACHRYMALIS.

Q. What is the caruncula lachrymalis?—*A.* It is a small eminence, lodged within the inferior canthus, between the eyeball and lids.

Q. What is its use?—*A.* It secretes a light yellow unctuous matter, with which the fine hairs on its surface being coated it detains any small foreign bodies that may float in the lachrymal secretion; it also directs the latter fluid into the puncta.

LACHRYMAL PUNCTA AND CONDUITS.

Q. What are the puncta lachrymalia? — A. Two small orifices situated on the inward margins of the two lids — superior and inferior — near the radix of the caruncle.

Q. What do the puncta terminate in? — A. The lachrymal conduits.

Q. What is their situation? — A. Within the substance of the eyelids.

Q. How are conduits formed? — A. A minute cartilaginous circle surrounds them, and they are lined by conjunctival membrane.

Q. What do the conduits terminate in? — A. The lachrymal sac.

LACHRYMAL SAC.

Q. Where is the lachrymal sac situated? — A. Within the depression which leads into the channel of the lachrymal bone, behind and below the small eminence upon the orbital ridge of that bone.

Q. Describe the sac and its connections? — A. It is an oblong membranous bag; its front is crossed by fibres of the orbicularis; it has also a connection with the tendon of that muscle. The posterior part of the sac adheres firmly to the lachrymal bone. It is composed of a dense, white, fibrous membrane, furnished with a lining from the conjunctiva.

Q. By what is this sac perforated? — A. By the lachrymal conduits.

Q. What does it open into? — A. Into the ductus ad nasum.

Q. What is the function of this sac? — A. It is a reservoir into which the tears flow from the lachrymal conduits, and from thence pass into the ductus ad nasum.

DUCTUS AD NASUM.

Q. What is the ductus ad nasum? — A. It is a long membranous canal, commencing at the contracted portion of the lachrymal sac, and running with the groove through the lachrymal bone; then along a canal in the superior maxillary bone, between it and the anterior turbinated bone; terminating at the inner and inferior part of the nostril.

Q. What is the organization of the ductus? — A. It appears to be a continuation of the membrane composing the lachrymal sac, which is strengthened by a fibrous sheath; its internal surface is probably mucous, which protects it from the action of the tears, or lachrymal secretion.

Q. Describe the course of the tears, or lachrymal secretion? — A. They are secreted by the lachrymal gland, and are poured by its excretory ducts over the surface of the eyeball; the puncta lachrymalia absorb them; they are then conveyed by the lachrymal ducts to the lachrymal sac; and through the ductus ad nasum pass into the nostril.

THE EYEBALL AND ITS COATS.

Q. What is the form of the globe of the eye? — A. Nearly of a spherical figure.

Q. Of what is the globe of the eye composed? — A. Of membranes, or coats, filled with humors or fluids, which preserve its form.

Q. How many coats has the eye? — A. Five: the sclerotic, choroid, retina, cornea, and iris.

Q. Does not the tunica conjunctiva enter into the composition of the membranes of the eye? — A. Yes; it may be considered as common to both.

Q. Where does it adhere most closely? — A. Over the cornea.

SCLEROTIC COAT.

Q. What is the use of the sclerotica? — A. It bounds the form of the eye, protects and supports the parts within.

Q. What are its perceivable boundaries? — A. It extends from the optic nerve to the cornea.

Q. What is inserted into it posteriorly? — A. The fleshy part of the retractor muscle.

Q. What is inserted into its anterior margins? — A. The tendons of the four recti.

CORNEA.

Q. What is the cornea? — A. The transparent, anterior part of the globe of the eye.

Q. How does its form compare with the sclerotica? — A. It is more convex.

Q. What covers its convex surface? — A. The conjunctiva.

Q. What is its structure? — A. Laminated.

IRIS.

Q. What is the iris? — A. It is a circular membrane, with an irregular central cavity, in the anterior chamber of the eye.

Q. What is its central perforation called? — A. The pupil.

Q. How is the periphery of the pupil bounded? — A. By several dark, colored, glandular bodies, termed corpora nigra.

Q. What is the use of the iris? — A. By contracting it excludes all superfluous rays of light, and by expanding admits through the pupil all that pass through the cornea.

Q. What is the structure of the iris? — A. It is a fibrous membrane, divisable into two layers, provided with blood-vessels and nerves.

CHOROID COAT.

Q. What is the choroid coat? — A. It is a dark-colored membrane of delicate texture, located immediately beneath the sclerotica.

Q. What are its boundaries? — A. It extends from around the termination of the optic nerve as far forward as the edge of the cornea, and ends in the ciliary circle.

Q. How is the choroid coat connected with the sclerotica? — *A.* By cellular membrane.

Q. What is remarkable at its outer edge? — *A.* It is thrown into folds, called ciliary circle and processes.

Q. What is the color of the choroides? — *A.* Externally, its whole surface is black; internally, the anterior parts are black, and the posterior half is of a brilliant variegated green.

Q. What is the black part termed? — *A.* Pigmentum nigrum.

Q. What name is given to the variegated part? — *A.* Tapetum lucidum.

Q. What difference do we observe in the pigment of the choroid surfaces? — *A.* The inner layer is thicker and more consistent than that found on the outer surface.

RETINA.

Q. What is the retina? — *A.* It is the third or innermost tunic of the eye. It cannot, however, be considered as a tunic, for it pervades the interior of the globular expansion without contracting any adhesions until it has reached the corpus ciliare.

Q. How is the retina formed? — *A.* The optic nerve, having reached the inner and inferior part of the globe of the eye, enters the sclerotic and choroid coats, and in its passage through them its diameter contracts; having arrived at the inner part of the globe, the nerve forms an eminence, from the circumference of which issues radiating fibres which form the retina.

Q. How is the retina sustained in this state of globular expansion? — *A.* By the humors of the eye, which keep it in contact with the choroides.

Q. The retina having radiated on the interior of the globe, where is it inserted? — *A.* Into the corpus ciliare.

HUMORS OF THE EYE.

Q. Of how many humors does the eye consist, and what are their names? — *A.* Of three; they are called aqueous, crystalline, and vitreous humors.

Q. What parts of the eye do they occupy? — *A.* They occupy in succession the spaces termed anterior, middle, and posterior chambers of the eye.

AQUEOUS HUMOR.

Q. What is the use of the aqueous humor? — *A.* It transmits the rays of light, and aids the free motions of the iris.

Q. What are its boundaries? — *A.* It fills the interval between the cornea and crystalline lens.

Q. Describe the aqueous humor. — *A.* It is a bright limpid fluid, and in properties bears some resemblance to the vitreous.

Q. What is the composition of both these humors? — *A.* They are composed of albumen, gelatine, and muriate of soda, held in solution by an aqueous menstruum.

Q. How is this fluid secreted? — *A.* By secretion from the transparent walls of its capsule.

CRYSTALLINE LENS.

Q. What is the use of the crystalline lens? — *A.* It concentrates the rays of light, so as to make a distinct image in the posterior chamber.

Q. Where is the crystalline lens situated? — *A.* Between the aqueous and vitreous humors.

Q. By what is the crystalline lens enclosed? — *A.* By a tunic, called tunica crystallina.

VITREOUS HUMOR.

Q. What is the vitreous humor, and where is it situated? — *A.* It is the most bulky humor of the eye; of a jelly-like consistence, yet quite transparent, and occupies that portion of the eyeball posterior to the crystalline lens.

Q. What is the use of the vitreous humor? — *A.* It supports the form of the eye, and maintains the other humors in their proper positions.

RESPIRATORY SYSTEM.

PHYSIOLOGICAL CONSIDERATIONS.

Respiration and Structure of the Lungs.—The organs of respiration are the larynx, the upper opening of which is named glottis, the trachea or windpipe, bronchiæ and the lungs.

The air is displaced out of the lungs by the action of the muscles of respiration; and, when these relax, the lungs expand to a certain calibre by their elasticity. This may be exemplified by means of a sponge, which may be compressed into a small bulk by the hand, but, upon opening the same, the sponge returns to its natural size, and all its cavities become filled with air. The purification of the blood in the lungs is of vital importance, and indispensably necessary to the due performance of all the functions. When the lungs, and muscles connected with them, are in a physiological state, the horse is said to be in *good wind*—a very desirable state for an animal to be in, whose usefulness depends on his being capable of a long continuance of quick motion. The trachea, or windpipe, after dividing into bronchiæ again subdivides into innumerable other branches, the extremities of which compose an infinite quantity of small cells, which, with the ramifications of the veins, arteries, nerves, lymphatics, and the connecting cellular membrane, make up the whole mass or substance of the lungs. The internal surface of the windpipe, bronchiæ and air-cell, is lined with a membrane, which secretes a mucous fluid: when, in consequence of an obstructed surface, this fluid becomes abundant, it is expelled by the nostrils. The whole is invested with a thin, transparent membrane, named pleura: the same membrane lines the internal surface of the ribs and diaphragm, and, by a duplicature of its folds, forms a separation between the lobes of the lungs.

RESPIRATORY SYSTEM.

The function of respiration is the conversion of venous into arterial blood. This arterialization of the venous blood is a process highly essential to the well-being of all animals; more important is it than the assimulation of aliment; for a horse may live several days without food, yet cannot exist many minutes unless his blood be arterialized.

In considering the function of respiration, our attention is first turned to the mechanical means by which the air is alternately admitted and discharged from the lungs. The mechanical act of respiration is divisible into two periods, that of inspiration, during which air is drawn into the lungs so as to increase its volume and distend its parenchyma and expiration, during which process the air which had been so received is expelled.

Inspiration is accompanied by enlargement of the capacity of the thorax in its various dimensions. This is effected by the action of different sets of muscles, operated on by the nervous system. The principal muscle of inspiration is the diaphragm.

Among the secondary muscles employed in inspiration are those which articulate the ribs, viz., the intercostales. Each rib is capable of a small degree of motion on the extremity by which it is articulated with the vertebræ. This motion is chiefly forward and backward; the intercostal muscles favor this motion, as they are disposed in two layers, each passing obliquely, but with opposite inclinations, from one to the adjacent rib. There are two ways in which the chest may be dilated: first, by the diaphragm; and secondly, by the intercostales, which elevate the ribs. In natural respiration, the horse breathes chiefly through the aid of the diaphragm. Should the respiration become quickened, the intercostales are employed, and, when the respiration is laborious, the auxiliary muscles of the abdomen, back, and sides, are brought into use.

The glottis is opened during inspiration by the muscles of the larynx.

The expulsion of the air from the lungs constitutes expiration. This takes place as soon as the air which has been expired has parted with its oxygen, and received in return a certain quantity of carbonic acid gas and vapor. In regard to the elasticity of the lungs, it is now demonstrated that they possess no inherent power of elasticity other than that common with all membranous textures. Hence, if an opening be made in the sides of the chest, the lobes on this side collapse in consequence of the pressure of air from without.

We have next to inquire what changes have, in the meanwhile, been effected in the blood by the action of the air to which it has been subjected in the lungs. A visible alteration, in the first place, is produced in its color, which, from being of a dark purple, nearly approaching to black, when it arrives at the air-cells by the pulmonary arteries, has acquired the bright, intensely scarlet hue of arterial blood, when brought back to the heart by the pulmonary veins. In other respects, however, its sensible qualities do not appear to have undergone any material change. Judging from the changes produced on the air which has been in contact with it, we are warranted in the inference that it has parted with a certain quantity of carbonic acid and of water, and that it has in return acquired a certain proportion of oxygen. Since it has been found that the quantity of oxygen absorbed is greater than that which enters into the composition of the carbonic acid evolved, it is obvious that at least the excess of oxygen is directly absorbed by the blood; and this absorption constitutes, no doubt, an essential part of its arterialization.

It has been much disputed whether the combination which seems to be effected between the oxygen of the air and the carbon furnished by the blood, occurs during the act of respiration, and takes place in the air-cells of the lungs, or whether it takes place in the course of circulation. On the first hypothesis, the chemical process would be very analogous to the simple combustion of charcoal, which may be conceived to be contained in the venous blood in a free state, exceedingly divided, and ready to combine with the oxygen of the air, and imparting to that venous blood its characteristic dark color; while arterial blood, from which the carbon had been eliminated, would exhibit the red color natural to blood. On the second hypothesis, we must suppose that the whole of the oxygen, which disappears from the air respired, is absorbed by the blood in the pulmonary capillaries, and passes on with it into the systemic circulation. The blood becoming venous in the course of the circulation, by the different processes to which it is subjected for supplying the organs with the materials required in the exercise of their respective functions, the proportion of carbon which it contains is increased, both by the abstraction of the other elements, and by the addition of nutritive materials prepared by the organs of digestion. The oxygen, which had been absorbed by the blood in the lungs, now combines with the redundant carbon, and forms with it either oxide of carbon, or carbonic acid, which is exhaled during a subsequent exposure to the air in the lungs. Many facts tend strongly to confirm our belief in the latter of these hypotheses.

OF THE LARYNX.[*]

The larynx is the organ producing the voice of the animal.

Situation.— It is joined to the top of the trachea (or windpipe), and is placed in the throat, between the posterior and broadest

[*] Percivall's Hippopathology.

parts of the branches of the lower jaw; having the pharynx and uppermost part of the œsophagus situated above it; the superior portions of the sterno-hyoidei and thyroidei below it; the tongue, with its muscles, and the os hyoides, in front of it; and the trachea issuing from below and behind it.

Attachment.— The larynx is retained in its place by its connection with the os hyoides and pharynx; by its muscles; and by its coalition with the trachea.

Conformation.— The larynx has so complete a fleshy covering, that it is not until it is divested of its muscles (which have been heretofore described) that it is discovered to be composed of five pieces of cartilage, so joined together as to be moveable on one another, and open both superiorly and inferiorly, to admit of the passage of air into and out of the trachea. These cartilages have received the names of *thyroid*, *cricoid* (two), *arytenoid*, and *epiglottis*.

The thyroid or *shield-like* cartilage, by much the largest of the five, forms the superior, anterior, and lateral parts of the larynx. It consists of two broad lateral portions, continuous and prominent at the upper and anterior part of the neck, the prominence corresponding to which in human anatomy has received the name of *pomum Adami*. Below this point of union the divisions recede from each other, leaving a triangular space between them, which is occupied by a ligament denominated the *ligamentum crico-thyroideum*. The four projecting corners from the posterior parts of the thyroid cartilage are named its *cornua*: the two superior are joined by capsular articulations to the body of the os hyoides; the two inferior are connected by very short capsular ligaments to the cricoid cartilages; the union of all which parts receives additional strength from expansions of membrane. At the roots of the superior cornua are two foramina, that give passage to nerves, of considerable importance, to the interior of the larynx. This cartilage not only constitutes by far the most extensive part of the larynx, but, as its name indicates, incloses and shields from external injury all the others.

The cricoid or *ring-like* cartilage is placed below the thyroid. In front it appears like part of the trachea; but it broadens so much behind, that it overlaps the first ring of the windpipe, somewhat after the form of a helmet. Upon its broad or posterior part are four surfaces of articulation: the two upper receive the hinder extremities of the arytenoid cartilages, the two lower are adapted to the inferior cornua of the thyroid cartilage; they are all furnished with capsular ligaments and synovial membranes. Furthermore, it is attached by ligamentous expansions to those parts, and likewise to the first ring of the trachea.

The two arytenoid, or *ewer-shaped* cartilages, triangular in their figure, lie over the upper and back part of the trachea, leaving an aperture between them leading into that canal, denominated, from its proximity to the tongue, the *glottis*. Their inward parts are everted, and form a triangular prominent border, over which is spread *the membrane of the glottis:* their outward surfaces are marked by concavities in which are lodged the arytenoid muscles. Posteriorly, they repose upon the cricoid cartilage, and are connected with them by capsular articulations: in front, they have a membranous connection with the cartilage next to be noticed.

The epiglottis, so named from being raised over the *glottis*, and occasionally covering it like the lid of a pot, is well adapted, from its heart-like shape, to the *rima glottidis;* whose margin is completed by two narrow slips of cartilage proceeding from the base of the lid to the arytenoid. By some, these slips of cartilage have been separately considered; but in my opinion improperly so; for they are, in reality, nothing more than prolongations or appendices of the epiglottis. The surface of this cartilage presented to the interior of the larynx is smooth and concave, and covered by an extension of membrane from the glottis; that part opposed to the tongue is unevenly convex, and is tied to that organ, as well as to the os hyoides, by a doubling of membrane infolding some muscular fibres; to this musculo-membranous ligature, which assists in retaining the cartilage in its elevated position, the name of *frænum epiglottidis* is properly given. The frænum receives co-operation in this function from strong elastic ligaments connecting the base of the epiglottis to the thyroid and arytenoid cartilages.

If we detach the epiglottis, or raise it forcibly in order to obtain a more complete view of the rima glottidis, the latter will be found to be stretched into an oblong quadrilateral figure, whose width gradually diminishes from the middle towards either extremity, and bears a ratio of about one to six when compared to its length. The sides turned forward are formed by the arytenoid cartilages; those directed backward by two prominent folds of membrane (which envelop the thyro-arytenoid muscles), commonly described as the *vocal ligaments*, from their being concerned in the formation and intonation of the voice. Immediately over them are slit-like apertures, opening into membranous sacs, each large enough to contain a walnut; these are the *ventricles of the larynx*, whose use is also connected with the production and modulation of the voice.

The *membrane lining the cavity* of the larynx is one of great susceptibility; on which account it is kept continually moist by a mucus, oozing from numerous *lacunæ*—the excretory orifices of small subjacent follicles whose situation is denoted by the little round eminences upon its surface. This is the common seat of that species of catarrh which is accompanied by cough.

OF THE TRACHEA.

The trachea, or windpipe, is a cartilaginous tube extending along the neck, from the larynx to the lungs, for the passage of air. In horses of ordinary size, it is from twenty-five to thirty inches in length.

Course.—The trachea commences from the inferior border of the cricoid cartilage, opposite to the body and transverse processes of the atlas; takes its course along the anterior and inferior part of the neck, inclining to the near side, between the sternomyloidei muscles (which by their approximation conceal the lower portion of it), and enters the chest between the two first ribs; wherein, under the curvature of the posterior aorta, it divides into two parts the *bronchial tubes*.

Structure.—From fifty to sixty annular pieces of cartilage enter into the composition of the windpipe; altogether constituting a structure so remarkable, for the inequality or asperity of its exterior, that the ancients, in order to at once distinguish it from all other vessels, called it the *aspera arteria*. No entire or undivided tubular substance could have partaken of the various motions of the head and neck, without having suffered more or less distortion, and consequent deformity and diminution of caliber, of some part of its canal, which would have been attended with frequent interruptions to the free passage of the air, dangerous, and even fatal, to the respiratory functions; whereas, constructed as it is, with the aid of its muscular power, no attitude into which the animal may naturally put himself will impede the freedom of passage through it. The cartilages, or, as they are commonly described, the *rings* of the windpipe, have all a close resemblance to one another: if there be any disparity between them worthy of notice, it consists in those that form the superior part of the pipe being somewhat larger and broader than those nearest to the bronchial tubes.[*] A ring is not uniform in its breadth, in consequence of having waving or scolloped borders; the advantage of which is, that a sort of dove-tailed connection is effected which materially contributes to the compactness and strength of the entire structure. Its front and sides measure, in the broadest places, half an inch in breadth, and nearly a

[*] Now and then we find, at the upper part of the tube, two or three or more of these rings accreted together: it gives rise to some prominence thereabouts generally, and may often be detected by taction in the living animal.

quarter of an inch in thickness—evidently made so substantial to resist external injury; whereas its posterior or unexposed parts grow suddenly thin and yielding, and taper to the extremities; which, instead of meeting and uniting, pass one over the other, and thus form a shield of defence behind, while they admit of a certain dilatation and contraction of the internal dimensions of the tube. These attenuated ends are joined together by a ligamentous expansion, mingled with a quantity of cellular membrane. The rings are likewise attached to one another by narrow ligamentary bands, strong and elastic; which, after they have been drawn apart in certain positions of the head and neck, have the power to approximate them; when the pipe is removed from the body, and suspended by the uppermost ring, these ligaments counteract the tendency its weight has to separate the rings, and still maintain them in apposition. The lowermost ten or twelve pieces of cartilage appear on examination but ill to deserve the name of rings; indeed, they are little more than semi-annular, the deficiences in them behind being made good by intermediate moveable pieces of cartilage. These pieces, whose breadth increases as we descend, are let into the vacuities in such manner as to overlap the terminations of the segments, and they are confined and concealed by the same sort of ligamentary and cellular investment as was before noticed.

Muscle.—Where the outward extremity of the ring suddenly turns inward, and degenerates into a thin flexible flap on either side, a band of muscular fibres is fixed and stretched across the canal, dividing it into two unequal semi-elliptical passages. The anterior one is the proper air channel; the posterior or smaller one is filled with a fine reticular membrane, connecting the band to the posterior part of the ring, and preventing it in action, from encroaching upon the main conduit. This self-acting band appears to me to have been added to the tube to enable it to *enlarge* its caliber—not to diminish it, as a superficial view of these parts might lead one to imagine; for, in consequence of the passage being naturally elliptical, and the muscle being extended across its long diameter, the contraction of its sides will give the tube a circular figure, by increasing the curvature of the ring anteriorly, and thereby, in effect, will expand and not contract the caliber of the canal. I would say, then, that the trachea was made muscular in order that it might have the power of increasing its capacity for the passage of air, whenever the lungs were called into extraordinary action: in addition to which, I think that this band may, in some degree, counteract any tendency certain positions of the head and neck have to alter its shape and diminish its circumference. This opinion is corroborated by the circumstance, that the muscle grows slender and pale as we approach the lower end of the pipe, where the canal itself is nearly circular, and where it is placed in the least moveable part of the neck.*

Membrane.—The trachea is lined by a soft, pale red membrane, which anteriorly has a close adhesion to the rings themselves, and presents a smooth, polished internal surface; but which, posteriorly, is loosely attached to the muscular band, and puckered into fourteen or fifteen longitudinal *plicæ* or folds, that extend with regularity from one end of the tube to the other. These folds were evidently made to allow of the contraction and elongation of this muscular band; for I cannot myself assign any reason why they should exist in its relaxed state, unless this fulness of membrane be given to admit of enlargement of the caliber of the tube during the contractions of that muscle; if this be plausible, I may adduce the corrugation of the membrane as another proof that the caliber of the trachea is susceptible of augmentation. This membrane is continuous with that which clothes the rima glottidis;

---

* In this opinion, says Mr. Percivall, I find I am at variance with Girard. The French professor ascribes to it the power of *contracting the caliber* of the trachea. "Cette couche, bien evidemment musculeuse, peut *retrecir le calibre* de la trachée, en rapportant les extremités des segmens."—*Anat. Vét.*, p.146 et 147, tom. II.

but it is paler than it, and not near so sensitive. Its arterial ramifications, also less abundant than upon the glottis, exhale a vapor from its surface; independently of which, it is kept continually lubricated by mucus, furnished from its numerous *lacunæ*, to defend it from anything acriminous that may be contained in the breath.

Bronchial Tubes.— The trachea having entered the thorax, bifurcates into the two *bronchial tubes*. Of them, the right is the more capacious canal, on account of having communication with the larger division of the lungs; the left the longer one, in consequence of having to cross under the posterior aorta, in its course to the left division of the lungs. The last cartilage of the main pipe has a spear-like or angular projection extending down between the bronchial tubes, filling up that space which would otherwise be left open from the divergent manner in which they branch off: it is quite loosely attached, in order that the branches may accommodate themselves to the motions of the neighboring parts. The bronchial tubes vary in structure from the trunk that gives origin to them: instead of their rings being formed of entire pieces of cartilage, they are constituted of several separate pieces, making up so many segments of the circle, overlapping one another, and united together and invested by an elastic cellular substance: they also differ in having no muscular band, another fact connected with the physiology of that part. The bronchial tubes, in penetrating the substance of the lungs, subdivide — the right into three principal branches, the left into two; from which spring innumerable others, that grow smaller and smaller, until the ramifications become so reduced that they are no longer traceable by the naked eye. In the larger branches, we may dissect out five and even six segments of cartilage, held together by a thin but dense and elastic cellular substance; in the smaller divisions, only two are found, and they are diminished in size; and, in the smallest visible ramifications of all, cartilage is altogether wanting, though, in many places, marks of the rings may be traced upon the continuation of the lining membrane, which in these intimate parts compose the entire parietes of the tube. In the larger branches this membrane (which is continuous throughout the bronchial system) assumes a plicated disposition — apparently, to admit the more readily of expansion.

THYROID GLANDS.

Two egg-shaped, apparently glandular bodies, attached just below the larynx to the sides of the trachea, and united in front of that tube by an intervening portion of the same substance, which, by way of distinction, is by some called the *isthmus*. They are enveloped and attached in their situation by cellular membrane; are larger and more vascular in the young than in the old subject; and exhibit a spongy texture, when cut into, which I am at present ignorant of the precise nature of. They are well supplied with blood-vessels, and have many small nerves going to them. Their physiology still remains obscure.

OF THE LUNGS AND PLEURA.

The lungs are the essential organs of respiration: *the pleura* is but the membrane by which they are invested.

PLEURA.

The pleura is a fine, semi-transparent membrane, lining the cavity of the chest, and giving a covering to the lungs. By that portion of it which is called the *mediastinum*, the cavity is divided into the right and left *sides* of the thorax.

General Conformation.— If the lungs be exposed, by breaking off one or two of the ribs, we shall perceive that their surface, as well as that of the cavity itself, is everywhere smooth, polished, and humid. This is owing to the extensive investment of the pleura, the surface of which is now presented; so that, in reality, without breaking the surface, nothing but pleura can be touched; although, from its extreme tenuity and pellucidity, the viscera appear, on a superficial view, to present their own bare exterior. Its other side, on the contrary, is rough, having numerous cellular flocculent appendages, by which it is united to the parts it invests; and so close and firm are these adhesions, that to cleanly detach it, in the recent subject, is a very difficult and tedious dissection.

The pleura is a *reflected* membrane; by which is meant, one that not only lines the cavity in which the viscera lie enclosed, but by duplicature, or what in anatomical language is called *reflection*, gives a partial or complete covering to the contained organs themselves. It is evident, therefore, that such a membrane admits of division into two portions — a *lining* or *parietal*, and a *reflected* portion; and these, with regard to the pleura, have, for the sake of more definite description, received the names of *pleura costalis* and *pleura pulmonalis*. They are both, however, continuous at all points, are precisely similar in structure and function, and, in fact, are still but one and the same pleura.

Mediastinum. — There is yet a third portion of this membrane to which a distinct appellation has been given, and that is *the mediastinum*, the membranous partition between the cavities or sides of the thorax; it differs from both the others in being composed of *two* layers, which are derived from the two pleuræ of the opposite sides. If we conceive the pleuræ of the two sides of the thorax to be perfect sacs or bags, with flattened sides turned inwardly, and closely applied and united together, in such a manner that the double membrane formed by their union extends through the middle of the chest, from the dorsal vertebræ to the sternum, we shall at once have a tolerably correct idea of the formation as well as situation of the mediastinum.

Structure. — The pleura, from the nature of its secretion, is one of those included in the list of *serous* membranes to which it has been demonstrated also to be similar in its intimate organization. Like them, it presents a shining secreting surface, of a whitish aspect, and considerable transparency, and is composed of little else than condensed cellular substance, whose texture is penetrated by blood-vessels, absorbents, and nerves: by long maceration in water, indeed, it may be entirely resolved into cellular substance. In most parts it is extremely thin, and by no means tough; but it is not so in all, for that portion which faces the diaphragm is much denser and stronger than the pulmonary or costal division of it.

Organization. — The *arteries* of the pleura, which come from the adjacent parts, are in the natural state exceeding small, admitting only the colorless parts of the blood — a circumstance that accounts for its pellucidity; under inflammation, however, they contain red blood, and such is the explanation of that arborescent vascularity upon the sides of the thorax in horses that die of pneumonia; than which state nothing can better demonstrate the comparative number and distribution of these blood-vessels. The majority of them terminate in exhalent orifices, from which is continually poured, upon the contiguous surfaces of the smooth interior of the membrane, a serous fluid, in the form of steam or vapor, which may at any time be rendered visible by opening the chest of an animal recently dead. The *absorbents* of this membrane are very numerous; and, though their extreme exility prevents us from demonstrating them in a state of health, yet may they often be seen in considerable numbers in horses that die of dropsy of the chest; we have also abundant proofs of their existence from various phenomena that occur in the diseases of the part. We know, for instance, that these vessels take up the serous fluid effused in hydrothorax, for they have been found full of it after death; and it is a fact that no longer admits of doubt, that blood, extravasated into the chest, is absorbed by the mouths of these minute vessels.

The *nerves* of the pleura are too small to be traced by dissection; but, though it is not possessed of much sensibility in a healthy state, we know, at least we presume from analogy, that it is highly sensitive in the diseased; for few diseases are more

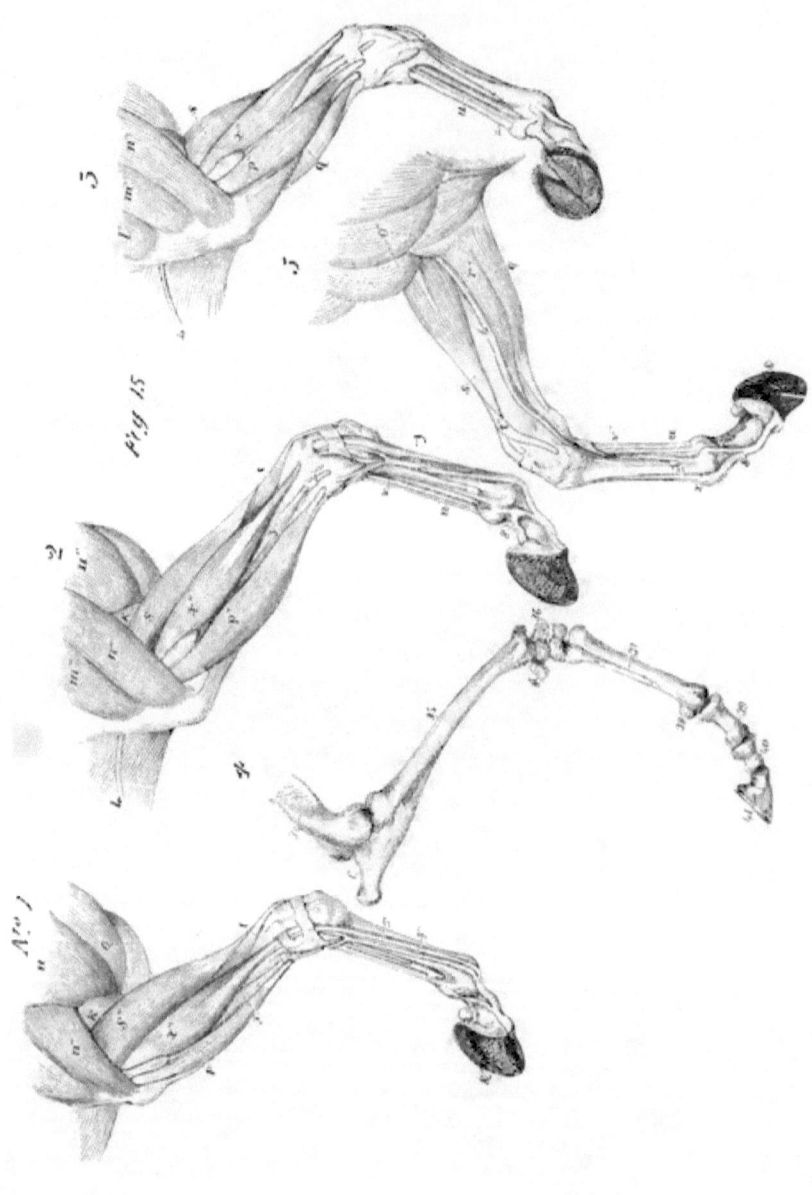

EXPLANATION OF FIGURE XV.

NO. 1. — FORE EXTREMITIES.

LATERAL VIEW OF THE OFF-FORE LIMB.

k. Humero cubital. — Flexor brachii.
n. Triceps externus.
o. Pectoralis transversalis.
p'''. Flexor metacarpi externus.
s''. Extensor metacarpi magnus.
t'. Extensor metacarpi obliquus.
u'. Levator humeri.
x. x. Extensor pedis.
y. y. Extensor suffraginis.
§. The hoof.

NO. 2.

(VIEW AS ABOVE.)

k''. Humero cubital, or flexor brachii.
m''. n''. Two of the triceps extensor brachii.
p''. Flexor metacarpi externus.
s. Extensor " magnus.
t. " " obliquus.
o. Levator humeri.
v'. u. Flexor tendons.
x''. Extensor pedis.
y'''. y. Extensor suffraginis.
8. Perforatus et perforans.
4. Subcutaneous thoracic vein.

NO. 3.

(THE SAME VIEW.)

l''. m''. n''. Triceps extensor brachii.
p''. Flexor metacarpi externus.
q. Extensor suffraginis.
S. Extensor metacarpi magnus.
u. Flexor tendons.
x''. Extensor pedis.
z. Suspensory ligaments.
4. Subcutaneous thoracic vein.

NO. 4.

OSSEOUS STRUCTURE.

34. Os humeri.
f. Os ulnaris.
35. Os Radialis.
g. Trapezium.
36. Ossa carpi.

EXPLANATION OF FIGURE XV. CONTINUED.

37. Metacarpi magnum.
z.   "   parvum
38. Ossa sesamoidea.
39. Os suffraginis.
40. Os coronæ.
41. Os pedis.

NO. 5.

INSIDE VIEW OF THE OFF-FORE LEG.

o". Pectoralis transversus.
p". Flexor metacarpi internus.
q". Flexor metacarpi medius.
s". Extensor metacarpi magnus.
t". Extensor metacarpi obliquus.
u", v". Flexor tendons.
x. Extensor pedis.
y". Suspensory ligament.
6. Radial vein.
8. Bifurcation of the suspensory ligament.
z. Splent bone.
&. Inferior border of the hoof.

acutely painful in the human subject than pleurisy, and we have every reason to believe that horses suffer much from the same malady.

Secretion.— It has been observed that the exhalents of the pleura secrete a serous fluid, which is emitted, in the form of an exhalation or vapor, into the cavity of the thorax; and that it may be rendered visible at any time, if an animal, recently dead, be opened while yet warm; or if an opening be made into the chest of a live animal. In either case, a whitish steam will be perceived to issue from the interior of the cavity. This vapor, shortly after death, becomes condensed and converted into a liquid; which accounts for the contiguous surfaces of the pleura being moist, and for a collection of more or less fluid, resembling water, existing in the most depending parts of the cavity. In consequence of every part of the membrane being bedewed in this manner, the lung itself may be said to be in an insulated state; for the pleura costalis does not, philosophically speaking, touch the pleura pulmonalis, nor is the latter in actual contact with the mediastinum: all friction, therefore, in the motions of these parts, is by this interfluent secretion effectually prevented. In this, then, consists the chief use of the pleura, viz., to furnish a secretion for the purposes of lubrication and facility of motion, which it further promotes by its extreme glibness of surface. It is said also to answer the purpose of *ligaments* to the contained organs, thereby confining and strengthening them. The use of the mediastinum is to divide the chest into two compartments.

LUNGS.

The lungs (by butchers called the *lights*) are two spongy bodies formed for the purpose of respiration.

Situation and Relation.— They are contained in the lateral regions or sides of the thoracic cavity; separated from each other by the mediastinum and heart, which occupy the middle region. Prior to any opening being made into the thorax, the lungs continue to fill up every vacuity; no sooner, however, is a perforation made into the thoracic cavity than they shrink in volume, and become in appearance too small for the spaces they occupy. This arises from their being during life—or rather during the unopened state of the thorax—in a constant state of inflation with atmospheric air, which preserves them expanded; and they suffer collapse of substance the instant air is admitted, in consequence of the pressure of the atmosphere upon them, from which they were protected before by the parietes of the thorax.

Division.— The lungs are two in number, the *right* and the *left lung*; partitioned from each other by the mediastinum. A further division of these organs has been made into *lobes*. That on the right side, the larger of the two, consists of three lobes; the left, only of two. These lobes, which are nothing more than partial divisions of the lung by fissures of variable extent through its substance, serve to adapt them more accurately to the thoracic cavities, and, at the same time, render them fitter for the purposes of expansion and contraction.

Volume.— The lungs of the horse, when inflated, are of great bulk;* and the right is the larger of the two: in consequence of the heart being inclined to the left side, less space is given for the left lung.

Attachment.— The lungs are attached, superiorly, to the spine (which attachment is sometimes called their *roots*) by blood-vessels, the divisions of the trachea, and the mediastinal portions of the pleura: everywhere else, in a healthy subject, they are free and unconnected.

Figure.— In form, the lungs of the horse are very like those of the human subject; and the latter have been compared to the foot of an ox, to which the injected lung of the fœtus bears indeed much resemblance; for, though the two lungs are not symmetrical, yet, both together, they put on this shape, which is the counterpart of that of

* I consider, in comparison with the body, that they exceed in magnitude those of the human subject.

the cavity they occupy. With regard to their general figure, however, the lungs may be said to be conical: being broad and concave posteriorly, where they are opposed to the convex surface of the diaphragm; narrow and somewhat pointed anteriorly, where they are received into the blind pouches of the pleura, in the space between the two first ribs.

Color.—In color, these organs vary somewhat, depending upon the age of the animal, and upon the quantity and distribution of the blood they contain. In the young subject, they are of a lighter and more uniform shade than in the adult. In perfect health they assume a pink hue; which, as age advances, becomes mottled with purple and grayish patches. Sometimes, in the dead subject, they are found of the color of the darkest venous blood, which arises from an inordinate congestion of that fluid within the pulmonary veins.

Structure.—The lungs are composed of the branches of arteries and veins, and of the ramifications of the trachea; all which vessels are connected together by an abundant intervening cellular substance, known by the name of *parenchyma*. Beneath the curve made within the chest by the posterior aorta, the trachea divides into the two bronchial tubes, of which the right is the larger, but the shorter; the left the longer, in consequence of having to pass under the aorta in order to reach the left lung. Having entered the substance of the lung, the right tube divides into four others; the left only into three; which difference arises from the right lung possessing an additional lobe. These branches may be traced for a considerable extent within the parenchyma, giving off in their passage numerous other smaller tubes of similar structure; but, as we prosecute our dissection of them, we shall find that, in growing smaller, they partake less and less of the nature of cartilage, and that the extreme ramifications are not only entirely membranous in their composition, but of so fine a texture as to be perfectly transparent. It will be remembered here, that, in speaking of the trachea, a membranous lining to it was described of the mucous kind, which, it was observed, thence passed into the bronchial vessels: now, it is of the continuation of this membrane in an attenuated state that the minute air-tubes appear entirely to consist; at the extremity of every one of which the membrane is prolonged into a kind of blind bag, or cul-de-sac, to which the name of *air-cell* has been given.

From the arborescent ramification and peculiar mode of termination of the bronchical tubes, some anatomists have compared them, and the cells at their extremities, to a bunch of grapes—supposing the stalks to represent the ramifications of the former, and the grapes connected with them the air-cells; others have described them as having resemblance to a honeycomb: and so far as the knife, with the aid of glasses, can develope their intimate structure, the first is an apt comparison, insomuch as it relates to the disposition of their cells; the last, insomuch as it conveys an idea of their ready inter-communication. For, though they do not communicate but through the ramifications of the bronchial tubes, this is a medium of intercourse at once so general and free, that numbers of them are inflated at the same time by impelling air into any one of the larger branches. With the parenchymatous substance, however, they have no communication whatever.*

The blood-vessels that enter into the composition of the lungs are denominated the *pulmonary*. The pulmonary artery, having taken its origin from the right ventricle of the heart, winds upward to the root of the left lung, and there divides into the *right* and *left* pulmonary arteries, which divisions

---

* If the substance of the lungs be lacerated or rent asunder, the surface will be found to present a lobulated aspect. Introduce a blow-pipe into one of these *lobuli*, and all the other lobules—the entire lung—may be inflated from this one; showing the free communication existing between them. The same may be effected by injecting quicksilver. You may do the same with the interstitial substance; but in this case you do not fill the lobules. In fine, the lungs with their cells resemble a sponge; only that the connecting tissue has no communication with the sponge.

enter their correspondent lungs. The ramifications of these vessels (which differ from other arteries in having no anastomotic communications one with another) accompany those of the bronchial tubes, and, like them, divide and subdivide, grow smaller and augment in number, as they approach the air-cells; upon the internal* surfaces of which they become capillary, and assume a texture of correspondent thinness and pellucidity with the cells themselves. Through these minute vessels every particle of blood is impelled every time it is circulated over the system, as was stated when on the blood: a remarkable change of color is thereby effected in it, and we have now an opportunity of seeing in what manner this fluid is exposed to the influence of atmospheric air for the purpose. It is evident that no immediate contact can happen between the air and the blood, for the thin, transparent side of the vessel, if not that of the air-cell likewise, must ever be interposed: so that, whatever this influence be, it must take effect through one or other or both of these membranes. We might conceive, indeed, that such minute vessels could not transmit through them such a body of fluid as the blood; but, when we look at the volume of the lungs, and consider the incalculable number of air-cells they must contain, the globular surface of every one of which is furnished with an expansion of pulmonary vessels, we shall feel more surprise and admiration at the extreme division and diffusion of this fluid in order to receive the necessary change, than that such a prodigious number of capillaries should be equal, in their united caliber, to the pulmonary artery itself.

From the extremities of the arteries, upon the surface of the air-cells, arise the pulmonary veins. These, by repeated union with one another, form themselves, first, into visible branches, which subsequently become branches of larger size, until at length they end in eight pulmonary venous trunks, which proceed to, and by four openings terminate in, the left auricle of the heart. The ramifications of these veins, unlike the generality of others, are not more numerous than those of their correspondent arteries: and the reason for this is obvious; for, here, one set of vessels are not more subject to compression than the other, nor does the heart (which is so proximate to them) require any such aid as an additional number of veins affords to carry on the circulation. The pulmonary veins have only to convey the blood back to the heart, after it has received its due change within the capillaries upon the air-cells.

Organization. — Besides the pulmonary blood-vessels, there are two others, named *bronchial* arteries. They come off, by one trunk, from the posterior aorta, and each of them enters a division of the lungs, in the substance of which it branches forth, and takes the course of the bronchial tubes. These tubes they supply, as well as the coats of the pulmonary vessels, and the parenchyma of the lungs, with blood: in fact, they may be regarded as the nutrient vessels of these organs. It has been, however, and still remains, a subject of dispute, whether these vessels do *exclusively* nourish the substance of the lungs or not; some say that they do; while others assert that they are assisted in this function by the pulmonary artery, with some of the branches of which they anastomose. The latter opinion certainly does not appear to be supported by facts of much weight; on the contrary, the blood which the pulmonary arteries contain is dark-colored, and unfit for the nutriment of any organ; and as for anastomosis, we have no demonstrative proof of its existence. The bronchial veins end in one trunk, which returns the blood into the vena azygos.

The nerves of the lungs are derived principally from a large plexus within the chest, constituted of the par vagum and sympathetic. They enter the pulmonary structure in company with the bronchial tubes and blood-vessels, and continue their course with them, to be dispersed upon the bronchial membrane and parietes of the air-cells.

---

* Some say, "upon the *external* surfaces."

The absorbents of the lungs are large and numerous, particularly the deep-seated: and of the superficial, we may often succeed in injecting considerable numbers, by introducing a quicksilver-pipe under the pleura pulmonalis. They all pass through the absorbent glands situated around the roots of the bronchial tubes.

Parenchyma. — The connecting medium of the various constituent parts of these organs, or, as it is termed, their *parenchyma*, appears to consist of little else than cellular tissue, without any intermixture of adipose matter: it admits of the free diffusion of any fluid that may be extravasated into it — of air that may have escaped from the air-cells, or of serous fluid poured out when the lungs become anasarcous; but, as was observed before, there is no intercommunication between it and the cells or vessels, as long as the organs preserve their integrity of structure.

Specific Gravity. — The lungs, when healthy, are exceeding light in comparison to their volume; so that, if they be immersed in water, unlike most other parts, they will float upon the surface, — a fact familiar to every one who has seen the liver and lights of an animal thrown into a pail of water to be washed: indeed, the name of *lights* itself seem to have been given to them from this very property. If the fœtal lungs, however, be so treated, they will instantly sink to the bottom of the vessel: and this experimental result at once shows why those of an animal that has once breathed should swim; for, in the one instance they contain air, in the other they are wholly free from it. They are not to be regarded as respiratory organs in the fœtus. It is evident, therefore, that the lungs owe their property of lightness to the air they contain; and, as a further proof of it, if that air be by any means absorbed or pressed from them, and their bulk diminished by collapse of the air-cells, like other viscera, they will prove heavier than an equal volume of water: hence it is that the lungs of a horse that has died of hydrothorax, even though they be sound, are of a greater specific gravity than those of one in health. It occasionally happens, however, that these viscera evince, in this particular, the properties of airless lung, while their natural volume and general appearance remain the same: there must be present interstitial deposition.

BRONCHIAL GLANDS.

Small, oval-shaped, glandular-looking bodies, situated about the roots of the lungs, adhering more particularly to the bottom of the trachea and the bronchial tubes. They exhibit a dirty French gray hue, interspersed with dark blueish spots, and are about the volume (though this varies much) of a tick-bean. For a long time the nature of these bodies remained obscure: of late, skilful injections have clearly shown them to be absorbent glands. They possess their capsules, and, when cut open, exhibit a cellular structure. They contain a dark fluid, which will soil anything it touches; whose principal ingredient chemists have found to be carbon.

CIRCULATORY SYSTEM.

PRELIMINARY REMARKS (ON THE BLOOD, ETC.).

The appearance of blood is familiar to most persons. It contains the elements for building up and nourishing the whole animal structure. On examining blood with a microscope, it is found full of little red globules, which vary in their size and shape in different animals, and are more numerous in warm than in cold-blooded animals; probably this arises from the fact that the latter absorb less oxygen. If the blood of one animal be transfused into another, it will frequently cause death.

When blood stands for a time after being drawn, it separates into two parts. One is called serum, and resembles the white of an egg; the other is the clot or crassamentum, and forms the red coagulum, or jelly-like substance: this is accompanied by whitish, tough threads, called fibrine. When blood has been drawn from a horse, and it assumes a cupped or hollow form, if serum, or buffy coat, remain on its surface, it denotes an impoverished state; but if the whole, when coagulated, be of one uniform mass, it indicates a healthy state of this fluid. The blood of a young horse generally coagulates into a firm mass, while that of an old or debilitated one is generally less dense, and more easily divided or broken down. The power that propels the blood into the different ramifications of the animal, is a mechanico-vital power, and is accomplished through the medium of the heart and lungs; the former is a powerful muscular organ contained in the chest. From certain parts of it arteries arise; in others the veins terminate; and it is principally by its alternate contractions and expansions, aided as already stated, that the circulation of the blood is carried on. The heart is invested with a membranous sac, called pericardium, which adheres to the tendinous centre of the diaphragm, and to the great vessels at the base of the heart. The heart is lubricated by a serous fluid within the pericardium, which guards against friction. In dropsical affections, the quantity of this fluid is considerably increased, and constitutes a disease called hydrops pericardii. The heart is divided into four cavities, viz., two auricles, named from their resemblance to an ear, and two ventricles, forming the body. The left ventricle is smaller than the right; but its sides are much thicker and stronger: it is from this part that the grand trunk of the arteries proceeds, called the great aorta. The right cavity, or ventricle, is the receptacle for the blood that is brought back by the veins after going the rounds of the circulation; which, like an inverted tree, become larger and less numerous as they approach the heart, where they terminate in the right auricle. The auricle on the left side of the heart receives the blood that has been distributed through the lungs for purification. Where the veins terminate in auricles, there are valves placed. The coronary vein, which enters the right auricle, has its mouth protected by a valve called semilunar, or half-moon shape, which opens only toward the heart, and prevents the blood taking a retrograde course. The different tubes coming from and entering into the heart are also provided with valves to prevent the blood from returning. For example, the blood proceeds out of the heart, along the aorta; the valve opens forward or upward, the blood also moves upward, and pushes the valve asunder, and passes through; the pressure from above effectually closes the passage. The valves of the heart are composed of elastic cartilage, which enables them to work with ease. In some diseases, however, they become ossified. This, of course, is fatal. The heart and its appendages are also subject to other diseases, called dilatation, softening, hardening, etc. Now, the blood, having been brought from all parts of the system by the veins, enters into the vena cava anterior and posterior, which empty themselves into the right auricle; and this, when distended with blood, contracts, and forces its contents into the right ventricle, which, contracting in its turn, propels the blood into the pulmonary arteries, whose numerous ramifications bring it in contact with the air-cells of the lungs. It then assumes a crimson color, and is then adapted to build up and supply the waste. Having passed through the vessels of the lungs, it continues on, and passes into the left auricle: this also contracts, and forces the blood through a valve into the left ventricle. This ventricle then contracts in its turn, and the blood passes through another valve into the great aorta, from which it is distributed into the whole arterial structure: after going the rounds of the circulation, it is again returned to the heart by the veins.

EXAMINATIONS ON THE NATURE AND PROPERTIES OF BLOOD.

Q. What are the properties of blood? — A. In health, it is a smooth homogeneous fluid, of unctuous adhesive consistence, of a slightly saline taste, and of a specific gravity somewhat exceeding that of water. It exhales a vapor which has a peculiar odor; this, however, differs in various animals.

Q. Does the blood always preserve the same density? — A. No. Its density is liable to great variations, under the states of rest, labor, disease, and health.

Q. What do you understand by the "*crassamentum*" of the blood? — A. It is supposed to consist chiefly of fibrin.

Q. How is it colored? — A. It owes its peculiar color to what is termed the red globules, which are entangled in it during its coagulation.

Q. How can this be demonstrated? — A. By long continued ablution in water, the red particles are liberated; and we have remaining a white, solid, and elastic substance, which has all the properties of fibrine, and is almost exactly similar to the basis of muscle.

Q. By what name was fibrine formerly known? — A. Coagulable lymph.

Q. What is the form of the red globules of the blood? — A. The Abbé de la Torré, who examined them under microscopes of considerable power, states that they obtained the appearance of flattened annular bodies, with a depression, sometimes perforation, in the centre, but they differ in size and shape in various animals.

Q. By what means is the blood colored? — A. By means of iron and oxygen.

Q. Describe the properties of the serum? — A. It is the yellow fluid part that is left after the separation of the crassamentum; it is of a saline taste, and homogeneous, adhesive consistence.

Q. What effect has a temperature of 160° on it? — A. The whole is converted into a firm white mass, perfectly analagous to the white of an egg which has been hardened by boiling.

Q. Can any liquor be extracted from the serum after having been coagulated by heat? — A. Yes. If the coagulum be cut into slices, and subjected to gentle pressure, an opaque liquor drains from it, which is called the serosity.

PERICARDIUM.

Q. By what is the heart surrounded? — A. The pericardium.

Q. What is the structure of this? — A. It is a fibroserous membranous bag, composed of two coats; one fibrous, the other serous; these are united by cellular tissue.

Q. What are its connections? — A. It is attached to the sternum, pleura, diaphragm, and to the roots of the large blood-vessels at the base of the heart.

Q. What is the function of the serous surface of the pericardium? — A. To secrete the liquor pericardii.

Q. What is the use of this liquor? — A. It serves to protect its own surface, and that of the heart, from friction.

Q. What office does the pericardium perform? — A. It sustains the heart in its proper situation.

---

* Dr. B. Babington is of opinion that the blood, whilst circulating in the vessels, consists of two parts only — a fluid which he calls *liquor sanguinis*, and red globules; and he is induced to believe, from his experiments, that fibrin and serum do not exist as such in the circulating fluid, but that the liquor sanguinis, when removed from the vessels, and no longer subjected to the laws of life, has then, and not before, the property of separating into fibrin and serum. — *M. J. Chossat. Transact.* vol. xvi. pt. 2. Lond. 1831. and art. Blood (morbid conditions of the) in Cyclop. of Anat. and Physiol. Lond. 1836.

HEART.

Q. What is the form of the heart? — *A.* Its form is conoid, yet somewhat flattened on the anterior surface and rounded on the other.

Q. Where is the heart situated? — *A.* Within the thorax, in the region of the fourth, fifth, and sixth dorsal vertebræ; bounded on the sides by the lungs and walls of the thorax; posteriorly, by the diaphragm; inferiorly and anteriorly, by the sternum.

Q. How is the body of heart divided? — *A.* Into a base and apex.

Q. What are the divisions internally? — *A.* It is divided into four cavities, viz: two auricles, or anterior cavities; two ventricles, or posterior cavities.

Q. What communications exist between the cavities of the heart? — *A.* Between the two auricles there is no communication, nor between the two ventricles; but the right auricle opens into the right ventricle, and a similar opening exists between the left auricle and ventricle.

Q. How do veterinarians describe the relative situation of the cavities of the heart? — *A.* The auricles are described as anterior and posterior, because the right auricle forms the upper and fore part, and the left is in a posterior direction; the ventricles being located under their respective auricles; thus we have the anterior and posterior ventricles.

Q. How is the exterior surface of the heart protected? — By a duplicature of the pericardium.

Q. What is the function of the auricles? — *A.* To receive the blood from the various vessels and transmit it to the ventricles.

Q. What is the function of the ventricles? — *A.* One propels the blood to the lungs, for purification; the other distributes it through the arterial ramifications.

Q. Name the venous vessels which terminate in the right auricle. — *A.* Three venous vessels terminate in it, viz: the vena cava, anterior and posterior, and the coronary vein; the vena azygos forms a junction with the anterior cava, just as the latter pierces the walls of the auricles.

Q. How are the auricles divided? — *A.* By the septum auricularum.

Q. Describe the internal mechanism of the right ventricle? — *A.* It has within it numerous fleshy pillars, longitudinally distributed; also, three fleshy prominences, termed carnæ columnæ, from which several tendinous cords proceed to the edges of those membranous and fibrous productions; these close the auriculo-ventricular opening; the apparatus altogether forms valvula tricuspis. Other cords, similar to the cordæ tendinæ, pass between the outer wall and the septum.

Q. Where is the origin of the right pulmonary artery? — *A.* It emerges from the upper and back part of the ventricle.

Q. How is the mouth of this artery protected? — *A.* By three semilunar valves, which present little pouches within its cavity; these valves consist of doublings of the lining membrane of the parts.

Q. Describe the left ventricle? — *A.* Its cavity is smaller than that of the right, and its wall is thicker. Its musculi pertinati appear mostly upon the septum, within the apex and under the valves; it has two, instead of three, carnæ columnæ; they are more bulky, and project more into the cavity than those of the right.

Q. From whence does the aorta arise? — *A.* From the upper and fore part of the left ventricle.

Q. What is remarkable about the mouth of the aorta? — *A.* It has three semilunar valves, similar to those at the origin of the pulmonary artery.

Q. By what are the ventricles divided? — *A.* They are divided by a fleshy partition called septum ventriculorum.

Q. How is the circulation of the blood effected? — *A.* By the alternate contraction of the auricles and ventricles, called the dyastole and systole of the heart.

Q. By what vessels is the heart itself supplied with blood? — *A.* By the coronary arteries.

ARTERIAL SYSTEM.

DISTRIBUTION OF ARTERIES.

The blood is propelled by the heart through the great aorta, which rises out of the base of the left ventricle, in the space between the left auricle and the pulmonary artery. The branches furnished by the main trunk are the coronary arteries. The right coronary artery emerges from between the pulmonary and right auricle, winds round the fissure separating that cavity from the right ventricle, and turns down under the termination of the vena cava; and distributes ramifications in its course, which penetrate the substance of the parietes, and end in spiral branches. The left coronary artery, in passing out between the pulmonary artery and left auricle, sends off a large branch, which encircles the other auricle; it then takes its course downward, and ends in spiral ramifications.

ANTERIOR AORTA.

This is a shorter division of the main trunk. The course of this vessel is under the windpipe; it gives origin to those large arteries which are distributed over the breast, head, neck, brain, and anterior extremities. It divides, at a short distance from the heart, into the right and left arteria innominata; the right is considerably longer than the left, and measures nearly as much again in circumference; it forms the trunk from which the two carotid arteries spring; the left terminates in the following vessels:*

1. The dorsal artery. 2. Posterior cervical. 3. Vertebral. 4. Internal pectoral. 5. External pectoral. 6. Inferior cervical.

* The vertebral artery, forming the basilar, gives off the posterior cerebellar, anterior cerebellar, posterior cerebral, and the circular arteriosus.

7. Axillary. Each of these arteries ramify and anastomose with others, and are distributed to muscular and adipose substance. From the axillary artery spring all the arteries of the fore extremity. This vessel can only be seen by detaching the shoulder from the body. It arises within the chest, from the arteria innominata; gains exit by making a sudden turn around the first rib, rather below its middle, crossing the lower border of the scalenus in the turn; it is first directed outward in this flexure, and then backward, and at length reaches the inner part of the head of the humerus, where it makes another turn backward, and afterwards takes the name of the brachial artery. Its branches are — 1. The external thoracic. 2. The internal thoracic, which runs to the point of the shoulder, and gives its branches to the levator humeri and shoulder joint. 3. The dorsalis scapulæ ascends, in a flexuous manner, to the shoulder joint, crossing the insertion of the subscapularis. It runs for a short distance along the ribs. 4. The subscapularis, a large artery, which also arises from the upper part of the trunk, but near to its termination. It passes along the ribs, screened from view by the edges of the subscapularis and teres major, to both of which muscles it detaches several small branches, and ends near the lower angle of the bone; it also gives off several branches to the triceps and panniculus. 5. The humeral.

The *humeral artery* descends from the inner and back part of the head of the os humeri, in an oblique direction on the body of the bone, where it divides into the ulnar, spiral, and radial arteries. On its inner side, it has the spiral and ulnar nerves; in front, the radial nerve; and behind, the humeral veins; and it is covered internally by the large pectoral muscle, to which it sends some small branches. But its principal branches are — 1. One near its origin, which crosses the bone to get to the flexor brachii, and sends twigs to the shoulder-joint. 2. A posterior branch, arising a little lower down, which enters the muscle called triceps. 3. Near its termination, another branch to the flexor brachii. Where the artery divides, it is covered by the humeral plexus of veins, and by the absorbent glands of the arm.

The *ulnar artery* consists of a common root, from which spring three or four vessels of considerable size, running in waving lines upon the inner side of the lower end of the humerus. The upper one is directed to the ulnar, splitting before it reaches the bone, and sending one branch upward upon the elbow, and another downward to the heads of the flexors; to which muscles the other branches of this vessel are distributed.

The *spiral artery*, the outermost division, turns round the os humeri, passing under the flexor brachii, and sending a recurrent branch to it, to arrive at the front of the radius, where it splits into several branches, of which — 1. Some run into the elbow joint. 2. Others, larger and more numerous, penetrate the heads of the extensors. 3. Two long, slender ones descend upon the radius, and give branches, in their course, to the extensor muscles as low as the knee, and there end in ramifications about and into the joint, joining with others coming from the radial.

The *radial artery*, the principal division humeral, continues its descent along the radius, about the middle of the arm; the nerve accompanies it first on its outer side, and subsequently behind it. A short way above the knee, it splits into the metacarpal arteries.

The *small metacarpal artery* descends, within a cellular sheath, along the inner and back part of the knee. It continues its descent along the metacarpal vein (which runs to its inner side), till it gets below the knee, and then transmits its divisions down the front of the suspensory ligament; between it and the canon bone, it sends off branches over the front of the knee, the canon, and suspensory ligament.

The *large metacarpal artery*, a continuation of the radial trunk, continues its course down the leg, by the side of the tendo perforatus, passing under the posterior annular ligament, approaches the fetlock just above the joint, and then splits into three vessels; from the middle division three recurrent arteries are given out; the side divisions become the plantar arteries. From the arch below come off two other branches, which descend into the joint. The plantar arteries, external and internal, in the fore extremity, result from the fork of the metacarpal; in the hind, from that of the metatarsal. (Their general distribution is the same, both in the hind and fore feet.) They descend the fetlock upon the sides of the sessamoids, in company with the veins which run in front of them, and with the plantar nerves which proceed behind them; the artery then passes down to, and into, the substance of what is called the "fatty frog;" it next passes the inner and upper extremity of the coffin bone, and afterwards to the foramen of the posterior concavity of the bone. The branches of the plantar artery are many and important. After detaching some small ramifications inwardly to the fetlock, posteriorly to the flexor tendons, and anteriorly to the extensor tendon, it then sends off — 1. The perpendicular artery. 2. The transverse artery. 3. The artery of the frog. 4. The lateral laminal artery. 5. The circulus arteriosus. From the latter arise two principal sets of vessels — 1. The anterior laminated arteries. 2. The inferior communicating arteries, "thirteen, and sometimes fourteen, in number." 3. The circumflex artery. Then, again, from this vessel spring the solar arteries, which may be so named from their radiated arrangement. These latter are destined for the supply of the sole, upon which they run in radii at equal distances, whose common centre is the toe of the frog, where they end in communications with the arteries of that body.

THE CAROTID ARTERY.

The right arteria innominata, having detached seven important branches, which vary but little in their mode of origin, general course, and distribution, from the several arteries into which the left division resolves itself, become the common carotid—a large vessel emerging through the upper opening of the chest; it divides, as it quits the chest, into two branches, called the right and left carotids. These arteries ascend, and having reached the top of the larynx, the carotid of either side branches into three divisions—the external and internal carotids, and the ramus anastomoticus: here, though the trunk itself becomes deeply lodged in soft parts, its situation is well indicated by the larynx, with which it is in contact. This vessel detaches—1. Several unimportant muscular branches in its progress up the neck. 2. The thyroideal artery, which furnishes the laryngeal, a small artery that perforates the ligament uniting the cartilages of the throat.

The external carotid artery is the large division, which may be regarded as the continuation of the carotid itself. This artery is imbedded in glandular substance, surrounded by venous and nervous trunks, and protected by bony prominences and muscles. The first branch of the external carotid is the submaxillary artery; it comes off behind the horn of the os hyoides, just as the carotid makes its second curve, and ranks next in size to the trunk itself. After reaching the lower jaw (about one-third of its length downwards), it arrives upon the face; here it becomes subcutaneous, ending in an equal division, called the facial and inferior labial arteries. Its branches are, the ascending laryngeal, pharyngeal: smaller branches go to parotid gland, and a large branch, called the lingual. The latter detaches a few twigs into the submaxillary space; it then branches into two arteries, the ranine and the sublingual. The ranine, apparently a continuation of the lingual, passes along the under part of the tongue, and transmits branches to the interior, and continues of large size even to the tip of the organ, wherever its extreme ramifications are expended. The sublingual artery winds along the under and outer border of the tongue, preserving a more superficial course than the former. It supplies the sublingual gland, and distributes branches over the membrane of the tongue. The submental artery leaves the submaxillary, follows the course of the branch of the jaw, and detaches twigs to muscles; it then transmits its ramifications into the gums internally. The anterior masseter branches pass on the external side of the jaw.

The inferior labial artery courses the side of the jaw, invested in the cellular and fleshy substance belonging to the buccinator. It gives off slender ramifications to the investing cellular substance, also the buccinator arteries; the buccal twigs bifurcate, sending their divisions respectively to the upper and under lips; these form the superior and inferior coronary arteries of the lips.

The facial artery ascends upon the side of the face, crosses the buccinator, then, having run as high as the bony ridge from whence the masseter arises, it detaches a large branch, and then expands upon the upper and fore part of the face; its terminating ramifications are in the cellular substance and skin covering the fore part of the face.

The posterior auricular gives branches to the parotid gland, and to the different muscles of the ears.

The temporal artery, the anterior auricular, and the internal maxillary, may be considered as the terminating branches of the external carotid. The internal maxillary gives off deep temporal branches, long slender twigs, to the soft palate, to the ear, and to the articulation of the jaw; the facial artery also gives off the inferior maxillary, the supra-orbitar, the ocular, the infra-orbitar, and the palate maxillary. The second and smallest division of the carotid is the

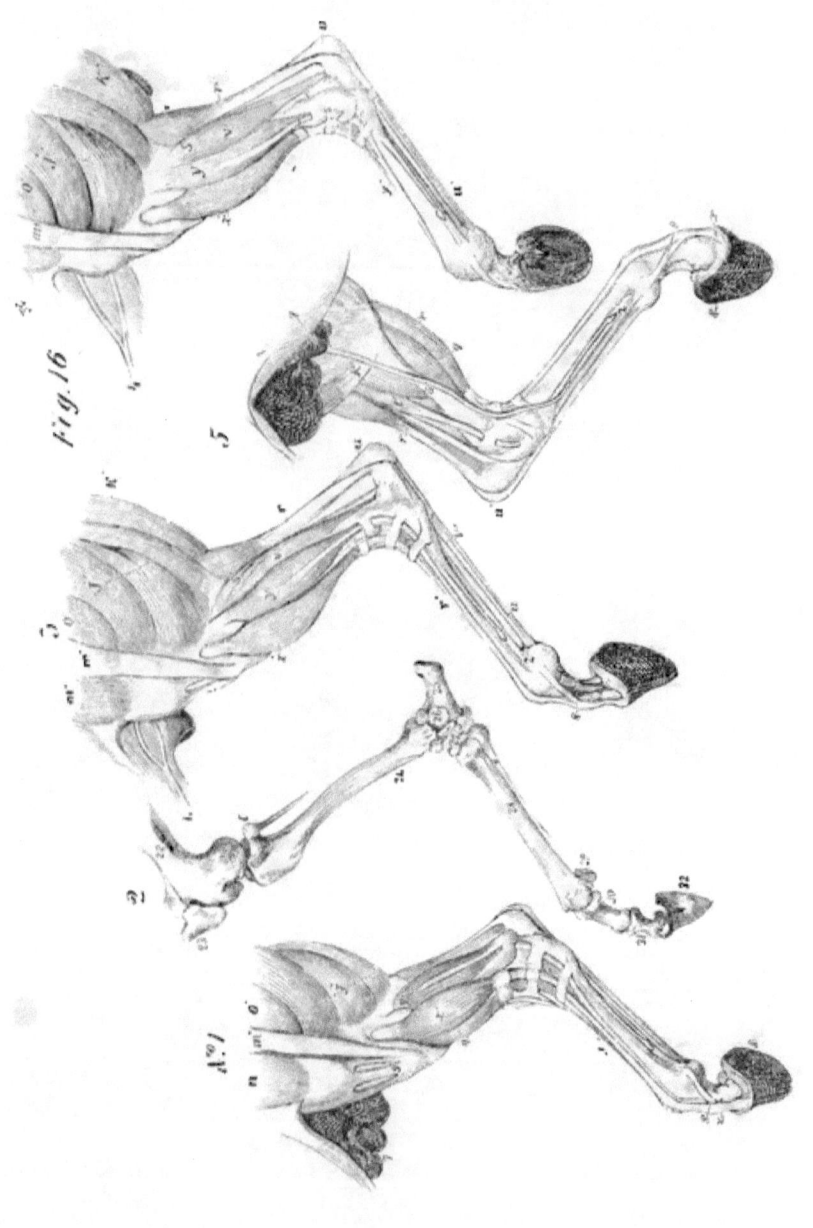

Fig. 16

EXPLANATION OF FIGURE XVI.

NO. 1.
FRONT AND SIDE VIEW OF THE NEAR-HIND LEG.

- g. Ligaments of the patella.
- J''. Triceps.
- m'. Tensor vaginæ.
- n'. Rectus.
- o'. Vastus externus.
- q''. y'. Extensor suffraginis.
- x. x. Extensor pedis.
- 7. Sheath and penis.
- 8. Bifurcation of the suspensory ligament.
- &. The hoof.

NO. 2.—OSSEOUS STRUCTURE.

- 22. Femur.
- 23. Patella.
- 24. Tibia.
- e. Fibula.
- 25. Os calcis.
- 26. Astragalus.
- 27. Inferior tarsus.
- 28. Metacarpi magnum.
- *. " parvum.
- 29. Sessamoids.
- 30. Os suffraginis.
- 31. Os coronæ.
- 32. Os pedis.

NO. 3.
OUTSIDE VIEW OF THE NEAR-HIND LEG.

- K. Abductors.
- J'''. Triceps.
- m'. Tensor vaginæ.
- m'. Rectus.
- o'. Vastus externus.
- r'. Gastrocnemius externus.
- r. s'. Plantaris.
- n'. Gastrocnemius externus.
- v'. u'. Flexor pedis.
- x. Extensor pedis.
- y. y. Peroneus.
- u. v. Flexor tendons.
- &. The hoof.
- 8. Bifurcation of suspensory ligament.
  No. 4 is nearly the same as No. 3, and therefore needs no further description.
- z. Suspensory ligament.

EXPLANATION OF FIGURE XVI. CONTINUED.

NO. 5.

k'. Abductor tibialis.
q. Glans penis.
r. Gastrocnemius externus.
t'. Flexor pedis accessorius.
u'. Insertion of the gastrocnemius.
x'. x. q. Extensor pedis.
8. Bifurcation of the suspensory ligament.
&. The hoof.
9. Flexor metatarsi.
5. The saphena vein.
z. Suspensory ligament.

RAMUS ANASTOMOTICUS.

It leaves the trunk of the carotid, joins the vertebral, and from it arises the occipital artery, which gives off twigs to be dispersed upon the dura mater, temporal muscle, and muscles of the occiput.

INTERNAL CAROTID.

This vessel, whose calibre is not more than half that of the external carotid, ascends to the base of the skull: at its entrance into the skull, a vessel comes off named the arteria communicans: after having given off this vessel, the internal carotid pierces the dura mater, takes its course up near the optic nerve, and branches into four divisions, which supply the cerebrum with blood.

The remaining vessels of the brain are derived from the vertebral artery, which gives off posterior arteries to the dura mater, and ramifications to the medulla oblongata.

The basilar artery sends off branches to the cerebellum.

THE POSTERIOR AORTA.

Considerably longer and larger than the anterior is the main trunk, from which are derived the artery of the abdomen, pelvis, and posterior extremities, in addition to the posterior intercostals, and some few of the thoracic arteries. It commences opposite the fourth dorsal vertebræ: from its origin it courses first upward, and then backward, having the pulmonary artery on its left, the termination of the windpipe on its right, then takes a course along the spine, inclining to the left side. From the inferior part of the curvature of the aorta arise the right and left bronchial arteries: these vessels penetrate the lungs in company with the bronchia, to the branches of which they cling in the course of their ramifications within the substance of the lungs.

The esophageal also spring from the concavity of the arch near to the former, and proceed backward to the esophagus, where it divides into an inferior and superior artery. The intercostal, the remaining branches, come off in pairs from the sides of the vessel, to supply all those intercostal spaces posteriorly to the last. These arteries run along the lower borders of the ribs, and end about the inferior parts of the chest and abdomen. They furnish, near their origin, small branches, which enter the vertebral canal. Having detached these small vessels, the posterior aorta continues its passage into the abdomen. In making its exit from the chest, it gives off the phrenic or diaphragmatic arteries.

Within the abdomen, the aorta continues to be firmly fixed to the spine, by its several cellular attachments, as far as the lumbar vertebra, under the body of which it branches into four large arterial trunks. Prior to this division, the abdominal aorta gives off the cœliac artery, which is nothing more than the common root of the splenic, gastric, and hepatic — arteries that in some instances have separate origins.

The splenic artery, after passing between the stomach and spleen, ends in the left gastric artery. In its course it gives off several branches to the pancreas, called pancreatic arteries.

The gastric artery, the smallest of the cœliac divisions, runs forward to the small curvature of the stomach, between the layers of the omentum, branching, before it reaches this organ, into two vessels, called inferior and superior gastric, which finally ramify upon the upper and under surface of the stomach.

The hepatic artery, the largest of the cœliac division, proceeds before the pancreas to the right side of the cavity, and passes over the pyloric end of the stomach, and gives off small branches to the pancreas. Near the pylorus, it sends a branch to the duodenum, which, as soon as it reaches the intestine, divides: one division — the duodenal — retrogrades along the gut, and ends in anastomosis, with branches coming from the anterior mesenteric; the other — the right gastric — crosses the gut, proceeds to the great curvature of the stomach, where it inosculates with the left gastric. The hepatic artery itself is continued forward to the porta of the liver, where it divides into the right and left hepatic; the right,— the larger and shorter one,— after giving off a considerable branch to the portio media, turns back to reach the right lobe; the left, after giving off a branch or two to the middle portion, penetrates the left lobe.

The anterior, or great mesenteric, is the next vessel to the cœliac, and arises from the under part of the posterior aorta. From its origin, it passes downward within the layers of the mesentery, detaching some small twigs to the pancreas; it then separates into larger vessels (commonly from eight to twelve in number), from which are derived a branch that runs to the duodenum; several other branches encircle and ramify on and around the intestines.

The renal or emulgent arteries leave the aorta at right angles just below the preceding vessel; they each pass into the respective kidneys, and therein divide into branches that penetrate the glandular substance.

The spermatic arteries, right and left, originate from the under part of the aorta; they pass out of the abdomen, at the abdominal ring, to the testicles. In the female, they pass to the ovaries, fallopian tubes, and horns of the uterus.

The posterior aorta also gives off the small mesenteric, and five or six pairs of lumbar arteries. Under the last lumbar vertebræ, the aorta gives off two pairs of arterial trunks, called the external and internal iliacs.

The internal iliacs give off a branch called the artery of the bulb, and afterwards branches into three divisions — the obturator, gluteal, and lateral sacral arteries.

The artery of the bulb passes to the bulb of the penis, where it terminates. In the female, this artery sends its terminating branches to the vagina. It gives off the fœtal umbilical artery. In leaving the pelvis, the prostatic artery, which detaches twigs to the vesiculæ seminales, also distributes its ultimate ramifications to the prostate gland. It also gives off divers branches, anal and perineal, to the posterior portion of the rectum, anus, and parts comprising the perineum.

The obturator artery is the lowest of the divisions of the internal iliac. Its branches are the arteria innominati, and ramifications to the obturator muscles and ligaments. Its divisions are the ischiatic, which distributes its branches to the triceps; next, the pubic: the internal pubic artery gives two sets of branches, which pass to the penis.

The gluteal artery is destined principally to supply the gluteal muscles.

The lateral sacral artery, having reached the coccyx, divides into two branches. It furnishes the sacro-spinal branches, five or six in number, and the perineal artery. It soon divides into several ramifications, of which many run into the gluteal muscles; others descend on the back of the thigh, and others are distributed to the anal muscles, and to the skin and cellular substance of the perineum. The lateral sacral also furnishes the lateral coccygeal, and the inferior coccygeal.

The external iliac artery, right and left, results from a branch of the posterior aorta, which takes place under the body of the last of the lumbar vertebræ, and passes into muscles, forming the inside of the thighs. The vessel gives off the circumflex artery of the ileum, the artery of the cord, and the arteria profunda: the latter, having reached the posterior quarters, it sends its ramifications into the biceps. Before this vessel dips into the substance of the thigh, it gives rise to a large branch called the epigastic artery.

The epigastric artery, in passing the margin of the internal ring, forms a branch which divides into several small arteries; of these a twig runs to the groin, and ramifies among the adipose membrane and absorbent glands; then, next, a slender branch to the cremaster, and subcutaneous twig to the thigh, and, lastly, the external pudic artery.

The femoral artery. — Regarding the profunda femoris as a limb of the external iliac, we descend to the femoral artery, the subsequent continuation of the same trunk. This artery proceeds in an oblique direction down the haunch, preserving nearly the line of its middle; opposite to the head of the tibia, it branches into the anterior and posterior tibial arteries; the anterior tibial gives off the inguinal artery, also three or four branches to the sartorius, and one to the side and front of the stifle. Its posterior sub-branches are a large artery to the gracilis (which detaches twigs to the long and short heads of the triceps), also one to the biceps. At the back of the stifle come off the popliteal branches, four or five in number, taking opposite directions, which are destined for the supply of the joint; one runs down upon the posterior tibial muscles; another—the recurrent branches—climbs the back of the os femoris, and anastomoses with the descending ramifications of the profunda femoris.

The tibial arteries are a continuation of the femoral trunk, which branch off into tibial arteries at the head of tibia.

The posterior tibial artery, the smaller of the two, passes along the posterior deep region of the thigh, to the hock, where it ends in bifurcation. Its branches are, one that runs into the flexor pedis; another to the upper and back part of the tibia; and small twigs to both the flexors. There are several terminating branches, some ramifying subcutaneously, others continuing down the leg internally over the tendon of the flexor pedis, and ending at the lower part of the canon in divers small ramifications.

The anterior tibial artery, after leaving the trunk, passes down the fore part of the thigh to the hock and metatarsal bone, where it becomes the metatarsal artery.

The metatarsal artery pursues its course downwards to about two-thirds the length of the leg; it then gains the posterior part of the latter; a little above the fetlock, it divides into three vessels: one forms an arc, (as in the fore extremity), from which come off the recurrents, and they anastomose with the posterior tibial artery; the lateral divisions become the plantar arteries.

REMARKS ON THE DISTRIBUTION OF ARTERIES.

The preceding is a brief sketch of the arterial structure, and the professional man will perceive that we have not named the whole of the arteries; therefore, in order to supply this deficiency, the author here introduces a table of the arteries, constructed by Mr. Percivall.

TABLE OF THE ARTERIES.

AORTA { Anterior Aorta. / Posterior Aorta.

ANTERIOR AORTA { Right Arteria Innominata. / Left Arteria Innominata.

LEFT ARTERIA INNOMINATA {
- Dorsal.
- Posterior Cervical.
- Vertebral, forming the Basilar. { Posterior Cerebellar. / Anterior Cerebellar. / Posterior Cerebral. / Circular Arteriosus.
- Internal Pectoral.
- External Pectoral.
- Inferior Cervical.
- Axillary.

Axillary {
- External Thoracic.
- Internal Thoracic.
- Dorsalis Scapulæ.
- Subscapular.
- Humeral.

Humeral {
- Ulnar.
- Spiral.
- Radial... { Small Metacarpal. / Large ditto.

Large Metacarpal { External Plantar. / Internal ditto.

Plantar {
- Perpendicular.
- Transverse.
- Artery of the Frog.
- Lateral Laminal.
- Circulus Arteriosus... { Ant. Laminal. / Inf. Communicating. / Circumflex.... { Solar.

The Right Arteria Innominata sends off branches correspondent to those on the left side; and, in addition, the

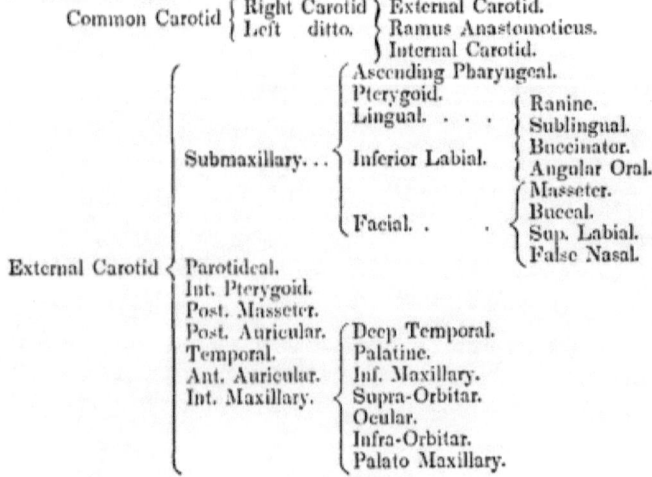

Common Carotid { Right Carotid / Left ditto. } External Carotid. / Ramus Anastomoticus. / Internal Carotid.

External Carotid {
- Submaxillary... {
  - Ascending Pharyngeal.
  - Pterygoid.
  - Lingual. ... { Ranine. / Sublingual.
  - Inferior Labial. { Buccinator. / Angular Oral.
  - Facial. . { Masseter. / Buccal. / Sup. Labial. / False Nasal.
- Parotideal.
- Int. Pterygoid.
- Post. Masseter.
- Post. Auricular.
- Temporal.
- Ant. Auricular.
- Int. Maxillary. {
  - Deep Temporal.
  - Palatine.
  - Inf. Maxillary.
  - Supra-Orbitar.
  - Ocular.
  - Infra-Orbitar.
  - Palato Maxillary.

Ramus Anastomoticus } Occipital. { Dura Matral. / Temporal. / Nuchal.

Internal Carotid { Arteria Communicans. / Anterior Dura Matral. / Anterior Cerebral... < Ophthalmic / Middle Cerebral. / Lateral Cerebral.

Posterior Aorta { Thoracic Division. / Abdominal ditto.

Thoracic Division { Bronchial { Right Bronchial. / Left ditto. / Œsophageal. { Superior Œsophageal. / Inferior ditto. / Intercostals. / Phrenic.

Abdominal Division {
  Cœlic { Splenic { Pancreatic Branches. / Splenic Branches. / Left Gastric. / Gastric { Superior Gastric. / Inferior ditto.
  Hepatic { Pancreatic Branches. / Duodenal. / Right Gastric. / Right and Left Hepatic.
  Anterior Mesentric { Duodenal Branches. / Small Mesenteric. / Cœcal Branches. / Anterior Colic Branches.
  Renal . . . . { Right Renal. } External Branches. / Left ditto. } Capsular Renal.
  Spermatic { Right and Left.
  Posterior Mesenteric { Posterior Colic. / Rectal.
  Lumbars — five or six pairs.

Bifurcation of the Posterior Aorta into External and Internal Iliac Arteries.

Internal Iliac {
  Artery of the Bulb { Umbilical. / Vesical Branches. / Prostatic. / Anal and Perineal Branches.
  Obdurator { Arteria Innominata. / Foraminal Branches. / Ischiatic. / Pubic / Int. Pubic { Branches to the Crus Penis. / Ditto Corpus Cavernosum. / Ditto Dorsum Penis. / Ditto Glans Penis. / Cutaneous Branches.

The Middle Sacral issues at the Bifurcation of the Trunk.

External Iliac { Circumflex of the Ileum. / Artery of the Cord. / Arteria Profunda. { Epigastric. { Branch to the Groin. / Branch to the Ring. / External Pudic. / Femoral.

Femoral
- Inguinal.
- Muscular Branches.
- Stifle Branches.
- Muscular Branches.
- Popliteal. { Recurrent.
- Anterior Tibial ....
  - Recurrent Articular.
  - Muscular Branches.
  - Cutaneous Branches.
  - Metatarsal Branches.
  - Metatarsal Artery. . { Recurrent. External Plantar. Internal Plantar.
- Posterior Tibia ....
  - Muscular Branches.
  - Medullary.
  - Tarsal.
  - Internal Metatarsal. { Recurrent.

DISTRIBUTION OF VEINS IN THE HORSE.

THE two main venous trunks, the *renæ cavæ*, anterior and posterior, correspond to the anterior and posterior aortic.

THE ANTERIOR VENA CAVA

Forms the main trunk of the veins, returning the blood from the head, neck, chest, and fore extremities. It is principally formed by the concurrent union of the jugular and axillary veins, and is situated at its formation in the space between the two first ribs, about midway between the sternum and vertebræ; it also receives the pectoral, vertebral, dorso-cervical, and inferior cervical veins, and the vena azygos.

THE JUGULAR VEIN.

It passes behind the condyle of the lower jaw, under the parotid gland, and joins the external carotid artery, and continues its course down the neck with the latter. It now receives the auricular veins, anterior and posterior, and also internal. The next is the temporal, the third is the internal maxillary; the latter in its course receives the blood of many small veins,— the palato-maxillary, infra and supra orbitar, ocular, inferior maxillary, and deep temporal; the fourth branch, received by the jugular vein, is the parotideal, and the last branches from the masseter muscles.

THE OCCIPITAL VEIN

Descends from the head, along with the occipital artery. It brings blood from the occipital sinuses, receives veins from the posterior lobes of the cerebrum and cerebellum; also from the dura mater.

The submaxillary vein is a large branch of the jugular. It is formed upon the side of the face by the concurrence of the facial, labial, and varicose veins. It joins the trunk by the side of the trachea, just below the parotid gland. In its course it receives a number of veins; the principal are — the submental, sublingual, lingual, pharyngeal, and superior laryngeal veins. The facial vein results from an expansion of small veins upon the side of the face, one of which is the varicose from the masseter. The labial vein is formed by the union of a plexus of venous branches, coming principally from the angle of the mouth, joined by others both from the upper and lower lips. The varicose vein is buried in the masseter.

The jugular trunk having received the submaxillary, proceeds down the neck, and terminates in the anterior vena cava, within the space between the two first ribs. Near the junction of the submaxillary the jugular receives the small thyroideal, cutaneous, muscular, and tracheal veins. Near its termination it receives a branch of the superficial brachial, and plait or plat vein.

The vertebral vein runs the same course as the artery, through the foramina, in the transverse processes of the cervical vertebræ, with the exception of the last. This vein has communications with the occipital sinus and posterior cerebral veins, medulla oblongata, and spinal marrow; it also receives vessels from the deep-seated muscles in the vicinity, and ends in the anterior vena cava, just behind the first rib.

The axillary vein returns the blood distributed by the axillary artery to the various parts of the fore extremity; there is a superficial and deep-seated set; the former run under the skin, the latter among the muscles. The plantar veins are an intricate network of small veins, and cover the foot with a venous netting. The veins of the sole pour their blood into the veins of the laminæ; the latter increase in size towards the coronet, and gradually unravel themselves, so as to collect in a great many branches; these run upward, through the substance of the coronary ligament, and form the superficial coronary vein; from them other branches proceed and join the deep coronary, and afterwards unite in a single vein opposite the pastern joint.

The veins of the frog, after ramifying in the form of network over that body, ascend into the heel, growing larger as they leave the foot; they make a single branch at the pastern joint, then unite with the vein coming from the laminæ, thereby forming the plantar vein. The plantar vein ascends, unites with other vessels, and becomes metacarpal.

The metacarpal veins, two in number, result from the union of the plantar; these veins pursue their course up the leg, one on either side, to the back of the knee, where they end in anastomosis. The internal metacarpal vein preserves the line of the splint bone. These vessels receive in their course cutaneous veins from the front of the canon, and one or two descending veins from the back of the leg; it afterwards forms the deep-seated veins of the arm.

The superficial brachial vein ascends along the inner side of the radius to the elbow-joint; here it crosses over to the front of the biceps and pursues its ascent upon that muscle toward the point of the shoulder, and then passes inward to the jugular vein. In its course to the latter, it receives numerous cutaneous and muscular branches, communicates with the humeral vein, and anastomoses with other veins of the arm.

The radial veins, two in number, arise from the junction of the metacarpal veins above the knee; they take the course of the radial artery, and receive anastomosing vessels as they ascend from the ulnar and superficial veins.

The ulnar veins (with one exception) end in the common trunk of the humeral vein.

The humeral vein accompanies the artery; it receives small veins from the muscles.

The axillary vein is the continuation of the humeral, augmented by the accession of the triceps vein. Its branches are, the subscapular vein, and dorsalis scapular; the latter terminates about midway between the chest and shoulder. The remaining branches of this vein are the humeral thoracic, and the external thoracic; it also receives other small veins, which contribute more or less to its volume.

The pectoral vein runs the course of the pectoral artery. It originates in branches from the abdominal parietes, continues to receive accessory vessels in its course, and ascends along the inner and lower border of the first rib.

The dorso-cervical vein consists of two divisions, ramifying with the dorsal and posterior cervical arteries; it receives the anterior intercostal vein.

The inferior cervical vein runs down the lower part of the neck in company with the artery; the principal branches are muscular, though some come from the skin and absorbent glands in the vicinity.

The vena azygos ends just as the trunk opens into the auricle; it returns the blood from the lower intercostal veins.

THE POSTERIOR VENA CAVA.

This is the corresponding venous trunk to the posterior aorta, returning the blood from the parietes of the abdomen and pelvis, the urinary and genital organs, and the posterior extremities. It takes its course under the bodies of the lumbar vertebræ, runs along the great fissure of the liver, perforates the cordiform tendon, and pursues its way directly across the cavity of the chest to the lower part of the right auricle; in its passage it is joined by the lumbar spermatic, renal, hepatic, and diaphragmatic veins.

The *common iliac veins* are formed under the sacro-iliac, symphysis, by the union of the external and internal iliacs; they receive a vein from the psoæ and iliacus, circumflex vein of the ileum, middle sacral, and azygos.

The *ischiatic vein*, situated upon the side of the pelvic cavity, midway between the external iliac and lateral sacral veins; external and internal branches unite to form it. The internal comprise veins coming from the bladder, anus, perineum, and, in the male, from the bulb and prostate; in the female, from the vulva and body of the vagina. The external come principally from the gluteal and obturator muscles.

The *lateral sacral vein* comes from the tail, formed by coccygeal veins; it runs forward to the sacrum, and receives in its course the perineal and sacro-spinal branches.

The *external iliac vein* takes the same course as the artery; as it departs from the belly, this vessel receives

The *inguinal vein* (coming from the groin), also a superficial or sub-cutaneous abdominal vein, known as the milk vein in cattle.

The *femoral vein* is the continuation of the iliac trunk below the brim of the pelvis; and is the main channel into which the deep-seated veins of the hind extremity pour their blood. We commence the description, as in the fore extremities, at the leg.

The *large metatarsal vein* ascends the canon by the side of the flexor tendons, and passes over the front and inner part of the hock; it sends out branches, from which result the

Anterior tibial veins, which run between the tibia and fibula to the back and lower part of the os femoris, and then are joined by the posterior tibial vein, and all three unite to form the femoral.

The *posterior tibial vein* is a continuation of the small metatarsal vein, and corresponds in size to the small metacarpal. It runs in company with the posterior tibial artery, receiving various muscular branches in its course, also the medullary vein of the tibia.

The *femoral vein* results from the two last-named vessels; runs behind the femoral artery, and ends in the external iliac vein. It receives muscular veins, as well as veins from the stifle joint, and the medullary vein of the os femoris; also, about two-thirds of its length upwards, it is joined by the saphena vein.

The *vena saphena major* results from the large metatarsal vein; at the hock it anastomoses with the anterior tibial vein; it also receives cutaneous and muscular branches in its course.

The *vena saphena minor* springs from the small metatarsal vein; it runs up the back of the hock, over the root of the os calcis, and ultimately reaches the femoral vein.

The *vena porta* circulates the blood through the liver, and is principally formed by the union of the splenic and mesenteric veins.

THE HORSE.

TABLE OF THE VEINS.

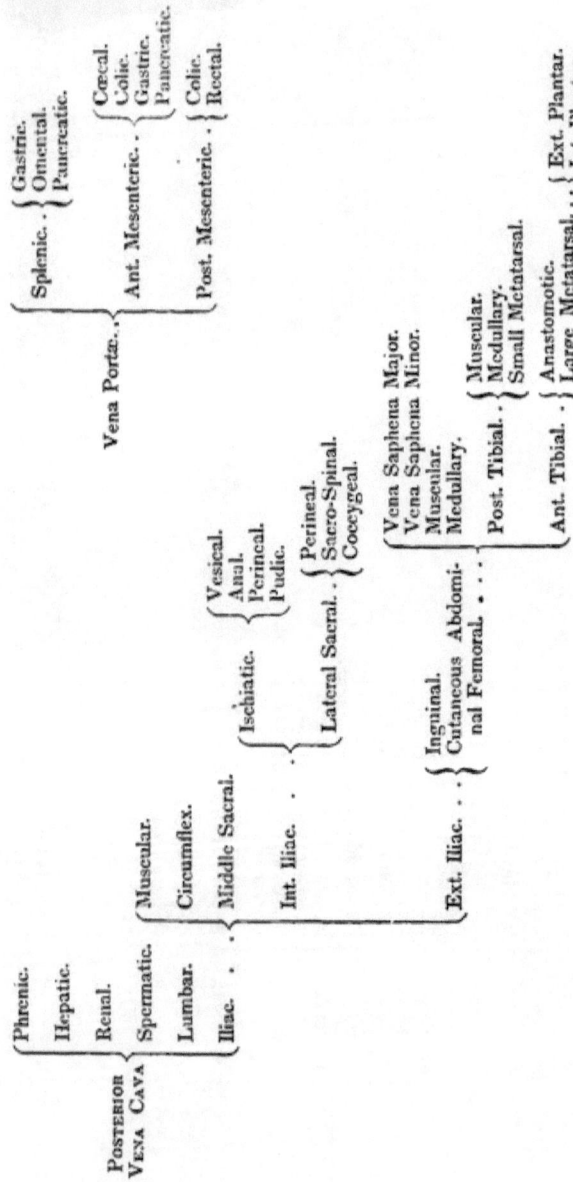

POSTERIOR VENA CAVA
- Phrenic.
- Hepatic.
- Renal.
- Spermatic.
- Lumbar.
- Iliac.
  - Muscular.
  - Circumflex.
  - Middle Sacral.
  - Int. Iliac.
    - Ischiatic.
      - Vesical.
      - Anal.
      - Perineal.
      - Pudic.
    - Lateral Sacral.
      - Perineal.
      - Sacro-Spinal.
      - Coccygeal.
  - Ext. Iliac.
    - Inguinal.
    - Cutaneous Abdominal Femoral.
      - Vena Saphena Major.
      - Vena Saphena Minor.
      - Muscular.
      - Medullary.
    - Post. Tibial.
      - Muscular.
      - Medullary.
      - Small Metatarsal.
    - Ant. Tibial.
      - Anastomotic.
      - Large Metatarsal.
        - Ext. Plantar.
        - Int. Plantar.

Vena Portæ.
- Splenic.
  - Gastric.
  - Omental.
  - Pancreatic.
- Ant. Mesenteric.
  - Cœcal.
  - Colic.
  - Gastric.
  - Pancreatic.
- Post. Mesenteric.
  - Colic.
  - Rectal.

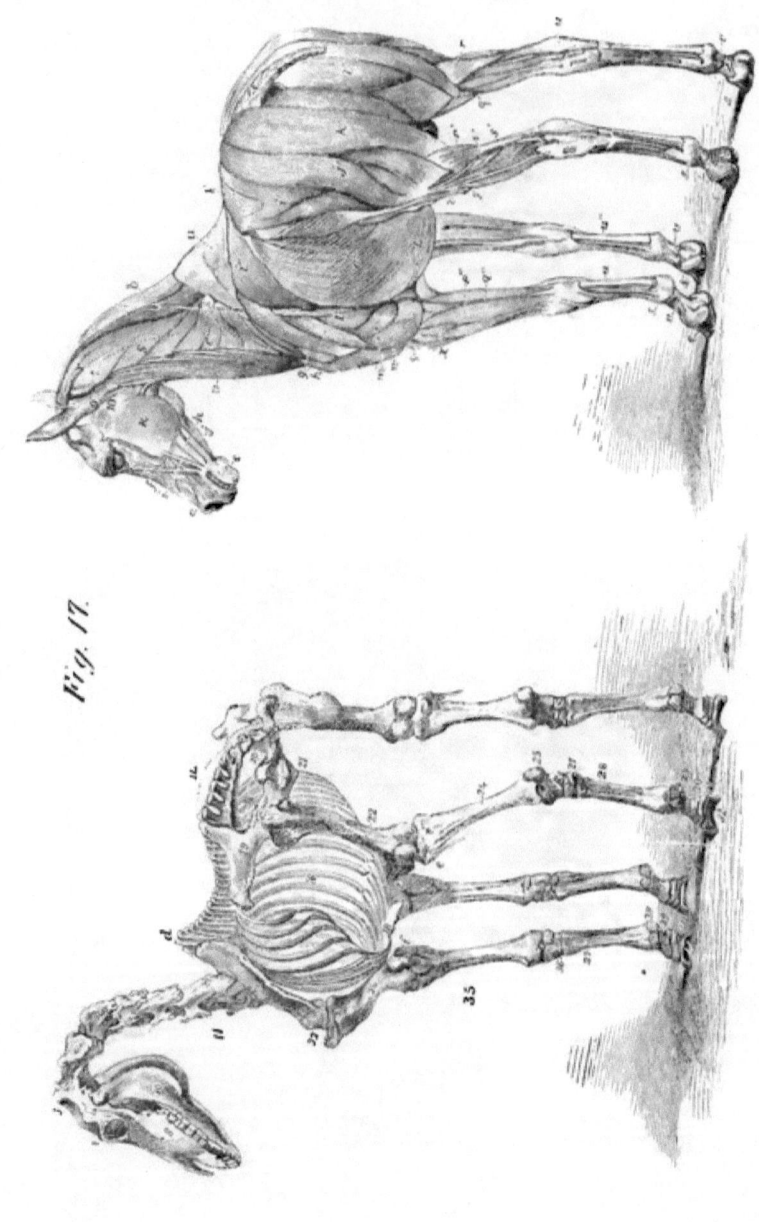

Fig. 17.

EXPLANATION OF FIGURE XVII.

OSSEOUS STRUCTURE.

1. Frontal bones.
3. Occipital.
4. Temporal.
8. Superior maxillaris.
10. Inferior   "
11. Cervical vertebræ.
14. The sacrum.
16. The false ribs.
18. The sternum.
19. The ileum.
20. The ischium.
21. Pubis.
22. Femur.
23. Patella.
24. Tibia.
e. Fibula.
25. Os calcis.
26. Tarsal bones.
27. The inferior tarsal bones.
28. Metatarsi magnum
29. Sessamoides.
30. Os suffraginis.
31. Os coronæ
32. Os pedis.
33. Scapula.
34. Os humeri
35. Radius.
f. Os ulnaris.
36. The carpal bones.
37. Metacarpi magnum.
38. Sessamoids.
39. Os coronæ.
40. Os suffraginis.
41. Os pedis.
d. Dorsal spines.

MUSCULAR STRUCTURE.

THE HEAD.

a. Orbicularis palpebrarum.
c. Dilator naris lateralis.
e. Orbicularis oris.
f. Nasalis longus.
h. Buccinator.
j. Depressor labii.
k. Masseter.
10. Region of the parotid gland.
o. Adducens vel deprimens aurem.

EXPLANATION OF FIGURE XVII. CONTINUED.

THE NECK.

b". Cervical ligament — Ligamentum colli.
c". Trachelo subscapularis (scalenus).
s. Splenius.
r, t. Tendon of the splenius and complexus major.
n. Levator humeri.

SUPERIOR PART OF THE SHOULDER AND BACK.

l". Latissimus dorsi.
u". Trapezius.

SHOULDER AND FORE EXTREMITIES.

g, k". Spinatus muscles.
l"', m', n'. Triceps extensor brachii.
s", x". Extensors metacarpi.
p", q". Flexors externus et internus.
u, n, v, n. Flexors perforans et perforatus.
s. The pastern.
&, &. Hoofs.

ABDOMINAL REGION.

c"'. Abdominis transversalis.
4. Subcutaneous thoracic vein.

POSTERIOR EXTREMITIES.

c'. Erector coccygis.
g'. Compressor coccygis.
i'. Glutei.
J'. Triceps.
k'. Biceps abductor.
l'. Abductor tibialis internus.
r', r, s'. Gastrocnemii.
v. Peroneus.
q'. Extensor metatarsi internus.
u. Insertion of the gastrocnemius.
r. Flexor perforans et perforatus.
x, y. Fleshy belly of the extensors.
s. The pastern.
&, &. Hoofs.

THE BRAIN AND ITS APPENDAGES.

The cerebrum, cerebellum, and medulla oblongata, and medulla spinalis are invested with three membranes: the dura mater, pia mater, and tunica arachnoides. Of these the exterior is the dura mater; which, though called a membrane, is of a dense, tough, and inelastic texture. It is so firmly adherent, by means of numerous little prominences, to the sutures of the cranium, that it is difficult to separate them; this membrane is to the internal cranium what the pericranium is externally. The inner surface of the dura mater is lubricated by a fluid furnished by its own blood-vessels.

The pia mater is that membrane which closely envelopes the substance of the brain, and dips down between its convolutions, and adheres to its surface by numberless minute blood-vessels. It differs in its appearance and texture from the dura mater; presenting a smooth surface exteriorly, but a rough and villous one next to the brain, and being composed of a beautiful network of blood-vessels united together by a delicate cellular tissue.

The third membrane has been compared to a spider's web, in allusion to which the name of *membrana arachnoides* has been given to it.

The arteries which supply the brain with blood are the two vertebrals, besides two other branches called the internal carotids. Its blood is returned from the sinuses of the dura mater by the vertebral and jugular veins.

It is on the supply of the vertebral arteries, however, that the brain mainly depends for its supply, for, if ligatures are placed on these arteries, the animal dies; whereas, both the carotids may be tied without occasioning any apparent ill effects. If our memory serves us, Dr. J. C. Warren, of this city, has performed the latter operation on the human subject with success.

In raising the bony covering of the brain, we meet with two processes, called the falx, or longitudinal process, and the tentorium, or transverse process. The former resembles the blade of a scythe, hence its name. These processes are formed from duplicatures of the dura mater; the first descends for a short distance between the lobes of the cerebrum. It takes its rise from the crista galii, and terminates on the os occipitis.

The tentorium is extended from the inner plate of the os occipitis along the sides of the cranium to its base, whence it may be traced to the sphenoid bone, and is lost in the common covering of the dura mater. It is composed of two laminæ: one is continuous with the falx; the other forms that portion of the membrane which covers the cerebellum. The tentorium is equally divided by the falx into two lateral portions.

The sinuses. — The superior, or longitudinal sinus, runs within the duplication of the falx, along its superior border.

The two lateral sinuses are formed within the duplicature of that part of the tentorium which is attached to the temporal and occipital bones; one extending to the right, and the other to the left. They receive veins both from the cerebrum and cerebellum.

The cavernous sinuses, so named from the cavernous appearance of their interior; they receive some important nervous trunks in their passage from the brain, and for lodging the terminations of the internal carotid arteries. They commonly communicate with the sub-occipital sinus; these are also of membranous formation, and are found upon the cuneiform process of the os occipitis, running longitudinally to the foramen magnum. They receive veins from the cerebellum and posterior parts of the cerebrum.

Cerebrum. — The largest portion of the cerebral mass, and that which presents itself to our view in raising the skull, is the cerebrum. It is equally divided by a longitudinal fissure along its middle, into which the falx cerebri descends; and its divisions, which are symmetrical, both internally and externally, are denominated hemispheres.

Cerebellum. — The cerebellum is at once distinguished from the cerebrum by its being only one-sixth the size of the latter. Its figure is irregular; it has two oval ends, and its lateral dimensions exceed its longitudinal. It is divided into three oblong lobes — a middle and two lateral.

Medulla oblongata, the smallest division of the cerebral mass. It rests on the cuneiform process of the occipital bone, and is continued upward and backward to the foramen magnum.

The upper surface of the medulla oblongata forms, with the tuber annulare, the floor of the fourth ventricle.

Pituitary gland, a red body, of an egg form, seated upon the sella tursica, within a fold of the dura mater. It has a membranous capsule, surrounded by cellular adhesions, by which it is firmly retained in its place.

Medulla spinalis. — The spinal marrow is that extended portion of brain-like substance which is continued from the posterior part of the medulla oblongata through the entire length of the spinal canal. It is inclosed in the same membranes that envelop the brain; but, in addition to them, the superior ligament of the spine serves as a covering and defence to it below. To this, and to the periosteum lining the canal, its proper theca is loosely attached by cellular, adipose, and gelatinous matter. Its dura mater is derived from that which covers the brain; in being continued through the foramen magnum, the membrane is contracted into a cylindrical sheath which loosely encases the marrow, and is generally described under the denomination of *theca vertebralis*. The arachnoid membrane and pia mater have the same relation to the marrow that the same membranes have to the brain, of which they may be considered prolongations.

ORIGIN AND DISTRIBUTION OF THE NERVES.

The nerves, being symmetrical in number and distribution on either side of the body, take their origin in pairs, and these pairs are numbered, and so distinguished from one another, according to the order in which they arise. There are forty-six pairs of nerves; ten, coming from the brain, are distinguished as the cerebral nerves; thirty-six, from the spinal marrow, denominated the spinal nerves.

CEREBRAL NERVES.

First pair, or olfactory nerves, arise from the corpora striata along the posterior borders of which bodies the medullary bands or roots of them may be traced as high up as the middle lobes of the cerebrum. These are the largest of the cerebral nerves, are bulbous at their origin, pulpy in texture, and exhibit, when cut into, comparatively to their size, large cavities, which are walled in by a layer of medullary matter, enclosed within a thinner one of cortical substance.

Second pair, or optic nerves, arise from the thalamia nervorum. They leave the cranium through the optic foramen, and pass to enter the globe of the eye, within the interior of which it expands, and forms the retina. In its whole course, it is enclosed within a sheath prolonged from the dura mater.

Third pair, or motores oculorum, take their origin by several filaments, from the inward parts of the crura cerebri. The trunk of the nerve first runs obliquely outward, across the back of the crus, then turns downward and enters the cavernous sinus, on through the foramen lacerum-orbitale. In entering the cavity, the nerve divides into two branches. The smaller is generally received by the levator oculi. The larger branch subdivides into several others; the longest of these runs round the eyeball, and penetrates the oblique muscle. Two or three others run to the abductor and depressor muscles.

Fourth pair of pathetic. — These take a filamentous origin, and pass the border of the tentorium, entering the cavernous sinus, from thence to the orbit. Its destination is the superior oblique muscle of the eye.

Fifth pair, or par trigemini. — These are

the largest nerves of the brain. They take their origin by filaments from the crura cerebelli, and pierce the dura mater. Each nerve appears to form a ganglion; from this ganglion we say that three nerves depart. One is called the ophthalmic; the second, the anterior maxillary; the third is the posterior maxillary nerve. The ophthalmic nerve is the smallest of the three divisions; as it emerges from the orbit, it divides into three branches, called the lachrymal, the super-orbitar, and the lateral nasal branch.

The second division, or anterior maxillary nerve, leaves the cranium through the hole called foramen rotundum, of the sphenoid bone, and takes its passage through the inferior orbital canal, whence it emerges, covered by the levator labii superioris, upon the face: here it splits into several large branches, denominated the facial nerves. But prior to its entering this canal it detaches several important branches to the eyelid, lachrymal duct; also several long filaments, which descend on the tuberosity of the anterior maxilla, penetrate the bone, and furnish twigs to the antrum, and the two superior molar teeth. The largest branch is the spheno-palatine, or lateral nasal nerve, to which the foramen sphenopalatine gives passage into the nose, wherein it divides into two sets of filaments. One of these is spread over the lateral parietes of the nasal cavity; the other ramifies over the sinuses, and sends a filament to the lower border of the septum. A branch also goes to the velum palati, and another branch accompanies the palatine blood-vessels, and ramifies over the soft palate.

The facial branches of this division terminate on the front and sides of the face, and receive communicating filaments from the anterior facial branch of the portio dura, and with them form a plexus.

The third division, the posterior maxillary nerve, gives off a branch which runs up in front of the parotid gland, and joins the portio dura; also branches called the buccal nerve, pterygoideus, and gustatory. The latter descends by the side of the tongue, penetrates that organ about its middle, and vanishes in its tip. It also sends ramifications to the roots of the incisive teeth, and to the under lip.

Sixth pair, or abducentes, arise by means of filaments from the medulla oblongata: this nerve gives off two or three filaments to the retractor oculi; but its principal destination is to the abductor, along the fascicula of which, its ramifications are equally distributed.

Seventh pair, or auditory nerves. — This pair includes two separate nerves on either side; one, from its remarkable softness, is denominated the portio mollis; the other, in contradistinction, the portio dura. The portio mollis enters the organs of hearing, and is distributed to the labyrinth.

The portio dura arises from the medulla oblongata, and passes to the internal part of the ear, the tympanum, and eustachian tube. It is also distributed to the temples, eyelids, nose, lips, cheek, and neck.

Eighth pair, or par vagum. — At its commencement it consists of two separate portions; the first called the glosso-pharyngeal nerve, and the second the true par vagum. They arise from the corpora olivaria, and make their exit through the base of the cranium. The glosso-pharyngeus gives off branches, which join the portio dura, to the constrictors of the pharynx, and form branches which ramify in the base of the tongue.

The proper par vagum, having disunited from the glosso-pharyngeal nerve, proceeds downwards to join the carotid artery, and takes its course along the neck to the chest. Its filaments are — 1. To the cervical ganglion. 2. The pharyngeal branch, whose filaments pass to the esophagus and larynx. 3. Two slender branches to the carotid artery, which form a plexus. 4. The laryngeal branch.

At the back part of the neck the par vagum inclines upwards, and is found above the carotid artery; it then passes between the two first ribs into the chest. Having entered the thoracic cavity, it runs within the superior mediastinum; the right nerve adheres to the trachea, crosses above the root of the right lung, alongside of the œsophagus and gains the under side of that tube before it leaves the chest. On the left side the nerve accompanies the anterior aorta, and crosses the root of the posterior aorta, and also reaches the œsophagus. Its branches within the chest are filaments to the tracheal and cardiac plexuses; also, a branch called the recurrent nerve; branches to the pulmonary plexus, and also two cords that branch out and penetrate the walls of the auricles. The recurrent nerve of the left side originates from the par vagum, by the side of the anterior aorta, and coils round the root of the posterior aorta.

The recurrent nerve, so denominated from its retrograde course, passes upwardly and outwardly, and is found between the carotid artery and the trachea; having reached the top of the latter, it spreads into fine terminating branches, several of which run to the muscles of the larynx and thyroid cartilage, and end in ramifications upon the membrane of the glottis. Its branches are filaments to the pulmonary plexus, cardiac plexus, posterior cervical ganglion, and branches to the œsophagus and trachea.

The par vagum runs to the stomach. The left nerve sends filaments to the heart, and others along the small curvature, which communicates with the ramifications of the right nerve; the other crosses to the left side, and joins the great semilunar ganglion. The right nerve, as soon as it reaches the heart, divides into numerous branches, which join the left, and spread their ramifications upon the under part of the heart; some run to the pylorus, and others join the hepatic plexus.

Accessory nerves to the eighth.—These nerves are considered as accessory to the eighth, in consequence of their being found in close connection in issuing from the cranium; it originates in the vertebral canal, by the union of several filaments. In its course into the cranium it receives many other fine threads, and in that cavity joins the par vagum. Beneath the atlas, the accessory nerve divides; the front division runs downward, and penetrates the belly, transmitting side twigs in its course. The posterior division turns round the transverse process of the atlas to the scapula, near which it is lost in muscular substance. The branches of the accessory pass to the par vagum, anterior cervical ganglion, and communicate with the sub-occipital nerve.

Ninth pair, or linguales, arise behind the eighth pair, from the corpora olivaria; it is found in company with the par vagum, near the coronoid process. The nerve passes down the lower jaw, between the muscles forming the root of the tongue, and ends in the tip of the latter. It sends branches to the lingual muscles and to the hyo-glossus longus.

Tenth pair, or sub-occipital nerves.—They arise from the medulla oblongata, and beginning of the spinal marrow; they pass out through a hole in the fore part of the body of the atlas. It then branches into a superior and inferior division. The superior is distributed to the extensor muscles of the head and neck. The inferior branch goes to the trachea, lymphatic glands, and muscles of the neck.

CERVICAL NERVES.

These consist of seven pairs, originating from the cervical portion of the spinal marrow. Each nerve, as soon as it issues from the spinal canal, forms two nervous filaments, one superior, the other inferior.

The first cervical nerve makes its exit between the first and second cervical vertebræ. It sends branches to different muscles, and communicates with the

Second cervical nerve, which makes its appearance between the second and third vertebræ. Its superior filament sends branches to the muscles of the neck, and levator humeri, communicates with the accessory nerve, and

Third cervical.—This also sends branches and twigs to the different muscles of the neck, and communicates with the fourth.

The fourth, fifth, sixth, and seventh pairs

pass from the spine, between their respective vertebræ, and send branches to the phrenic nerve, and ramifications to the muscles, sympathetic nerve, and unite with the dorsal.

THE DIAPHRAGMATIC OR PHRENIC NERVE.

This is formed by branches from several of the cervical nerves. It takes its course down along the inferior border of the scalenus muscle. It terminates by numerous ramifications on the tendinous parts of the diaphragm.

DORSAL NERVES.

These consist of eighteen pairs. They pass from the vertebral canal in the same manner as the cervical, having superior and inferior branches. The inferior branches follow the course of the intercostal blood-vessels, and are called intercostal nerves. The superior branches are distributed to the back and loins.

LUMBAR NERVES,

Consist of five pairs (corresponding to the number of the lumbar vertebræ).

The first nerve ends in ramifications near the stifle, and gives off branches to the last dorsal nerve, to the sympathetic, and to the second lumbar nerve.

The second nerve has communication with the first nerve, and sympathetic; also the crural. It sends one division to the fore part of the haunch, where it becomes subcutaneous, and ramifies over the stifle. The other division crosses the ilio-lumbar artery, just below its origin, and takes nearly a similar course to the inward part of the haunch, and then ramifies upon the skin; in its way it detaches a considerable branch, called the spermaticus externus, which passes through the abdominal ring, and sends twigs, in the male, to the scrotum and testicle; in the female, filaments go from it to the uterus, udder, and external labia.

The third nerve contributes to form the crural and obturator. It sends small branches to the sympathetic, psoas, and obturator nerves.

The fourth nerve sends a branch to the sympathetic, contributes to the production of the crural; and also sends a branch to the obturator.

The fifth nerve communicates with the sympathetic, crural, and sciatic plexus.

SACRAL NERVES,

Consist of five pairs; a superior and inferior fascicula. The superior make their exit through holes upon the upper part of the sacrum, and are there buried under a thick mass of muscle, and become cutaneous upon the outer part of the haunch.

The inferior fascicula. — The first nerve largely contributes to the origin of the sciatic plexus, and sends a branch to the gluteal nerve; also to the sympathetic and second lumbar nerves. The second nerve communicates with the third and sympathetic, and sends branches to the surrounding muscles and sciatic plexus. The third and fourth have similar connections. The fifth passes into the coccygeal muscles.

COCCYGEAL NERVES.

These issue from the spine, in the same manner as the last described. They communicate with one another, are distributed to muscles in the vicinity, and end in filamentous ramifications at the end of the tail.

NERVES OF THE FORE EXTREMITY.

The fore extremity receives its nerves from the axillary or humeral plexus, and this plexus is formed by the union of portions of the sixth and seventh cervical nerves, and a division of the first dorsal nerve.

The external thoracic nerves, six or seven in number, arise from the humeral plexus, and are distributed to the pectoral, triceps, and other muscles; they finally ramify into the skin.

The scapular nerves are called anterior, posterior, and sub-scapular. The former sends its ultimate filaments to the triceps.

The posterior scapular nerve sends branches to the sub-scapularis, triceps, teres minor, and shoulder joint, and ends in the insertion of the levator humeri.

The *subscapular nerves* run upward between the shoulder and chest, and enter the subscapularis.

The *spiral or external cutaneous nerve* is furnished by the axillary plexus; arises behind the humeral artery, and passes between the os humeri and the head of the triceps, through the extensors, to the external flexors of the canon. It gives off several branches to the triceps, ramifies on the fore and outward part of the knee, and sends branches to the heads of the extensor muscles.

The *radial nerve* descends with the humeral artery to the inward side of the elbow joint, and runs along the back part of the radius to the knee; passing under the annular ligament, it descends to the leg, and takes the name of the internal metacarpal nerve. It gives off numerous twigs to the muscles, and finally becomes subcutaneous.

The *ulnar nerve* originates from the humeral plexus. It passes down the radius, under the annular ligament, to the tendo perforans, and there becomes the external metacarpal nerve. It gives off internal cutaneous and subcutaneous branches, ramifies into cellular substance, penetrates the heads of the flexors, and finally disperses its ramifications in front of the leg.

The *metacarpal nerves* continue down the leg, over the fetlock joint, where they become the plantar nerves; these pursue their course behind their corresponding blood-vessels to the back part of the foot, which they penetrate to the inner side of the lateral cartilages.

The *plantar nerve* detaches a branch from the fetlock to the lateral cartilage; another passes to the fatty frog. The final branch enters a hole in the back and lower part of the coffin bone, in company with the plantar artery, and there divides and distributes its ultimate branches around the edges of the sole.

NERVES OF THE HIND EXTREMITY.

The *crural nerve* is derived partly from the second, third, fourth, and fifth lumbar nerves. It makes its appearance under the transverse process of the loins, and proceeds in a line with the external iliac artery. It gives off filaments to the psoas magnus, iliacus, rectus, and vastus internus muscles. It also gives off cutaneous filaments; one runs to the stifle, and ends in ramifications upon the fore part of the thigh. The other continues down the leg, and can be traced as low as the fetlock.

The *obturator nerve*, contributed to by third and fourth lumbar nerves, sweeps round the brim of the pelvis, and detaches twigs to the obturator muscles. Its ultimate filaments are expended on the triceps and gracilis.

The *gluteal nerve*, after leaving the cavity of the pelvis, accompanies the gluteal artery, and passes into the substance of the gluteal muscles.

The *sciatic nerve* derives its origin from the sacral and last of the lumbar nerves; after leaving the cavity of the pelvis, passes between the hip joint and the tuberosity of the ischium, and plunges into the substance of the haunch. Here it divides into branches called the popliteal nerves. At the hock its principal branch separates into the external and internal metatarsal nerves; the former runs over the flexor pedis to the os calcis. Their subsequent course and ultimate distribution are the same as those of the plantar nerves of the fore extremity. The second popliteal nerve passes between the bellies of the gastrocnemii, above the first, detaching twigs to them in its passage, and then spreads into many branches, which penetrate the heads of the flexor muscles of the foot, and send filaments into the stifle joint.

SYMPATHETIC NERVE.

This nerve derives its name from the universal influence which it has on the nervous system. It communicates with the head, neck, chest, pelvis, and abdomen, by its frequent intercourse and connection with their respective nerves. It is supposed by some writers to be a nervous system of itself. It has, at different distances, a great number of gangliform tubercles, from which ramifications proceed forward, as well as filaments backward, to the ganglia of the nerves of the medulla spinalis. It is considered generally as beginning from a branch of the fifth and sixth pair, given off at the base of the cranium. The ganglionic structures and the different plexuses are named from their form, location, and distribution; hence we have the cervical ganglion, semilunar, sacral, etc. From the semilunar ganglion nervous filaments shoot in various directions, which, from their being compared to the rays of the sun, are denominated the solar plexus. From the divergent filaments of the latter, the several smaller plexuses of the abdomen may be said to derive their formation, taking names according to the viscera they are particularly designed to furnish with nerves; hence we have the splenic plexus, that sends filaments to the spleen, the hepatic plexus, mesenteric, aortic, hypogastric, and renal plexuses. The sympathetic nerve in the abdomen travels over the sides of the bodies of the lumbar vertebræ, below the articulations of the ribs, and pursues its course into the pelvis. Here, also, it forms ganglia, which correspond in number to those of the lumbar nerves; and from every ganglion come off two filaments: one which runs to the corresponding lumbar nerve; the other crosses the aorta, and, by joining the aortic plexus, communicates with nerves coming from the sympathetic of the other side.

From the loins, the sympathetic descends into the pelvis, and takes its course along the side of the sacrum, and forms five ganglia, corresponding to the sacral nerves; it finally terminates by forming a union with its fellow.

EXAMINATIONS ON NEUROLOGY.

Examinations on Neurology, which will include the names of parts not alluded to in the preceding summary of the nervous system.

NERVES.

Q. What are nerves? — *A.* Long, firm, and white chords, which ramify after the manner of blood-vessels, and are distributed to all parts of the horse's body.

Q. Where do they arise? — *A.* From the brain, medulla oblongata, and medulla spinalis.

Q. What communications have the different nerves with each other? — *A.* They anastomose: forming sometimes a plexus; at others, a knot or ganglion, from which other branches arise.

Q. What is the structure of nerves? — *A.* They consist of fasciculi, or bundles, of distinct longitudinal fibres, closely connected together by cellular substance.

Q. What are the coverings of nerves? — *A.* Continuations of those which envelop the brain and spinal marrow, termed neurilema.

Q. What is the structure of ganglions? — *A.* They are formed by a close intermixture of filaments.

BRAIN AND ITS MEMBRANES.

Q. Where is the brain situated? — *A.* It occupies the cranial cavity.

Q. How is the brain divided? — *A.* Into cerebrum, cerebellum, and medulla oblongata.

Q. By what membranes is the brain enveloped? — *A.* By three membranes, or meninges: 1st, The dura mater; 2d, Pia mater; 3d, tunica arachnoides.

DURA MATER.

Q. What is the situation of the *dura mater?* — *A.* It is the external covering of the brain.

Q. How does it differ from the other coverings of the brain? — *A.* It is more dense, tough, and inelastic.

Q. How is it retained within the cranium? — *A.* It is firmly adherent to the interior of the cranium, more particularly to the depressions between the teeth of the cranial sutures.

Q. How does the internal differ from the external surface? — *A.* It has a smooth, polished, and lubricated surface.

Q. Is the dura mater supplied with nerves? — *A.* Being composed of tendinous fibre, it is supposed to be destitute of nerves.

Q. How are the processes of the dura mater formed? — *A.* By duplicatures.

Q. What are the use of the processes? — *A.* They steady and protect the various divisions of the brain.

Q. By what names are the processes known? — *A.* The longitudinal process is called falx cerebri, and the transverse ditto is called tentorium.

Q. What is the situation of the falx cerebri? — *A.* It forms a partition under the anterior and superior parts of the cranial cavity extending from the crista galli to the occiput, and ends in continuity with the tentorium.

Q. What is the situation of the tentorium cerebelli? — *A.* It is extended, after the manner of an arch, from the cerebral plate of the occipitis along the sides of the cranium to its base; whence, greatly diminished in breadth it continues onward to the os sphenoides.

SINUSES OF THE DURA MATER.

Q. What are the names of the principal sinuses of the dura mater?—*A.* The superior or longitudinal sinus; two lateral, cavernous, and sub-occipital sinuses.

PIA MATER.

Q. What is the situation of the pia mater?—*A.* It surrounds and closely invests the convolutions of the brain, and passes into the ventricles, furnishing them with an internal membrane.

Q. What is the structure of the pia mater?—*A.* It presents a smooth exterior surface; next the brain it is rough and villous, and is composed of a network of blood-vessels, which are united together by a delicate cellular tissue. Being highly vascular, it is supposed that the blood-vessels of the brain ramify in it before entering the latter.

ARACHNOID MEMBRANE.

Q. Where is the tunica arachnoidea situated?—*A.* It is a delicate and transparent membrane, spread uniformly over the surface of the brain.

CEREBRUM.

Q. Where is the cerebrum situated?—*A.* It occupies the superior part of the cranium.

Q. What is its form, and how is it divided?—*A.* It is oval, convex above and concave below, and is divided by a longitudinal fissure along its middle, into which the falx cerebri descends. Its divisions are denominated hemispheres.

Q. What is the appearance of the surface of the cerebrum?—*A.* It is covered with eminences called convolutions.

Q. Of what is the substance of the brain supposed to consist?—*A.* Of two kinds of matter; the external is called cortical or cineritious, and the internal is termed medullary.

Q. What is the color of the cortical?—*A.* Reddish-ash.

Q. What is the color of the medullary portion?—*A.* Of a milk-white hue.

CORPUS CALLOSUM.

Q. What is the situation of the corpus callosum?—*A.* It is an oblong white body, located at the bottom of the fissure which divides the two hemispheres of the brain.

Q. What does the corpus callosum join on each side?—*A.* Its edges blend with the medullary substance of the two hemispheres of the cerebrum.

Q. What name is given to the medullary substance of both hemispheres, together with the corpus callosum, when the usual anatomical section is made?—*A.* By cutting off the hemispheres of the cerebrum nearly even with the corpus callosum, there is seen a large oval mass of medullary substance, called the centrum ovale.

LATERAL VENTRICLES.

Q. What are the lateral ventricles?—*A.* Two cavities situated beneath the corpus callosum and medullary arches of the cerebrum.

Q. What divides the lateral ventricles from each other?—*A.* The septum lucidum.

Q. Name the parts which are generally considered as the contents of the lateral ventricles.—*A.* They are the corpora striata, the hippocampi, plexus choroides, fornix, and the thalami nervorum opticorum.

Q. What is the situation and form of the corpora striata?—*A.* They are found on the lower and back parts of the ventricles, projecting into the centre of the cavities, where they expand as they approach the septum; grow narrower and recede from each other above; below, they extend to the anterior cornua.

HIPPOCAMPI.

Q. What is the situation of the hippocampi?—*A.* They occupy the superior spaces of the ventricles in contact with the septum.

Q. From whence do they originate?—*A.* From the centres of the hemispheres.

Q. What is their structure?—*A.* They consist of alternate laminæ of medullary and cortical matter.

PLEXUS CHOROIDES.

Q. What is the situation of the plexus choroides?—*A.* They are situated in the channel between the corpus striatum and hippocampus.

Q. Describe the appearance of the same?—*A.* It is a soft vascular substance, consisting of a plexus of minute blood-vessels; it makes its appearance from behind the fornix, and ends abruptly in a round bulbous mass.

FORNIX.

Q. Describe the fornix and its situation?—*A.* The fornix is that part which receives the posterior border of the septum lucidum. It is extended after the manner of an arch, between the corpora stratia below and the heads of the hippocampi above, where it forms a junction with the corpus callosum.

Q. Describe the processes or crura of the fornix?—*A.* The two inferior crura spring from the corpus albicantium, at the base of the brain, and finally unite; thus united, they appear within the ventricles and constitute the body of the fornix. The superior crura proceed from the upper end of the fornix, and descend into the superior cornua of the lateral ventricles, and end in sharp, pointed extremities.

THALAMIA.

Q. What is the situation of the thalamia nervorum opticorum?—*A.* They form the upper and back parts of the lateral ventricles.

Q. Describe the thalamia.—*A.* They have a white

appearance, conoid in form, narrow and approximated inferiorly; broad superiorly; they finally contract into medullary bands, the tractus optici, which turn round the crura cerebri to the base of the brain.

Q. How are the thalami distinguished from the corpora striata? — A. They are more dense and firmer in composition.

TÆNIA.

Q. What is the situation of the tænia? — A. They are located in the groove between the thalamus and corpus striatum, partly covered by the plexus choroides.

COMMISURES.

Q. Name the commisures of the brain. — A. 1st, commissura mollis; 2d, commissura inferior cerebri; 3d, commisura superior cerebri.

Q. How is the commisura mollis formed? — A. By contiguous parts of the thalami, which are united by cortical matter.

Q. How is the commisura inferior cerebri formed? — By a connection between the hemispheres of the brain.

Q. Where is the superior commissure located? — A. Above the commissura mollis; it has the appearance of a short medullary chord.

FORAMEN.

Q. What is the foramen? — A. It is a triangular depression under the arch of the fornix, into which the lateral ventricles open.

VENTRICLES.

[*Remarks.* — Having put the usual questions regarding the lateral ventricles, which may be numbered 1 and 2, we now come to the third ventricle, which is not so well marked as in the human subject.]

THIRD VENTRICLE.

Q. How is the third ventricle formed? — A. By a mere fissure existing between the thalami.

[*Remarks.* — The fourth ventricle, being located in the cerebellum, will be considered under this head.]

INFUNDIBULUM.

Q. Where is the infundibulum located? — A. At the inferior part of the third ventricle.

PINEAL GLAND.

Q. Where is the pineal gland located? — A. Between the summits of the thalami, over the third ventricle, and above and before the superior commissure.

Q. Describe the pineal gland. — A. It is a small conoid body, of grayish color, marked by a slight depression along its centre.

Q. What are its attachments? — A. It is attached by means of the pia mater to the thalamia and tubercula quadragemina.

Q. What is the internal structure of the pineal gland? — A. It consists of cortical and granular matter.

NATES AND TESTES.

Q. Where are the nates and testes situated? — A. Above the third ventricle, behind the pineal gland, and immediately over and within the third and fourth ventricles.

Q. How do the nates differ from the testes? — A. The former are larger than the latter, and are separated by a groove from the testes, and by a deep perpendicular fissure from each other.

Q. What is their form? — A. Semi-oval.

Q. What is their composition? — A. They are composed of cineritous and medullary matter.

CEREBELLUM.

Q. What is the situation of the cerebellum? — A. In the inferior and posterior parts of the cranium.

Q. How does the cerebellum compare in size with the cerebrum? — A. The former is only about one-sixth the volume of the latter.

Q. Describe the appearance of the cerebellum? — A. Its surface is lobular and convoluted; its form is irregular, having two oval ends placed transversely, united in the centre by a broad vermiform belt; its lateral dimensions exceed its longitudinal.

Q. How is the cerebellum divided? — A. Into three lobes, a central and two lateral.

Q. How does the composition of the cerebellum differ from that of the cerebrum? — A. In the former the cortical substance exceeds the medullary, and, instead of forming the bulk of the outer parts, as is the case in the cerebrum, it pervades the inner.

FOURTH VENTRICLE.

Q. What is the situation of the fourth ventricle? — A. It is situated between the cerebellum, tuber annulare, and medulla oblongata.

Q. Where is the choroid plexus of the cerebellum situated? — A. Within and across the posterior part of the fourth ventricle, between the cerebellum and medulla oblongata.

Q. How is the choroid plexus of the cerebellum distributed? — A. It is distributed into three divisions: one lies in the middle of the calamus; the two latter are found within fissures in the cerebellum, occupying the spaces between it and the tuber annulare.

BASE OF THE BRAIN.

Q. How is the base or posterior part of the cerebrum divided? — A. It is divided into six lobes.

Q. Describe their divisions. — A. There are two anterior or inferior, resting upon the wings of the ethmoid bone; two middle, upon those of the sphenoid; and two superior or posterior, lodged in the fossa of the squamous portions of the temporal bones.

Q. What name is given to two broad, smooth prominences which appear over the middle lobes at the base of the brain? — A. These are the corpora striata.

Q. What nerves originate from this vicinity? — A. The olfactory nerves.

Q. What lobes rest on the wings of the sphenoid bone? — *A.* The crura cerebri.

Q. From whence do they arise? — *A.* From the inferior and middle lobes of the cerebrum, and are continued into an ovoid protuberance above them, named tuber annulare.

Q. What is observable between the crura cerebri? — *A.* A small hemispherical medullary eminence, called corpus albicantium.

Q. Where are the tractus optici situated? — *A.* They wind obliquely downward around the crura.

Q. Where do they proceed from? — *A.* From the terminations of the thalamia.

Q. What is the situation of the crura cerebelli? — *A.* They are located higher up and in a more outward direction than the crura cerebri.

Q. Describe the crura cerebelli. — *A.* They are two cylindroid, medullary chords, which join the lateral lobes of the cerebellum to the tuber annulare.

Q. What does the tuber annulare rest upon? — *A.* On the cuneiform process of the posterior occipital bone.

Q. Where are the foraminæ cæca situated? — *A.* Above and below the tuber.

Q. What is their appearance? — *A.* They are described as little, round depressions, or blind holes.

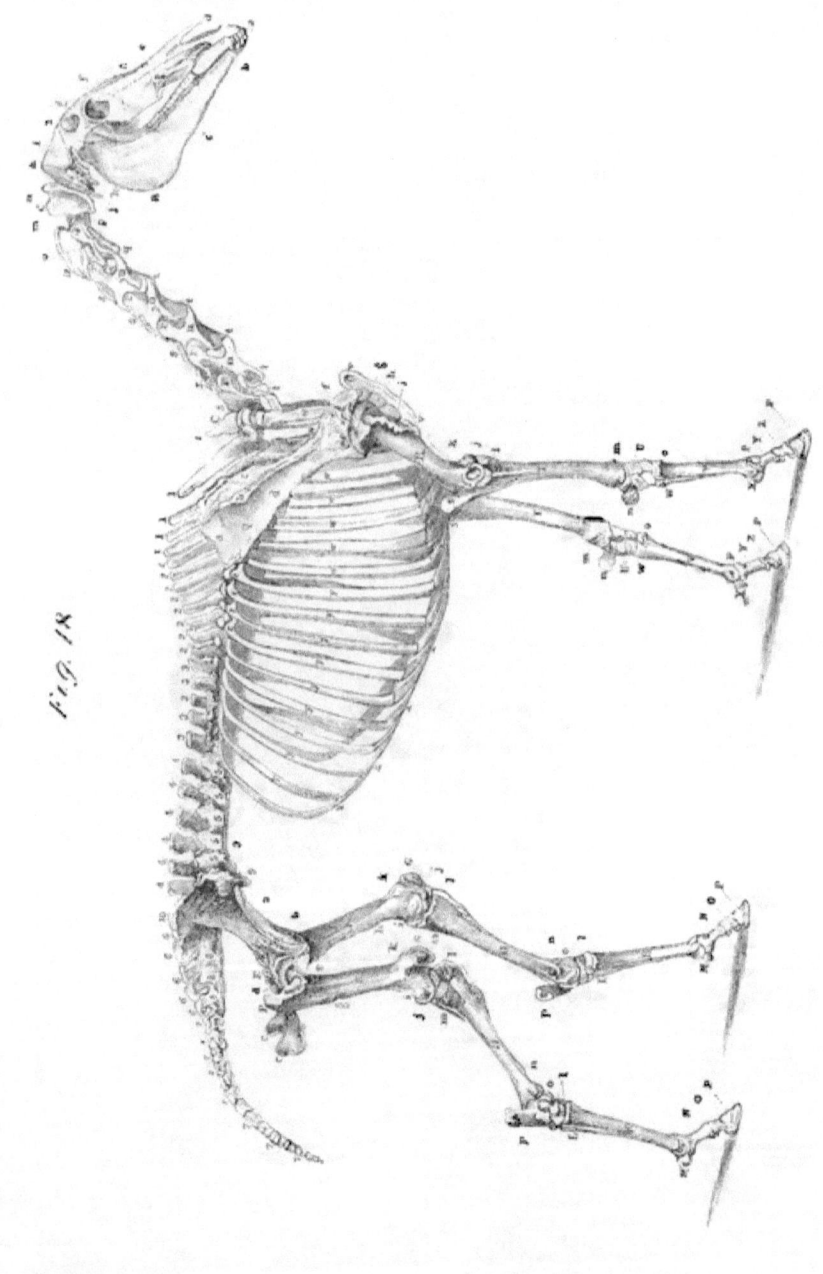

EXPLANATION OF FIGURE XVIII.

[FROM BLAINE'S "OUTLINES."]

THE HEAD.

A. The skull, face, and upper jaw, in one piece.
B. Lower jaw.
a. Incisor teeth.
b. Tushes.
c. Molares, or grinders.
d. Peak formed by the extremities of the nasal bones.
e. Zygomatic spine, to the bottom of which the masseter takes its origin.
f. Orbit.
g. Cavity above the orbital arch.
h. Pole.
i. Zygomatic arch.
j, j. Styloid processes for the attachment of the muscles.
k. Joint formed by the upper and lower jaws.
l. Meatus auditorius, or opening to the internal ear.

THE NECK.

C, C. Marks the extent of the cervical vertebræ.
D. Dentata.
m. Atlas.
n. Wing of the atlas.
o. Large superior spine of the dentata.
p. Body of the dentata.
q. Inferior spine of the dentata.
s, s, s, s, s. Superior spines of the five remaining cervical vertebræ.
r, r, r, r, r. Oblique processes of the five last cervical vertebræ.
u, u, u, u, u. Transverse processes of the same bones.
t, t, t, t, t. Inferior spines of the five last cervical vertebræ.

THE THORAX.

v, v. Cariniform process of the sternum.
w, w, w, w, w, w, w. Costæ or true ribs.
y, y, y, y, y, y, y, y, y. Ribs as distinguished from the costæ.
x, x, x, x, x, x, x. Cartilages by means of which the ribs are attached to the sternum.
z, z, z, z, z, z, z. Heads of the ribs.
1, 1, 1, 1, 1. Superior spines of the first five dorsal vertebræ, the fifth being generally the longest spine in the body.
2, 2, 2, 2, 2, 2, 2, 2. Superior spines from the sixth to the thirteenth, towards which they slope downward; the thirteenth is generally the most upright spine in the dorsal region.
3, 3, 3, 3, 3. Last five of the superior of the back spines, which have an inclination forward.

THE LOINS, OR LUMBAR REGION.

4, 4, 4, 4, 4. Superior spines of the lumbar region, thicker than the dorsal spines, and having a decided inclination forward.
5, 5, 5, 5. Projecting transverse processes of the loins.

THE SACRUM.

6, 6, 6, 6, 6. Superior spines of the sacrum leaning decidedly backward, thus leaving a large space between the points of the last lumbar and the first sacral spine, at which place occurs the great hinge of the back.
8, 8, 8, 8. Bodies of the sacral vertebræ.

THE TAIL.

7, 7, 7, 7, 7, 7, 7, 7. Coccygeal bones.

THE PELVIS.

E. Ossa innominata, consisting of three bones upon each side.
a. Ilium.
b. Pubis.
c. Ischium: the three bones unite at the cavity which receives the head of the thigh bone.
9, 9. The inferior spines of the ilium.
10. Superior spine, which partly covers the first sacral spine.
e, e. Ischiatic spines.

EXPLANATION OF FIGURE XVIII. CONTINUED.

THE THIGH AND STIFLE JOINT.

F, F. Femurs.
d. Round head of the bone.
e. Short neck of the femur.
f. Great trochanter.
g. Small external trochanter.
h. Small internal trochanter.
i, i. Sulcus whence the gastrocnemii muscles originate.
J, J. Posterior condyles of the femur.
k, k. Anterior trochlea over which the patella glides.
G, G. Patellas: the interarticular cartilages of the stifle joint, as well as the cartilages tipping the dorsal lumbar sacral spines, and the superior margin of the blade bone or scapula, are necessarily omitted in this delineation, which is admirably drawn from a macerated skeleton.

THE TIBIA AND FIBULA, OR LEG BONES, AND THE HOCK JOINT.

H, H. Tibias.
l, l. Heads of the bones.
m, m. Fibulas.
n, n. Inferior head of the tibia.
I, I. Hock joint.
o, o. Astragalus.
p, p. Calcis forming the point of the hock.

THE POSTERIOR SHANK BONES.

K, K. Canons, metatarsals, or shank bones.
L, L. Splint bones.

THE BONES OF THE PASTERNS, AND FEET, OF THE POSTERIOR LIMBS.

M, M. Sessamoids.
N, N. Large pastern bone.
O, O. Smaller pastern bone.
p, p. Pedal bones.

BONES OF THE ANTERIOR EXTREMITY.

Q. Scapula or blade bone.
a. Superior margin whence the cartilage has been removed.
b. Spine of the scapula.
c. Anterior fossa of the scapula.
d. Posterior fossa.
e. Shallow cup which receives the head of the humerus: the cartilage, which is situated around the margin of this cup, and which serves to deepen it, has been destroyed by maceration.
f. Tuberosity terminating the spine of the scapula, whence the flexor brachii originates.
R. Humerus or arm bone.
g. Head of the bone.
h. Smooth cartilaginous and synovial pulley over which the tendon of the flexor brachii plays.
i. External trochanter of the humerus.
j. Inferior head of the humerus.
k. Pit into which the ulna is received.
S, S. Ulna, the top of which is termed the olecranon.
T, T. Radius.
l. Head of the bone.
m. Inferior head of the bone.
U, U. Carpus or knee joint, consisting of two rows of bones.
n. Trapezium, which gives security to the great flexors, and attachment to several of the lesser flexors of the fore leg.
V, V. Canon or shank bone.
o, o. Head of the bone receiving the lower row of the bones of the knee.
W, W. Splint bones.
p. Inferior head of the canon bone.
X, X. Sessamoid bones.
Y, Y. Large pastern bone.
Z, Z. Small pastern bones.
P. Pedal or coffin bone.

DISTRIBUTION OF THE LYMPHATICS.

Mr. Percivall remarks, in his lectures, that "no English veterinarian has, up to the present day (1820), been at the pains to demonstrate, practically, the particular distribution of the absorbing vessels of the horse. Professor Girard, whose 'Traité d'Anatomie Vétérinaire' does no less credit to the talent and industry of its author than honor to the veterinary school over which he presides, has presented us with an article on the ramification of the lymphatics, which I shall translate.

"THE THORACIC DUCT.

"The largest, longest, and most remarkable of the lymphatic vessels, in which terminate the majority of the lymphatics of the body, is situated within the thorax, on the right side of the dorsal vertebræ, between the aorta and vena azygos: it receives the lymphatics from the posterior extremity, pelvis, parietes, and viscera of the abdomen, head, neck, withers, and left anterior extremity.

"It takes its origin under the loins, in a dilation or sinus situated at the root of the great mesenteric artery, and is named the receptaculum chyli: it directs its course forward, enters the thoracic cavity by the aortic perforation through the diaphragm, extends along the bodies of the dorsal vertebræ, until it arrives opposite the base of the heart, where it curves downward to cross over to the left side in its way to the anterior opening of the thorax; as it leaves the spine for this purpose, it runs over the trachea and esophagus; having reached the left side, it stretches forward to the beginning of the anterior vena cava, and terminates in the base of the left axillary vein. Not unfrequently, it ends in the right axillary; in some instances, even in the beginning of the anterior cava. At its termination, it dilates and forms a sinus, whose mouth opens into the vein, is guarded by a broad valve, so disposed as to prevent any reflux of blood into the duct.* It has also a ligamentous band around it, at this part, which confines it to the vein receiving its contents.

"THE RECEPTACULUM CHYLI.†

"This reservoir forms the point of general confluence of all the lymphatics of the posterior limbs and abdomen, and from which originates the thoracic duct. It is maintained by the aorta on one side, the vena cava posterior on the other, and is formed by the union of five or six large lymphatics, of which two or three come from the entrance of the pelvis, two or three others from the mesentery, a single one from the environs of the stomach and liver."

The Professor here makes a classification of the lymphatics of the body.

LYMPHATICS DISCHARGING THEIR CONTENTS INTO THE ABDOMINAL PORTION OF THE THORACIC DUCT.

"1. *Lymphatics of the Posterior Extremities.*—These are distinguishable into the superficial and deep-seated. The first originate from the skin and subcutaneous cellular tissue. They form divers ramifications, which accompany the superficial veins; of which the most remarkable attend the vena saphena major, frequently anastomosing with one another, and forming an anastomotic network. All these lymphatics run to the subcutaneous ingui-

---

* Notwithstanding this valve, blood often gains admission into the canal; this is observable in all cases of violent death, or in which struggles and convulsions attend expiration.

† Percivall's Lectures.

nal glands, which are lodged upon the superior and anterior part of the thigh.

"The deep-seated lymphatics take their rise from the foot, ascend along with the plantar veins, continue upward among the muscles, in company with the deep-seated veins, corresponding in their principal divisions to those vessels, and proceed to the inguinal glands.

"All the lymphatics of the posterior limbs assemble at these glands, and here form a plexus, from which several large branches depart and traverse the iliac glands, clinging to the sides of the iliac vessels, and discharge their contents into the pelvic branch, contributing to the receptaculum chyli.

"2. *Lymphatics of the Pelvis.* — The vessels coming from this cavity run in part to the inguinal glands, and in part to the internal pelvic glands. The superficial lymphatics about the pubes and the outlet of the pelvis run and join those of the extremities; those of the perineum and anus enter the cavity, and are accompanied by those coming from the croup and tail, both proceeding to the glands within the interior of the pelvis. All the deep-seated lymphatics accompany the veins, make for the pelvic glands, form union with the others, and run and empty themselves into the main pelvic branch, wherein their lymph mixes with that coming from the inguinal glands.

"The lymphatics of the urinary and genital organs, included in the pelvic cavity, also traverse the glands lodged therein, and unite with those of the parietes of the pelvis. Those of the scrotum enter the inguinal glands, as also do those belonging to the sheath and penis. The ramifications derived from the testicle and spermatic cord take the course of the veins, and penetrate one or two of the lumbar glands lodged at the entrance of the pelvis. The lymphatics of the mammæ, which are also divisible into superficial and deep-seated, run to the inguinal glands, and anastomose with the superficial set belonging to the inferior parietes of the abdomen; but, before they reach these last glands, they pervade those of the mammæ.

"3. *Lymphatics of the Parietes of the Abdomen.* — These vessels, in general but little developed, for the most part run to the inguinal glands. The superficial set of the lower parietes accompany the cutaneous inguinal vein, anastomose with the lymphatics of the scrotum and mammæ, and traverse the glands in the groin: some of them direct their course forward, along with the cutaneous external thoracic veins of the thorax, unite with the superficial lymphatics of that part, and proceed to the axillary glands. The deep-seated vessels of the belly run in company with the epigastric vein, and go to the inguinal glands, or else they accompany the pectoral vein, and pervade the glands in front of the thorax.

"The superficial or subcutaneous lymphatics of the loins join either those of the croup or those of the flanks: the deep-seated, which spring from the peritoneum, muscles, or spinal canal, perforate one of the lumbar glands, and pass onward to terminate in the main pelvic branch.

"4. *Absorbents of the Mesentery.* — The mesenteric branches, ordinarily two or three in number, the most considerable of which is constantly united to the great mesenteric artery, receive all the vessels continued from the mesenteric glands, as well as those coming from the mesentery and intestines.

"The mesenteric absorbents, extremely numerous, are sustained between the layers of the mesentery, where they form a vascular network: many of them issue from the exhalent surface of the mesentery and intestinal tube; others take their rise from the interior of the intestines, from which they imbibe chyle. All these vessels converge towards the lymphatic reservoir, clinging in their passage around the mesenteric veins; some, however, taking a solitary course at a greater or less distance from any blood-vessel. Having arrived at the root of the mesentery, they pass through one or two, sometimes three, of the mesenteric glands, and afterwards join the principal lumbar lymphatics. The absorbents of the colon and cœcum caput coli run to the glands set at intervals along the intestinal tube, whence they proceed to the receptaculum chyli.

"5. *Lymphatics of the Liver, Stomach, Spleen, and Omentum.*— The hepatic trunk comprises the lymphatics issuing from the above viscera. This branch of the receptaculum chyli not uncommonly consists of two divisions, and receives in addition to the above-mentioned vessels many ramifications from the crura of the diaphragm.

"The lymphatics of the pancreas, like the above, also run with the divisions of its veins, and join either those of the liver or those of the spleen: some proceed directly to the common hepatic trunk.

"II. RAMIFICATIONS TERMINATING IN THE THORACIC PORTION OF THE MAIN COMMON DUCT.

"1. *Lymphatics of the Parietes of the Thorax.*— The superficial absorbents of the chest take their rise either from the surface of the skin or else from the subcutaneous muscles; they form several large branches which accompany the thoracic cutaneous vein, unite with the superficial lymphatics coming from the anterior parietes of the abdomen, and proceed to the axillary glands.

"The deep-seated set take divers directions, and pass through the different sets of glands. The pectoral, which anastomose with ramifications from the abdomen, follow the pectoral vein, and reach one or two glands at the entrance of the chest. The intercostal spring from the pleura and intercostal muscles, accompany the intercostal veins, pervade the internal dorsal glands, and terminate by several branches in the thoracic duct.

"The lymphatics of the fleshy part of the diaphragm unite, some with the posterior intercostal, others with pectoral; those coming from the crura run to the dorsal glands, where they anastomose with the intercostal: those from the cordiform tendon anastomose with the deep hepatic, run forward between the layers of mediastinum, nearly to the heart, and enter the cardiac glands.

"2. *Lymphatics of the Thoracic Viscera.*— The absorbents of the different organs contained within the thorax traverse one or several of the bronchial or cardiac glands, and afterwards form divers branches, which end in the thoracic duct. The pulmonary lymphatics, very numerous, are distinguished into superficial and deep-seated. The first take their rise from the surface of the lungs, creep along under their enveloping membrane, and make for one or more of the bronchial glands. The deep set, which originate from the air-cells and from the parenchymatous tissue, follow the divisions of the pulmonary veins, run to the roots of the bronchiæ; there unite with the superficial, and perforate one or two of the bronchial glands.

"The cardiac lymphatics derive their origin either from the surfaces (both exterior and interior) of the heart, or from the muscular substance of the organ; they mount upon the curvature of the posterior aorta, and disappear in the cardiac glands.

"The lymphatics of the superior part of the mediastinum, and of the œsophagus, join, some the intercostal, and others the bronchial; those coming from the anterior part of this membranous partition, from the thymus, trachea, and œsophagus, unite, either with the pectoral, or close with the cardiac and anterior intercostal.

"3. *Lymphatics of the Head.*— The lymphatics of the head form two planes, a superficial and a deep one. The superficial pursue the course of the cutaneous veins, and run in part to the sublingual and gutteral glands. The deep vessels, which come from the nostrils, fauces, palate, etc., also run to the gutteral and sublingual, in which they unite with the superficial. From these two groups of glands, through which pass the lymphatics of the head, depart several large branches, two or three of which descend upon the anterior face of the trachea; others follow the course of the deep-seated and cutaneous veins, unite with those of the neck, and descend to the front of the chest. Almost all these vessels terminate in the thoracic duct; some few alone, on the right side, ending in the right axillary trunk.

"4. *Lymphatics of the Left Fore Extremity.*—The lymphatics of this member present the same disposition as those of the posterior limbs, and are divided into superficial and deep-seated. The former, consisting of diverse ramifications, accompany the superficial veins; the more considerable of them forming a plexus, which accompanies the cutaneous (superficial brachial) vein of the limb. The deep vessels originate from the foot, muscles, and bones, pursue the divisions of the deep veins, and plunge into the axillary glands, wherein they unite with the superficial, and whence they extend to the thoracic duct.

"*The Right terminating Trunk of the Lymphatics.*—This very short lymphatic canal is obliquely situated at the entrance of the thorax, upon the transverse process of the last vertebræ of the neck, extending in a direction from above downward, and from without inward, and terminating most commonly in the right axillary vein; though, in some instances, it joins the thoracic duct. This trunk is formed by the lymphatics coming from the right axillary glands, and some from the right lung, and right side of the neck and trachea."

(See Appendix.)

EXAMINATIONS ON THE PHYSIOLOGY OF THE LYMPHATICS.

Q. What is the character of the fluid found in the lymphatics?—*A.* It resembles dilute, liquor sanguinis, or the liquid portion of the blood in which the corpuscles float.

Q. What finally becomes of the lymphatic fluid?—*A.* It was formerly supposed that the lymphatic fluid was eliminated from the system; but Carpenter and other physiologists now contend that this is not the case; that the same is poured into the common receptacle with the nutrient materials newly imbibed from the food, whence both are propelled together into the general current of the circulation; and thus, instead of being eliminated, the lymphatic fluid is employed in the formation of new tissues.

Q. From whence is the lymphatic fluid derived?—*A.* 1st, from the residual fluid, which, having escaped from the blood-vessels into the tissues, has furnished the latter with the materials of their nutrition, and is now to be returned to the current of the circulation. 2d, from the particles of the solid frame-work which have lost their vital powers, and are therefore unfit to be retained as components of the living system; they therefore reënter the circulation, to be again submitted to the assimilating process, so that nothing shall be lost.

Q. By what process do fluids enter the cutaneous lymphatics?—*A.* By a process of imbibation.

Q. What fluid is more readily absorbed than some others?—*A.* Milk.

Q. What authority have you for this?—*A.* Schoeger, in the course of his experiments, found that the lymphatics of a limb, long immersed in milk, became tinged with it, while none of it could be detected in blood drawn from the veins.

Glossary of Veterinary Technicalities.

A GLOSSARY OF VETERINARY TECHNICALITIES.

A.

Abdomen. — The posterior part of the body of the horse.
Abdominalis. — Pertaining to the abdomen.
Abdominal Regions. — The divisions of the exterior of the abdomen.
Abductor. — Muscles are named *abductors* which draw parts from the axis of the body, or given centres.
Abnormal. — Unnatural, irregular.
Accelerator. — A muscle of the penis.
Acetabulum. — A name given to the cavity in which the head of the thigh bone articulates.
Achillis Tendo. — The tendon of the muscle inserted into the heel of man.
Acuminated. — Pointed, like a needle.
Adductor. — Muscles which draw parts toward the axis of the body.
Adipose. — Fatty matter.
Adventitious. — Accidental.
Afferent. — A term used to designate the structures which convey fluids to different parts.
Alæ. — Wings.
Albumen. — An element which constitutes the chief part of the white of an egg.
Alimentary Canal. — The passage which commences in the œsophagus and ends in the anus.
Alveolus. — The bony sockets of the teeth.
Anal. — Relating to the anus.
Anatomy. — To cut, with a view of displaying the structure, relations, and uses of parts.
Animus. — The principle of vitality.
Annular. — A ring-like ligament, found at the posterior part of the knee of the horse.
Antagonist. — A term applied to counteracting muscles or tendons.
Anterior. — A term applied to what may be situated before another part of the same kind.
Anti. — A prefix, signifying against.
Antilabium. — Against the lips.
Antrum. — Cavity in bones.
Anus. — The posterior extremity of the rectum.
Aorta. — The largest artery of the body.
Aortic. — Pertaining to the aorta.
Apex. — The pointed end of an organ.
Aponeurosis. — A tendinous expansion of fibre.
Arachnoid. — A membrane of the brain.
Arch of the Colon. — Transverse portion of that intestine.
Areola. — The spaces between fibres composing an organ.
Arterial. — A property belonging to arteries.
Arterialization. — The change which occurs in venous blood when brought in contact with air in the lungs.
Artery. — The name of blood-vessels which distribute arterial blood.
Articular. — Belonging, or relating, to joints.
Articulation. — (From *articulus*.) A joint.
Asperity. — A roughness.
Astragalus. — The bone beneath the os calcis.
Atlas. — The anterior bone of the neck.
Attollens. — A name given to muscles which lift, or raise, the parts.
Auditory. — Muscles and parts connected with the ear.
Auricular. — Relating to the ear.
Auricles. — The anterior cavities of the heart.
Axilla. — The part between the superior region of the arm and the chest.

B.

Biceps. — (From, *bis* — twice, and *caput* — a head; two heads.) The term is applied to muscles, having two distinct heads, or origins.
Bifurcate. — (Bifurcus; from *bis*, twice, and *furca*, a fork.) A blood-vessel or muscle is said to *bifurcate* when it divides into two branches.
Biliary. — Relating to the bile.
Brachial. — Of, or belonging, to the arm.
Bronchiæ. — Bifurcations of the windpipe.
Bronchial. — Relating to the bronchiæ.
Buccal. — (From *bucca*, the cheek.) Belonging to the cheek.
Buccinator. — A muscle of the cheek.
Bulb. — A dilated portion of the tube at the base of the penis.
Bursæ. — Sacs, or bags.
Bursæ Mucosæ. — Sacs found in the region of joints.
Bursal. — Relating to bursæ.

C.

Cæcum. — The blind gut.
Cæcal. — Pertaining to the cæcum.
Calcis Os. — The prominent bone of the hock.
Cancelli. — Cellular structure of bones.
Canine Teeth. — The eye teeth, cuspidati.
Canthus. — The angle of the eye.
Capillary. — Hair-like vessels which are found between the arterial and venous vessels.
Capsule. — A membranous sac.
Capsular. — A term applied to ligaments which surround articulations.
Caput. — The head.
Cardia. — The heart.
Cardiac. — Pertaining to the heart.
Carotid. — The name of the principal arteries of the neck.
Carpus. — The bones of the knee.
Caruncle. — A small fleshy excrescence.
Carunculæ Lacrymalis. — Small fleshy bodies found in the angle of the eye.

(187)

Cauda. — The tail.
Cava. — The largest vein in the body of the horse.
Cavity. — A hollow part; the abdominal cavity, for example.
Cellular. — Composed of cells.
Centrum Ovale. — The appearance of the brain, when a horizontal section is made on a level with the corpus callosum.
Centrum Tendinosum. — Tendinous centre of the diaphragm.
Cephalic. — Pertaining to the head.
Cerebellum. — Inferior lobe of the brain.
Cerebrum. — Superior lobe of the brain.
Cerebral. — Relating to the brain.
Cerebrospinal. — Pertaining to both the brain and spinal cord.
Cervical. — Pertaining to the neck.
Cervix. — The neck or contracted portion of an organ.
Chorœ Tendinœ. — Part of the internal structure of the heart.
Choroid. — The inner tunic of the eye.
Chyle. — A fluid found in the thoracic duct and lacteals.
Chyme. — A name given to the food after it has passed the pylorus.
Cilia. — The eyelids, hair of the same, etc.
Cineritious. — A term applied to that part of the brain which is of an ash color.
Circulus. — A ring.
Clitoris. — A part of the pudendum of the mare corresponding to the glans penis of the horse.
Coccyx. — The bones of the tail.
Cochlea. — The spiral cavity of the ear.
Cœcum. — (Sometimes spelt cæcum.) The blind gut.
Cœliac. — Prolongation of the solar plexus, an artery and vein of the abdomen.
Colon. — The largest and most dilated portion of the intestines.
Columna Carnœ. — A muscular arrangement within the cavity of the heart.
Commissure. — A suture, junction, or joint.
Complexus. — To embrace or surround.
Concha. — External cavity of the ear.
Conduit. — A canal.
Condyle. — An irregular process or enlargement.
Condyloid. — A tubercle, wart-like.
Conglobate. — Ball-shape.
Conglomerate. — An assemblage of glands.
Conjunctiva. — External coat of the eyeball, and internal lining of the eyelids.
Conoid. — Cone-like.
Constrictor. — Muscles that are bound together are thus named. The office is to close an outlet.
Continuity. — Identity of parts, having direct connection.
Convolute. — Rolled up.
Coracoid. — Like a crow's beak; a process of the scapula.
Cornea. — Anterior coat of the eye.
Cornu. — A horn.
Corona. — A crown, the superior pastern is thus named: os coronæ.
Coronal Suture. — The uniting medium between the frontal and parietal bones.
Coronary. — Arteries and veins, proper to the heart, are thus named.

Coronoid. — Processes of bones are thus named when they form an eminence.
Corpora. — A term applied to numerous prominences in the brain and elsewhere.
Corpus. — A body.
Corpora Striata. — Striped eminences in the brain.
Corpuscle. — A minute body.
Corrugator. — A muscle which wrinkles the surrounding parts.
Cortical. — Resembling bark.
Costa. — A rib.
Costal. — Pertaining to the region of the ribs.
Costalis Pleura. — That portion of the pleura which lines the interior of the chest.
Cotyloid. — Cup-shaped.
Cranium. — The skull.
Cruassmentum. — The clot, or red globules, of the blood.
Cremaster. — A muscle of the testicle.
Crest of the Ileum. — The anterior, superior parts of the pelvis.
Cricoid. — Ring-like.
Crista. — A crest.
Crucial. — In the form of a cross.
Crural. — Belonging to the thigh.
Crystalloid. — Resembling a crystal.
Cuboides. — One of the bones of the knee, which resembles a cube, or die.
Cuneiforme. — A bone of the knee, in form resembling a wedge.
Cuspidata. — The tushes of the horse are thus named.
Cutaneous. — Belonging to the skin.
Cuticle. — The scarf skin.
Cyst. — A bladder or sac.

D.

Dartos. — A name given to the muscle which corrugates the scrotum.
Deferens. — The excretory canal of the testes.
Dentatus. — A tooth-like process on the second cervical vertebra.
Dentes Incisors. — The twelve front teeth of the horse.
Dentes Molares. — The twenty-four grinders.
Depressor. — A muscle is so named when it depresses the part on which it acts.
Diaphragm. — The muscle which separates the thorax from the abdomen.
Diastole. — Periodic dilation of the heart.
Dilator. — A name given to muscles which dilate certain parts.
Diploe. — The cellular structure, which separates bony tables.
Diverticulum. — A blind tube, diverging from the course of a larger one.
Dorsal. — Pertaining to the back.
Ducts. — Orifices of various canals.
Ductus. — A canal for conveying fluids.
Duplicate. — Doubled.
Duplicature. — Reflection of a membrane upon itself.
Dura Mater. — The outermost tunic of the brain.

E.

Efferent. — Vessels are thus named which convey fluids from glands.
Elevator. — A muscle is so called when it lifts or elevates the parts to which it is attached.

Encephalon. — The brain.
Ensiform. — Sword-like.
Epididymis. — An appendage to the testicle.
Epigastrium. — Region of the stomach.
Epiglottis. — Cartilage at the root of the tongue.
Epiphysis. — A union of bones by means of cartilage.
Epithelium. — A transparent membrane covering various internal parts.
Erector. — A name given to certain muscles, which raise or erect the parts.
Eroded. — Rough and jagged.
Esophagus. — The gullet.
Ethmoid. — Sieve-like.
Exito-Motory. — The true spinal nerves.
Excretory. — Vessels and ducts are thus named which discharge fluids.
Expiration. — The act of expelling air from the lungs.
Extensor. — To stretch out; a name given to several muscles and tendons.
Extremity. — The end.

F.

Facial. — Belonging to the face.
Falciform. — Scythe-shaped.
Falx. — The scythe process of the dura mater.
Fascia. — The tendinous expansion of muscles.
Fascicular. — Fibres arranged in bundles.
Fauces. — Posterior part of the mouth.
Femoral. — Of, or belonging to, the thigh.
Fenestra. — Part of the internal ear.
Fibre. — A thread or filament.
Fibrous. — Composed of fibres.
Fibula. — A small bone attached to the lateral part of the tibia of the horse.
Filament. — A minute fibre.
Filiform. — Thread-like.
Fimbria. — A fringe.
Fissure. — A crack or groove.
Flavus. — Yellow.
Flexor. — A name given to numerous muscles and tendons which bend the limbs.
Foliatus. — Leaf-form.
Follicle. — A minute sac or bag.
Foramen. — An opening.
Fornix. — Arch or vault; one of the structures of the brain.
Fossa. — A shallow cavity or depression.
Frenum. — A ligament which restrains motion.
Frontal. — Belonging to the anterior part of the cranium.
Function. — Any action by which vital phenomena are produced.
Fundus. — The base or bottom.
Funis. — The umbilical cord.

G.

Ganglion. — A knot or enlargement in the course of a nerve.
Gastric. — Pertaining to the stomach.
Gastric Juice. — A secretion, peculiar to the walls of the stomach.
Gastrocnemii. — The tendinous portion of muscles inserted into the os calcis, or point of the hock, are thus named.
Gemini. — Twins; two organs precisely alike are thus named.

Gestation. — Pregnancy.
Gland. — An organ of secretion.
Glandula. — A small gland.
Glandular. — Resembling a gland.
Glenoid. — The name of articulating cavities.
Glissons Capsule. — The fibrous envelope of the liver.
Globate. — Globe-like.
Globules. — Red particles of the blood.
Globuline. — Albuminous constituent of the blood.
Glomerate. — Congregated.
Glossa. — The tongue.
Glottis. — Upper opening into the windpipe.
Gluteal. — Belonging to the haunch.
Gracilis. — A muscle on the inner part of the thigh.
Granule. — A small grain.
Gutteral. — Belonging to the throat.

H.

Hepatic. — Belonging to the liver.
Hiatus. — An aperture or foramen.
Hippo. — A horse; a prefix.
Hippocampus. — Two eminences in the lateral ventricles of the brain.
Homo. — A prefix designating similarity.
Humerus. — The bone beneath the shoulder blade.
Hyo. — Names compounded with this prefix relate to muscles situated near the root of the tongue.
Hyoides. — Bone at the root of the tongue.
Hypochondrium. — A region of the abdomen.
Hypogastric. — Relating to the hypogastric region of the abdomen.

I, J.

Ileo. — A prefix, the ileum or bone of the pelvis.
Ileum. — A portion of the intestinal tube.
Iliac. — Region of the flanks.
Incisors. — The twelve front teeth, or nippers, of the horse are thus named.
Incus. — A bone of the ear.
Infra. — Under; a prefix to the name of several muscles.
Infra. — Without.
Infundibulum. — A funnel or duct.
Inguinal. — Pertaining to the groin.
Inguinal Ligament. — Pouparts ligament.
Innominatum. — Without a name.
Innominatum Os. — Union of the ileum, ischium, and pubic bones.
Inosculation. — Union of the extremities of vessels.
Insertion. — Attachment of a muscle or tendon to the part which it moves.
Integuments. — The skin and subtissues.
Interarticular. — Between the joints.
Interosseous. — Muscles and ligaments situated between bones are thus named.
Interseptum. — The uvula.
Interstitial. — A term applied to substances occupying the spaces between contiguous parts.
Interspinales. — Between the spines of bones.
Intertransversales. — Muscles located between the transverse processes of bones.
Intervertebral. — The articular cartilages between the vertebrae are thus named.
Intestinal Canal. — The interior of the duodenum, jejunum, ileum, caecum, colon, and rectum, compises the intestinal canal.

Intra. — Within.
Intestines. — The bowels.
Invertebrata. — Animals without internal bony structure.
Ischium. — A part of the bones of the pelvis.
Isthmus. — A narrow passage.
Jejunum. — A term applied to that portion of the intestine which is generally found empty.
Jugular. — Belonging to the neck.

L.

Labarium. — Relating to the lips.
Labia. — The lips.
Lachryma. — A tear.
Lachrymal. — Structures concerned in the secretion and transmission of tears.
Lacteals. — Absorbent vessels of the lymphatics.
Lactiferous. — Vessels conveying milk.
Lacunæ. — Ducts issuing from small glands.
Lamella. — Thin plates.
Laminæ. — A series of plates.
Laminated. — Leaf-like.
Laryngeal. — Relating to the larynx.
Larynx. — The superior part of the windpipe.
Lata. — Broad.
Lateral. — Pertaining to the side.
Latissimus. — A term applied to a muscle in consequence of its great breadth.
Lens. — A crystalline body; a lentil.
Lenticular. — Shaped like the lens.
Levator. — A term applied to muscles which raise the parts to which they are attached.
Ligament. — A tendinous cord.
Linea. — A white line; thread-like.
Linea Aspera. — A rough projection.
Linea Semilunares, or semi-circularis. — Lines on each side of the linea alba, formed by the termination of the fibres of the abdominal muscles.
Linea Transversalis. — Lines crossing the recti muscles of the abdomen.
Lingual. — Pertaining to the tongue.
Lingualis. — A muscle of the tongue.
Liquor Sanguinis. — The fluid element of the blood.
Lobi. — A division of an organ.
Lobus. — A lobe.
Local. — Confined to a part.
Loins. — The posterior part of the back.
Longissimus. — The longest.
Longus. — Long, lengthy.
Lumbar. — Belonging to the loins.
Lumbrici. — Worms.
Luna. — The moon.
Lunare. — A bone of the knee.
Lymph. — A fluid found in the lymphatics.
Lymphatic. — Of the nature of lymph.

M.

Major. — The greater.
Malar. — Belonging to the cheek.
Malleus. — A hammer or mallet.
Malpighian Bodies. — Dark points of the kidneys.
Mamma. — The udder.
Mammary. — Belonging to the udder.
Masseter. — A muscle of the jaw.
Mastoid. — Processes of bones presenting the form of a nipple are thus named.

Mater. — A mother.
Maxilla. — Bone of the jaw.
Maxillary. — Pertaining to the jaw.
Meatus. — A passage.
Meatus Auditorius. — The internal auditory passage of the ear.
Meatus Urinarius. — The orifice of the urethra.
Median. — Central, the central line.
Mediastinum. — The partition which divides the thorax.
Medulla. — The medullary substance of the brain is thus named. It signifies marrow or pith.
Medulla Spinalis. — The spinal marrow.
Membranes. — Tissues.
Membranous. — Having the texture of membranes.
Meninges. — Membranes of the brain.
Meningial. — Relating to the membranes of the brain.
Mesenteric. — Pertaining to the mesentery.
Mesentery. — Membranes uniting the intestines.
Mesian Line. — The middle line.
Meso. — Words compounded of meso signify the middle.
Metacarpal. — Relating to the knee of the horse.
Metacarpus. — The bones of the knee.
Molar. — The grinders.
Motor. — To move; the nerves of voluntary motion are thus named.
Mucus. — Animal mucilage.
Mucous. — A term applied to the mucous tissues.
Muscular. — Belonging to a muscle.
Muscle. — Flesh; a bundle of muscular fibres.
Mylo. — Names compounded of this word relate to muscles located in the region of the root of the tongue.
Myology. — A description of the muscles.

N.

Nares. — The anterior cavity of the nostrils.
Nasal. — Belonging to the nose.
Navicular. — Boat-shaped.
Neurilema. — The sheath investing the nerves.
Nudus. — Naked.
Nucha. — A part of the superior region of the neck.
Nucha Ligamentum. — A ligament of the spine.
Nymphæ. — Internal labiæ of the vulva.

O.

Oblique. — A term applied to muscles that have an oblique direction.
Obturator. — Name of muscles, foraminæ, etc.
Occipital. — Connected with the occiput or posterior part of the cranium.
Occipito Atloid. — That which has reference to the occiput and atlas.
Occipito Frontalis. — A muscle which reaches from the occiput to the forehead.
Ocular. — Belonging to the eye.
Odontoid. — Tooth-like.
Œsophagus. — The gullet.
Olecranon. — Point of the arm, formed by the ulna.
Olfactory. — Relating to the sense of smell.
Olivaris. — Resembling the olive.
Omentum. — The caul.
Omo. — Names compounded of this word signify muscles which are attached to the scapula.
Opaque. — Not transparent.
Optic. — Relating to vision.

Orbicular. — Spherical-circular.
Orbicularis Oris. — Muscle of the lips.
Orbicularis Palpebrarum. — Muscle of the eyelids.
Orbiculare. — The smallest bone of the internal ear.
Orbit. — The bony socket of the eye.
Orbitar. — Pertaining to the orbit.
Organ. — A part having a distinct office to perform.
Organism. — Vital organization.
Organized. — Possessed of organs; endowed with life.
Orifice. — An aperture.
Origin. — The fixed point or commencement of a muscle.
Os. — A bone.
Osseous. — Bony.
Ovaria. — The female testes.
Ovum. — An egg.

P.

Palatine. — Relating to the palate.
Palate. — The roof of the mouth.
Palate Os. — Bone of the palate.
Palpebræ. — The eyelids.
Paries. — A wall.
Parietes. — The walls of the abdomen and thorax, etc.
Parotid. — Name of the gland beneath the ear.
Parotid Duct. — Opening into the cheek from the parotid gland.
Patella. — The stifle bone.
Pectinated. — Shaped like the teeth of a comb.
Peduncle. — A stalk.
Pellicle. — A thin membrane.
Pelvis. — The cavity formed by the innominata and sacrum.
Penis. — The principal organ of generation in the male.
Perforans. — Perforating; the name of part of the flexor tendons.
Perforatus. — Perforated for the transmission of the preceding tendon.
Peri. — Around; an envelope.
Pericardium. — The sac containing the heart.
Pericranium. — The membrane investing the skull.
Perineum. — The part between the anus and organs of generation.
Periosteum. — Membrane investing bones.
Periphery. — The circumference.
Peritoneum. — The serous membrane which lines the interior of the abdomen and is reflected on its contents.
Peroneal. — Relating to the fibula.
Petaloid. — Shaped like a petal.
Petrosum Os. — Rough portion of the temporal bone.
Peyer's Glands. — Clustered mucous glands of the intestines.
Pharyngeal. — Relating to the pharynx.
Pharynx. — Superior part of the gullet.
Phrenic. — Belonging to the diaphragm.
Pia Mater. — A thin membrane investing the brain.
Pigmentum Nigrum. — Black pigment upon the choroid coat of the eye.
Pilus. — Hair.
Pineal. — Shaped like the fruit of the pine.
Pisiform. — Shaped like a pea.
Pituitary Membrane. — The schneiderian membrane of the nostrils.
Placenta. — The afterbirth.
Plantar. — Relating to the feet.

Pleura. — The serous membrane which lines the cavity of the chest, and is reflected on the contents of the same.
Plexus. — A network of nerves or vessels.
Plica. — A fold.
Pons. — A bridge.
Pons Varolii. — A part of the brain.
Popliteal. — Muscles, nerves, and vessels in the region of the hock.
Pores. — Extremities of the exhalents of the skin.
Porta. — A door or gate.
Portio. — A portion or branch.
Posterior. — Behind.
Posticus. — Situated behind.
Præcordia. — The anterior part of the chest.
Primæ Viæ. — First passages of the alimentary canal.
Process. — The projecting eminence on a bone.
Profundus. — Deep-seated.
Pronatus. — Muscles of the fore limbs.
Prostate. — A gland near the neck of the male bladder.
Pseudo. — False; a term applied to spurious membranes.
Psoæ. — The loins.
Psoas. — Belonging to the loins.
Pterygoid. — Shaped like a wing.
Pubes. — The junction of the pelvic bones at their inferior parts.
Pubic. — Pertaining to the pubes.
Pudendum. — The external parts of the female organs of generation.
Pudic. — Belonging to the pudenda.
Pulmonary. — Belonging to the lungs.
Puncta. — Lachrymalia. The tear-ducts within the eyelids.
Punctum. — A point.
Pylorus. — The outlet of a horse's stomach.
Pyriform. — Shaped like a pear.

Q.

Quadratus. — Square in form or figure.

R.

Radial. — Belonging to the radius.
Radiated. — Diverging from the centre like the sun's rays.
Radicles. — Germs of the roots.
Radius. — Bone of the fore extremity.
Ramify. — To branch out, or from.
Ramose. — Branched.
Ramus. — A branch.
Ranine. — Vessels under the tongue.
Raphe. — The central line of the scrotum.
Rectum. — The posterior termination of the intestines.
Rectus. — Straight.
Recurrent. — Running in a backward direction.
Reflection. — A duplicature.
Regions. — Divisions of the body.
Renal. — Belonging to the kidneys.
Rete. — Net-work.
Retiform. — Net like.
Retina. — Expansion of the optic nerve.
Retractor. — Muscles thus named draw backwards.
Retrahens. — Drawing back.
Rima. — An opening or fissure.
Rotator. — A name given to muscles that rotate or revolve a part.
Rotundus. — Circular, round.

Ruga. — A wrinkle.
Rugose. — Wrinkled.

S.

Sac. — A bag or cyst.
Sacculated. — Encysted.
Sacral. — Belonging to the sacrum.
Sagittal. — Arrow-shaped.
Salivary. — Relating to the saliva.
Sanguis. — Blood.
Saphena. — A vein of the hind extremities.
Scaphoid. — Shaped like a boat.
Scapula. — Shoulder blade.
Scrotal. — Relating to the scrotum.
Scrotum. — The sac in which the testicles are contained.
Sebaceous. — Resembling suet.
Secernent. — Secretory.
Semen. — Secretion peculiar to the testes.
Semi. — One-half.
Septum. — A partition or division.
Serrated. — Resembling the teeth of a saw.
Serum. — The fluid portion of the blood.
Sesamoid. — Like seeds.
Sigmoid. — Flexure.
Sinew. — A tendon.
Sinus. — A long cavity.
Spermatic. — Belonging to the testicles.
Sphenoid. — Wedge-like.
Sphenoidal. — Belonging to sphenoid bone.
Sphincter. — Circular muscles, which close an opening, are thus named.
Spinal. — Belonging to the spine.
Spinal Marrow. — Medulla spinalis.
Spine. — The vertebral column.
Spinus. — Thorn-like.
Splanchic. — Belonging to the intestines.
Squamous. — Resembling scales.
Stapes. — A stirrup; bone of the ear.
Sternal. — Belonging to the breast bone.
Sternum. — Breast bone.
Striated. — Marked with long lines.
Styloid. — Shaped like a pointed pencil.
Sub. — Under; beneath.
Sublimus. — This term is applied to a muscle when seated more superficially than another of the same kind.
Sublingual. — Beneath the tongue.
Submaxillary. — Under the inferior jaw.
Subscapular. — Inner side of the shoulder blade.
Super. — Above.
Superficial. — Upon or near the surface.
Superior. — The upper part.
Suture. — Junction or union.
Sympathetic. — Associated in function.
Symphysis. — A connection of bones by an intervening substance.
Synovia. — The lubricating fluid of joints, sometimes called joint-oil.
Systole. — Contraction of the heart.

T.

Tabula. — An extended surface.
Tarsus. — The hock.
Tegumentary. — Relating to the skin.
Temporal. — Relating to the temporal regions of the cranium.
Tendon. — The extremity of a muscle.
Tendo Achillis. — The tendon of the gastrocnemii, inserted into the hock or heel of man.
Tensor. — A name given to muscles which stretch or extend parts.
Tentorium. — A membranous partition of the brain.
Teres. — Round; cylindrical.
Testes. — The testicles.
Thalami Nervorum Opticorum. — Supposed origin of the optic nerves.
Thalamus. — A bed or origin of certain parts.
Theca. — A sheath.
Thoracic. — Belonging to the thorax or chest.
Thoracic Duct. — The trunk of the absorbents.
Thorax. — The chest.
Thyro. — Names compounded with this word belong to muscles which are attached to the thyroid cartilage.
Thyroid. — Resembling a shield.
Tibia. — The bone beneath the femur.
Tibial. — Belonging to the tibia.
Tinca. — The name of a fish; the tench.
Tissue. — An organized structure.
Trachea. — The windpipe.
Tracheal. — Pertaining to the windpipe.
Trachelo. — Names compounded with this word belong to muscles located in the region of the neck.
Transversalis. — Having a transverse direction.
Transversus. — Placed across.
Trapezium. — A four-sided figure, bone of the horse's knee.
Trapezoides. — A bone which in figure somewhat resembles the preceding; it also enters into the composition of the horse's knee.
Trapezius. — Four square; a muscle placed over the region of the withers.
Triangularis. — Triangular.
Triceps. — Three-headed.
Tricuspid. — Having three points; a name applied to a valve in the right ventricle.
Trifid. — Three-cleft.
Trigastric. — Having three bellies.
Trisplanchic Nerve. — The great sympathetic or ganglionic nerve.
Trochanter. — Eminences or tuberosities on the bones.
Tuba. — A tube.
Tuber. — A solid roundish substance.
Tuberosity. — Protuberance or projection.
Tubular. — Tube-like.
Tunic. — A membranous covering.
Turbinated. — Shaped like a sugar-loaf.
Turgid. — Swollen.

U.

Ulna. — Bone of the fore extremity, termed point of the elbow.
Ulnar. — Pertaining to the ulna.
Umbilicus. — The navel.
Unciform. — Shaped like a hook.
Ureter. — A tubular connection between the kidneys and bladder.
Urinal. — Pertaining to the urine.
Uterine. — Relating to the womb.
Uterus. — The womb.
Uvula. — A pendulous body, posterior to the soft palate.

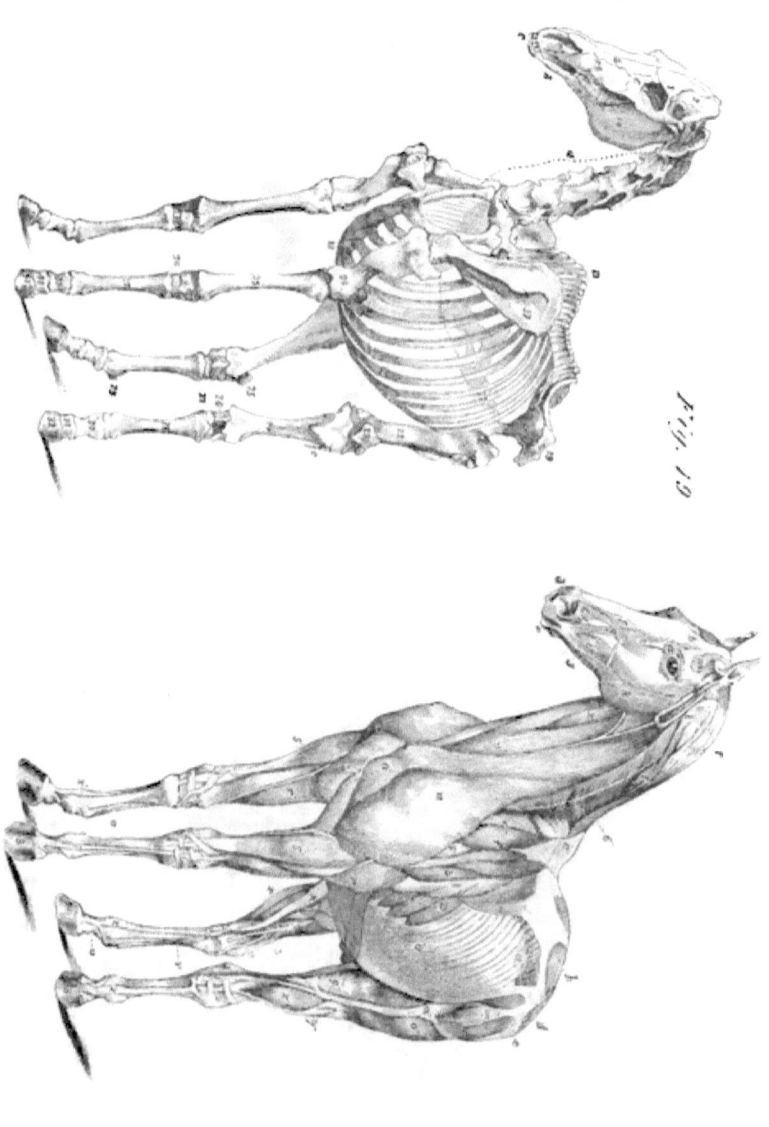

Fig. 19.

EXPLANATION OF FIGURE XIX.

OSSEOUS STRUCTURE.

1. Frontalis.
2. Parietalis.
3. Occipital.
5. Nasal.
6. Lachmyral.
8. Superior maxillary.
9. Anterior "
10. Inferior or lower jaw.
11. Cervical vertebræ.
16. True ribs.
17. False ribs.
18. Sternum.
19. Ileum.
22. Femur.
23. Patella.
24. Tibia.
25. Os calcis.
26. Astragalus.
27. Tarsal bones.
28. Metatarsi magnum.
29. Sessamoids.
30. Os suffraginis.
31. Os coronæ.
32. Os pedis.
33. Scapula.
34. Os humeri.
35. Radius.
36. Carpus.
37. Metacarpi magnum.
39. Os suffraginis.
40. Os coronæ.
41. Os pedis.
D. Dorsal spines.

MUSCULAR STRUCTURE.

FORWARD PARTS. — THE HEAD.

a. Orbicularis palpebrarum.
b. Levator palpebræ.
c. Dilator naris lateralis.
d. " " anterior.
e. Orbicularis oris.
f. Nasalis longus.
g. Levator labii superioris.
i. Buccinator.
J. Retractor labii inferioris.
k. Masseter.
m. Attolentes et abducens aurem.
1. Temporal vein.
2. Facial vein.

THE NECK.

c". Trachelo subscapularis. — Scalenus.
6. Rhomboideus longus.

EXPLANATION OF FIGURE XIX. CONTINUED.

f. Splenius.
o. Abducens vel deprimens aurem.
r. t. Tendon of the splenius and complexus major
v. Sterno maxillaris.
x. Subscapulo hyoideus.

THE SHOULDER, ANTERIOR MUSCLES, AND FORE EXTREMITIES.

a. Trapezius.
b'. Teres.
c". Pectoralis transversus.
f'. Antea spinatus.
g". Postea spinatus.
l'. r. Triceps extensor brachii.
o". Pectoralis magnus.
r". Flexor metacarpi internus.
s". s". Extensor metacarpi magnus.
t. t. Extensor metacarpi obliquus.
u. Tendons perforatus and perforans.
v. (At the humeral region.) Levator humeri.
x". x". Extensor tendons.
8. The hoof.

THE ABDOMEN AND POSTERIOR PARTS. — ABDOMINAL REGION, AND OF THE COSTA.

a". Levatores costarum.
o". Obliquus externus abdominis — (beneath the dotted line).
D. Serratus magnus.

POSTERIOR PARTS.

g". Ligaments of the patella.
h. d. e. Glutei.
k. Extensor metatarsi internus.
m. Tensor vaginæ.
m". Rectus.
o". Vastus externus.
n. Gastrocnemius internus.
e. Flexor pedis.
u. Flexors perforatus and perforans.
z". x". Fleshy belly of the extensor.
x. x. Extensor tendons.
y. Peroneus.
8. The hoof.

V.

Vagina. — A sheath; the cavity between the pudenda and womb.
Vaginal. — Pertaining to the vagina.
Valvular. — Valve-like.
Vas. — A vessel.
Vas Deferens. — Excretory duct of the testicle.
Vasa. — The plural of vas; vessels.
Vascular. — Highly organized with blood-vessels.
Vascular System. — The heart and its vessels.
Vastus. — Relates to size; large, thick and fleshy muscles of the thigh.
Vena. — A vein.
Vena Cava. — The great vein.
Vena Porta. — The largest vein of the liver.
Venter. — The belly.
Ventricles. — A term applied to the cavities of the brain and heart.
Vermiform. — Shaped like a worm.
Vertebræ. — Bones of the spinal column.
Vesical. — Formed like a bladder; pertaining to the bladder.
Vesicles Graafian. — Small bladders or cysts found in the ovaria (female testes).
Via. — Way or passage.
Villous. — Velvet-like, applied to the villous coat of a horse's stomach.
Viscera. — Internal organs.
Visceral. — Relating to a viscus.
Viscus. — An organ within the body.
Vital. — Life-like.
Vitreous. — Glassy; transparent.
Vivisection. — Surgical operations on living subjects.
Vivus. — Living; life-like.
Vulva. — The pudendum.

Z.

Zoology. — The science of animals.
Zootherapeutics. — Relates to the curative action of medicines.
Zootomy. — Comparative anatomy.
Zygoma. — An arch or yoke.
Zygomatic. — Belonging to the zygoma.

Veterinary Toxicological Chart.

A VETERINARY TOXICOLOGICAL CHART,

CONTAINING THOSE AGENTS WHICH ARE KNOWN TO CAUSE DEATH IN THE HORSE; WITH THE SYMPTOMS, ANTIDOTES, ACTION ON THE TISSUES, AND TESTS.

BY W. J. T. MORTON.

Lecturer on Veterinary Materia Medica, etc.

"Poisons are substances which are capable of altering or destroying, in a majority of cases, some or all of the functions necessary to the support of the vital principle."—*Fœdere*. They are derived both from the organic and inorganic kingdoms; and their action is either local or remote. Local action is referrable to, 1st, Chemical Decomposition; 2d, Irritation and Inflammation; 3d, Nervous Impression. Remote action is effected by, 1st, Absorption; 2d, Sympathy. Animal Poisons rank first in potency; next to these, the Mineral; and lastly, the Vegetable. Aerial poisons are, perhaps, the most insidious.

The manner in which poisons are introduced into the System varies. The Alimentary Tube, the Skin, the Circulation, and the Lungs, are the media. 1st, They may be taken into the Stomach inadvertently with the food, or they may be maliciously or accidentally administered. They may also be thrown up as Enemata. 2d, They may be placed underneath the Skin; or injected into the Circulation; or they may be absorbed from Wounds. 3d, If gaseous, they may be inhaled, and enter the blood during its transit through the Lungs. They are generally arranged according to the effects which they produce upon the Animal Economy. The great end of Toxicological Science is to counteract their influence, which may be accomplished by chemically decomposing them, by their expulsion from the System, and by restoring the Function of the Organ of which they have caused derangement. As comparatively large quantities of the Poisons are required to destroy Life in the Horse, the niceties of chemical manipulation in the application of Tests are uncalled for. It will generally be sufficient to collect some of the contents of the Stomach and Intestines, add distilled Water to them, filter and to the Solution apply the Test or Re-agent. Sometimes they require the influence of heat; and, when the contents are not attainable, portions of the Alimentary Tube which have been most acted upon by the Agent are to be boiled in Distilled Water, and similarly treated.

I.—IRRITANT POISONS.

These produce their action upon some part of the Alimentary Canal, particularly the Stomach and Intestines; and by absorption they are often carried to other Organs. The principal Symptoms are those of Irritation and Inflammation.

AGENTS	
ACIDUM SULPHURICUM.	are the most powerful of all local irritants. Indications of their action are uneasiness, frequent pawing and shifting of the position, increased secretion of saliva, which is sometimes viscid and fetid, the mouth inflamed, difficulty in swallowing from corrosion of the lining of the esophagus, acute gastric irritation extending to the intestines,
Sulphuric Acid.	
ACIDUM NITRICUM.	
Nitric Acid.	
ACIDUM HYDROCHLORICUM.	
Hydrochloric Acid.	

Symptoms.—The liquid mineral acids and giving rise to symptoms resembling a most violent attack of colic; pain on pressure being applied over the abdomen; frequent attempts to dung and stale; and, after the fæces have been voided, a discharge of mucus streaked with blood takes place: tenesmus, pulse quick and feeble, prostration of strength, profuse perspiration, coldness of the body, and death, after the animal has endured excruciating agonies.

In one case related to me, nitric acid was poured into the ear, and death took place from inflammation extending to the membranes of the brain.

Treatment. — As the general symptoms of poisoning by the liquid mineral acids do not materially differ, neither will the general treatment. This will consist, 1st, In diluting the agent by throwing into the stomach large quantities of water by means of Read's pump. 2d, In neutralizing it, by suspending in the water chalk, magnesia, or soap; or, in the absence of these, the plaster from the walls. 3d, In allaying the supervening inflammation by means of bloodletting, should the urgency of the symptoms demand it; and also by the administration of opium, and a free use of demulcents. The subsequent nervous debility and prostration of strength are to be combatted by the milder vegetable tonics, and a gradual return to liberal diet.

Morbid Appearances. — The mouth, pharynx, and esophagus, present traces of the action of the peculiar acid. The stomach is distended with gas, and occasionally lined with its disorganized tissue, which is eroded in patches, and so deeply ulcerated as to form perforations. Intense inflammation often exists in this viscus, which extends throughout the whole of the intestinal tube, involving its peritoneal tunic; this last circumstance has been thought to be distinctive between poisoning by acids and metallic compounds; this cannot, however, be relied upon. The blood in the larger vessels sometimes forms a firm clot. These appearances will not be so marked when an acid has been given in small doses for some time, or if much diluted we may then expect to find the coats of the stomach and intestines thickened and contracted, the result of chronic inflammation, with here and there eroded spots, but not of any depth.

Tests. — *General.* — Sour taste — neutralization by the alkalies — effervescing with the carbonates — reddening of litmus paper.

Particular. — *Sulphuric Acid.* — The parts with which it comes in contact are first whitened, and then changed to a brownish color. By macerating them or the contents of the stomach in distilled water, filtering, and adding a solution of the *nitrate of barytes*, an insoluble precipitate, *the sulphate of barytes*, is obtained.

Nitric Acid. — The tissues changed of a yellow color, which is heightened by ammonia. The filtered solution boiled on *copper filings* in a test tube emits orange-colored fumes of *nitrous acid*. Potassa being added to it, by evaporation a salt is obtained, which deflagrates; or a piece of bibulous paper may be saturated with the solution, dried, and inflamed.

Hydrochloric Acid. — Tissues blanched. Its fumes are rendered more manifest by a rod dipped in ammonia being held in them. This test, however, we are rarely able to avail ourselves of. On the addition of *nitrate of silver* to the solution, it gives a white precipitate, *the chloride of silver.*

AGENT.

ACIDUM OXALICUM.

Oxalic Acid.

Symptoms. — Instances are recorded of horses having been poisoned by this acid, but whether maliciously given, or administered by mistake for the sulphate of magnesia, I cannot say. The symptoms attendant on its action, when a concentrated solution is given, will not be dissimilar to those produced by the mineral acids. When diluted, however, it is said to cause death by palsying the heart and nervous system, or by inducing tetanus or narcotism; but I am not aware that such action has been observed in the horse.

Treatment.— Avoid large quantities of water, as it favors the absorption of the acid. Throw into the stomach a mixture of chalk, or of magnesia and water, particularly the former; or lime from the walls may be used; either of which will form an insoluble salt. The alkalies are inadmissible, because they form soluble salts. Demulcents to be freely employed, and the remaining irritation to be allayed by opium.

Morbid Appearances.— None recorded in the horse. In other animals the stomach has been found to contain black extravasated blood, its inner coat being of a cherry-red color; in some places the surface is brittle, and the subjacent stratum gelatinized. The intestines are usually inflamed throughout. When its influence has been through the medium of the blood on remote parts, the heart has been found to have lost its contractility, and to contain arterial blood.

Tests.— Acid reaction on litmus paper. A concentrated solution with *ammonia* forms a salt whose crystals radiate, *the oxalate of ammonia.*

Hydrochlorate of Lime throws down a white precipitate which is soluble in nitric acid, *the oxalate of lime.*

Sulphate of Copper yields a blue or greenish-white precipitate, *the oxalate of copper.*

Nitrate of Silver causes a dense white precipitate; also an *oxalate* which, when dried and heated, fulminates.

AGENT.

ACIDUM ARSENIOSUM.

Arsenious Acid.
White Arsenic.

Symptoms.— Intense pain, resembling acute enteritis; belly tympanitic, with a rumbling noise in the intestines; the dejections offensive, and mixed with mucus; pulse quick and feeble, becoming scarcely perceptible at the jaw; respiration laborious; surface of the body covered with an extremely cold, clammy sweat; extremities cold; efforts to vomit; countenance anxious, and indicative of great torture; mucous tissues injected; mouth hot; increased secretion of saliva, which is singularly fetid; delirium from pain which has become continuous; exhaustion; death. The action of this poison is not merely as a local irritant, it being often conveyed to remote parts through the medium of the circulation, thus causing death. Even as an external applicant it has been known to produce much general derangement of the system, independent of its influence as an escharotic, which is powerful. On this account, when the methods usually resorted to have failed to demonstrate its existence in the contents of the stomach and intestines, Orfila has succeeded in detecting it in the organic tissues, particularly the liver.

Treatment.— A free use of diluents, or of lime water; avoid blood-letting, as this promotes the absorption of the poison; give large doses of the hydrated peroxide of iron precipitated by ammonia from a solution of the sulphate of iron, so as to form an insoluble arsenic of iron, which may be expelled by the action of active purgatives. The subsequent inflammation is to be combatted by the ordinary antiphlogistic remedies; while the debility which supervenes, and which is often great, is best counteracted by the vegetable tonics and judicious dieting.

Morbid Appearances.— The stomach and intestines, especially the latter, highly inflamed and ulcerated in patches. The cæcum and colon present the most marked action, the villous coat being black from an effusion of altered blood, and the peritoneal tunic involved. Congestion of blood in the lungs, liver, and kidneys; redness of the lining membrane of the windpipe, extending to the air-passages generally; conjunctival membrane highly injected, and the blood in a fluid state throughout the body. Ecchymosis in the heart.

Tests.— 1st, *by Reduction.*— The suspected powder, being dried, is to be mixed with twice its weight of newly-burnt and pulverized charcoal, and introduced into a test-tube: the heat of a spirit lamp is now to be applied; first to the upper part of the mixture, and afterwards steadily to the bottom of the tube, when, if arsenious acid is present, the metal arsenicum will be sublimed, and, encoating the tube, form a ring of a polished-steel lustre, the inner surface of which is crystalline. The little watery vapor, which will be condensed within the tube before the metallic crust begins to appear, is to be removed by a roll of bibulous paper.

2d, *by Liquid Re-agents.* — The contents of the stomach, or such parts of that viscus as have been acted upon, being boiled in distilled water, the solution is to be filtered. The *ammoniacal sulphate of copper* added to this gives an apple-green precipitate, *the arsenite of copper.* The *ammoniacal nitrate of silver,* a lemon-yellow precipitate, changing to a dark brown on exposure to light, *the arsenite of silver.* *Sulphuretted Hydrogen,* — generated by the action of dilute sulphuric acid on sulphuret of iron, in a flask, having an emerging tube bent at a double right angle,— passed up through the solution for ten or fifteen minutes, gives a sulphur yellow precipitate, *the sulphuret of arsenicum.* Water impregnated with this gas affords the like compound. The solution for this test must be perfectly neutral. This precipitate may be afterwards subjected to reduction.

3d, *by Nascent Hydrogen.* — This is effected in Marsh's tube. The fluid contents of the stomach, or the filtered solution before spoken of, being introduced into it, zinc and sulphuric acid are added, and the *arseniuretted hydrogen* as it escapes from the jet inflamed, when *water* and *metallic arsenic* will be condensed upon the glass disc held above it. The former will be dissipated by the heat, and around the latter a ring of arsenious acid may be seen. In the absence of a Marsh's tube, a common two-ounce wide-mouthed vial, with a cork perforated by a piece of glass tube or even tobacco-pipe, may with care be made to answer all the purpose.

AGENT.

HYDRARGYRI BICHLORIDUM.
Bichloride of Mercury,
Corrosive Sublimate.

Symptoms. — The effects which follow the administration of large doses of this salt, resemble those which supervene when the mineral acids have been given, except that, generally, super-purgation is present, and the fœcal matter is profuse and highly offensive. Its solubility renders it more energetic than arsenious acid, although it is not so frequently had recourse to for poisoning.

The protochloride of mercury, *calomel,* when incautiously given, has also caused death, by inducing inflammation of the mucous lining of the intestines, accompanied with violent purging and tenesmus.

Treatment. — The white of eggs suspended in water, the albumen of which renders the bichloride of mercury insoluble; or large quantities of wheat-flour, or milk. Iron filings have also been advocated, which, reviving the metallic mercury, may be expelled by purgatives; a free use of dilutents. The treatment of the salivation, which sometimes supervenes, consists in exposure to cool air, the exhibition of saline purgatives, and nourishing diet.

Morbid Appearances. — These would closely resemble the effects produced by the above agent, the mucous lining of the alimentary canal being intensely inflamed throughout, its texture destroyed, and in parts corroded. The disorganized tissue often contains the poison, which it yields by analysis.

Tests. — *Lime-water,* which throws down an orange yellow precipitate, *the hydrated binoxide of mercury.*

Iodide of Potassium, which gives a beautiful scarlet compound, *the biniodide of mercury.*

Protochloride of Tin, which first affords a whitish precipitate, *the protochloride of mercury;* and, on adding more of the test a grayish-black powder is formed, which consists of minutely divided *metallic mercury*.

Sulphuretted Hydrogen, which gives a blackish compound, *a sulphuret of mercury*.

To these may be added the test *by reduction*, the reducing agent being the protochloride of tin, assisted by heat.

Albumen is not now relied on as a reagent.

AGENT.

ANTIMONII POTASSIO-TARTRAS.
Potassio-Tartrate of Antimony.
Emetic Tartar.

Symptoms. — Violent gastric irritation; nausea; efforts to vomit; profuse perspiration; catharsis, accompanied with colicky pains and much flatus; increased secretion of urine; the heart's action at first much quickened, and afterwards scarcely perceptible; labored respiration; injection of the mucous tissues; extreme distress; death.

Treatment. — The yellow bark, or any other astringent vegetable that contains tannin, to be administered both in powder and decoction; a free use of dilutents, oleaginous purges, and opium, should then be had recourse to for the purpose of allaying the irritation.

Morbid Appearances. — The stomach highly inflamed, and eroded patches on the mucous coat, which are of a deeper color than the surrounding parts; intestines reddened, encoated with slimy mucus, and thickened; lungs gorged with blood; and general inflammation of the whole system consequent on its absorption.

Tests. — *Caustic Potass* and *Lime-water*, which precipitates the *oxide of antimony*. The *carbonate of potass* acts with still greater delicacy.

Hydrochloric and *Sulphuric Acids* also afford the like precipitate. A strong infusion of the gall-nut gives a dirty, yellowish white precipitate, *the gallate of antimony*.

Sulphuretted Hydrogen throws down an orange-red precipitate, *the red sulphuret of antimony*, which is so peculiar as to be always distinguished; and the reduction of this precipitate by hydrogen at once dissipates all doubt.

AGENT.

SALTS OF COPPER.

Symptoms. — The salts of copper are rarely employed as poisons to the horse. Large doses of the sulphate improperly given have sometimes caused much intestinal irritation, followed by colicky pains and diarrhœa; and, in one instance, death from gastro-enteritis. Doubtless inordinate quantities would always destroy life, when symptoms similar to those caused by any other erodent would be manifested, it being a local irritant. The same, perhaps, may be said of the impure acetate of copper.

In the neighborhood of works for smelting of copper, horses are frequently attacked with diseases of the joints, indicated by swelling, bursal distension, exostosis, and, ultimately, anchylosis, arising either from the state of the herbage or the impregnation of the air by the vapors disengaged.

Treatment. — Give a solution of the ferrocyanide of potassium, or of soap. Albumen is also an antidote, and metallic iron, which latter precipitates the copper. This is to be expelled by oleaginous purgatives. Tepid water rendered slightly alkaline may also be freely given; and opium, to allay irritation. A free use of demulcents, as gruel, infusion of linseed, etc., is indicated.

For the local affection, puncture the bursal distention, and, after the evacuation of the cyst, apply an elastic bandage, giving gentle compression. Remove the animal to another locality.

Morbid Appearances. — Stomach ulcerated where the agent has adhered, and general inflammation of its mucous lining and that of the intestines, with here and there spots of erosion. In the instance adverted to of the sulphate of copper causing death, there was an engorgement of the blood-vessels of the lungs.

Copper is with extreme difficulty detected in any of the secretions. It, however, has been found in the organic tissues, particularly the liver after incineration, and also in the blood.

Tests. — *Water of Ammonia* affords an azure-colored precipitate, or a violet-colored solution, *the ammoniuret of copper.*

Ferrocyanide of Potassium causes a brown precipitate, *the ferrocyanide of copper.*

Sulphuretted Hydrogen throws down a blackish compound, *the sulphuret of copper.*

A *piece of polished iron* introduced into the solution is soon covered with *metallic copper.*

AGENT.

SALTS OF LEAD.

Symptoms. — Of these, like the above, comparatively large quantities are required to cause death. Violent spasms, tremors, obstinate constipation of the bowels, followed by paralysis, partial or complete, are the usual indications.

In the neighborhood of lead works animals are often thus affected, when, in addition to these symptoms, there is a depraved appetite present: the stomach after death being found filled to repletion with strange and incongruous substances.

Treatment. — Solutions of the sulphate of magnesia or soda, combined with croton or linseed oil; afterwards allaying the irritation by means of opium.

The phosphate of soda has also been extolled as an antidote.

The treatment for the paralysis which remains consists in judicious dieting and exercise.

Morbid Appearances. — The lining membrane of the stomach and intestines is sometimes inflamed, sometimes blanched; the caliber of the latter is diminished, and its coats corrugated; the muscular tissue throughout the body has lost its power of contractility; the buccal membrane is pale, and the blood of a brighter color than natural.

Tests. — *Chromate of Potass* throws down a yellow precipitate, *the chromate of lead.*

Iodide of Potassium likewise gives a yellow precipitate, *the iodide of lead.*

Sulphuretted Hydrogen causes a black precipitate, *the sulphuret of lead.*

A *rod of Zinc* introduced into it causes a deposition of *metallic lead* in a crystalline form.

The alkaline carbonates and sulphates, although they give white precipitates with lead, have been objected to as tests.

AGENT.

POTASSÆ NITRAS.

Nitrate of Potash, Nitre.

Symptoms. — Much uneasiness; cholicky pains; pulse feeble, quick, and irregular; respiration accelerated; mouth hot; mucous lining of the eyelids and nostrils highly injected; increased secretion of urine; frequent voiding of fæces. If the quantity given is very great, the abdominal pains are more intense; the breathing more labored; the pulse quicker; ineffectual efforts to stale are made; the extremities are cold; and the prostration of strength is extreme. If not, after manifesting much uneasiness, the fæces are frequently voided; diuresis supervenes; and relief is obtained.

Treatment. — A free use of demulcents; oleaginous purgatives; hot rugs to the abdomen, and over the loins; opiate enemata; if necessary, the abstraction of blood; with hand-rubbing, and bandages to the extremities. Possibly a mustard cataplasm or a sheepskin over the loins will be of service. Such counter irritants as are likely to be carried to the kidneys are to be avoided.

Morbid Appearances. — The villous coat of the stomach highly inflamed and studded with dark spots resembling ecchymosis, varying in size and running into patches; they are easily scraped off, and contain a deposit of serum. The cuticular coat is also inflamed, but not so highly, and its texture is much weakened; the small intestines are pervaded with an inflammatory blush; the lungs and heart are congested, and the venous blood is of a brighter color than usual. Constriction and inflammation of the neck of the bladder have also been observed.

Tests. — From the fluid contents of the alimentary tube, or from the urine, the salt may be obtained by evaporation and crystallization. It is known by deflagrating when thrown on a piece of ignited charcoal, of which it animates the combustion; and by yielding nitric acid when distilled with sulphuric acid. Heat also disengages oxygen from it.

Chloride of Platinum added to its solution gives a yellow precipitate.

AGENT.

CANTHARIS.

The Blistering Fly.

Symptoms. — Much uneasiness; intestinal irritation; frequent attempts to stale; strangury; bloody urine; accelerated pulse and respiration; continual pain, with much constitutional disturbance. These symptoms increase in urgency, until death closes the scene.

Treatment. — Expulsion of the agent from out of the alimentary tube by brisk purgatives; oil is objectionable, on account of the solubity of the active principle of the cantharides in it. A free use of diluents; opium may be given, and oleaginous and demulcent injections thrown into the bladder. Sheepskins over the loins; hot rags over the abdomen.

Morbid Appearances. — The mucous lining of the alimentary canal throughout in a high state of diffused inflammation; but the urinary organs are principally affected; the blood-vessels of the kidneys, bladder, and urethra, are much engorged, and the lining membrane of the latter has in some places a sphaceletic appearance.

Tests. — Washing of the contents of the stomach and alimentary tube will develop portions of the beautiful green case-wings of the fly, which appear not quickly to undergo decomposition; Orfila having detected them in a body some months after death.

AGENTS.

BITE OF THE VIPER,

STING OF THE HORNET, WASP, ETC.

Symptoms. — Subcutaneous inflammation, indicated by swelling and increased heat, with effusion into the cellular tissue, which sometimes goes on to gangrene; pain; constitutional excitement; quickened and irregular pulse; rigors.

The bite of a viper has been known to cause cerebral derangement and death, by its influence on the nervous system.

Treatment. — Removal of the stings; counter-irritants, as liniment of ammonia or turpentine, which, if sufficient, are to be followed by scarification, the use of emollients, and the general antiphlogistic remedies, such as bleeding, fomentations, and laxatives, with opium to allay the general irritation. The virus of the tooth of the viper may be extracted by cupping, or, which is preferable, let the part be excised, and the nitrate of silver freely applied afterwards.

Morbid Appearances. — But few instances of death are recorded. If it takes place, it is probably the result of sympathetic and general excitement of the whole system; the usual indications, therefore, of increased nervous and vascular action may be expected to be met with.

Tests. — None.

II.—NARCOTIC POISONS.

These produce Stupor, Delirium, and other affections of the Brain and Nervous System, followed by Death.

AGENT.
OPIUM.
Opium.

Symptoms.— The horse will bear large doses of this drug; the quantity necessary to destroy life is consequently great. Supposed instances are recorded of its causing death when given in doses of a few drachms in order to check superpurgation; also when the animal has been debilitated by disease, when symptoms of enteritis have shown themselves, accompanied with a torpidate of the bowels, and much suffering previous to death. Much larger quantities have, however, often been given with impunity, and frequently with advantage.

Treatment.— Expulsion of the agent from out the alimentary tube by means of oleaginous purgatives, enemata, venesection, and a free use of demulcents, with fomentations to the abdomen, and counter-irritants to the extremities.

Should a state of stupor prevail, exercise may be given, and cold water dashed over the head.

Morbid Appearances.— The mucous lining of the stomach and intestines inflamed, and easily torn asunder, the inflammation being diffused. This has been thought to be a distinctive between the effects induced by mineral and vegetable poisons; but it cannot be relied upon, as repeated small doses of an erodent will induce the like appearances, only there will be more thickening of the tunics.

Tests.— Odor, which is characteristic. To the suspected matter add distilled water acidulated with acetic acid; agitate for a few minutes, filter and evaporate to the consistence of syrup; boil this in alcohol, and again filter and evaporate; dissolve the residuum in distilled water, and add to the solution acetate of lead, which leaves morphia in solution: this being heated with sulphuretted hydrogen, any remaining lead will be precipitated. On nitric acid being added to the morphia obtained by evaporation, it dissolves with effervescence, and becomes of an orange-red color. Suspended in water and treated with a drop or two of the permuriate of iron, it is also dissolved, and forms a greenish-blue solution.

AGENT.
TAXUS BACCATA.
The Yew.

Symptoms.— Effects variable; large quantities have sometimes failed to cause any injury, while at others comparatively small quantities have destroyed life. It would appear to be very insidious in its influence, as the animal generally drops down dead without indicating any previous suffering. In some instances slight convulsions have preceded death. The partially dried leaves appear to be more energetic than the green leaves, probably from greater quantities being partaken of.

Treatment.— Usually no opportunity is afforded for the employment of remedies. Should it, however, be the case that the leaves of yew are suspected to have been eaten, I am not aware of any method which could be adopted but that of endeavoring to expel them from the system as quickly as possible, which may be effected by active purgatives. The after-treatment will depend upon the symptoms which may present themselves.

Morbid Appearances.— The alimentary tube distended with fæcal matter in a semi-fluid state, and highly fetid gases.

The mucous lining inflamed throughout, particularly of the larger intestines, with here and there patches of extravasated blood.

In some few cases scarcely a trace of diseased action in the tissues could be found.

Tests. — Portions of the vegetable in the stomach and intestines mixed with the ingesta.

The active principle of the poison is unknown, hence the difficulty in the treatment.

AGENT.

ACIDUM HYDROCYANICUM.

Hydrocyanic Acid,
Prussic Acid.

Symptoms. — Its influence is sudden, and that of a powerful sedative to the system, and, when the quantity is not large, evanescent: otherwise it is followed by marked cerebral derangement, manifested by giddiness and coma; the breathing becomes laborious; the nostrils expanded; the pulse quickened and fluttering; much debility is present, with loss of power: to these succeed tetanic spasms; the muscles become rigid; the jaws locked, and the membrana nictitans is forced over the eye, which is prominent, and has a glassy appearance; profuse perspiration covers the body, accompanied by violent convulsions and intense suffering. These effects are succeeded by a remission for a time, during which the animal appears to be in a state of partial insensibility; but the exacerbations return again and again, and then the paroxysms become less and less powerful, until at length, all action disappearing, the animal is left in a state of exhaustion, the vital powers being much depressed. In whatever way the agent is introduced into the system, the effects are similar. The most active form of the poison is that of vapor.

When the dose is sufficiently large to cause death, it is unaccompanied with suffering.

Treatment. — Cold affusions over the body; the inhalation of dilute ammoniacal and chlorine gases, particularly the latter.

The coma may be removed by bloodletting; and diffusible stimulants, such as ammonia, may be administered, combined with tonics to rouse the depressed vital powers.

Of course this treatment will only be of service when the drug has been too frequently given, or administered in too large quantities.

Morbid Appearances. — The inner tunic of the stomach and intestines slightly inflamed; the vessels of the lungs gorged with blood; the parenchyma natural; the lymphatics containing red blood; the heart inflamed, and spots of ecchymosis on its lining membrane; the vessels of the brain highly injected, particularly those of the medullary portion, in which organ the odor of the acid is easily recognized, as well as throughout the body, and particularly in the halitus from the blood. The eyes are glistening and prominent.

Tests. — Render fluid the contents of the stomach, and distil an eighth part over, when the following tests will be available: *The odor,* which resembles that of bitter almonds, and impresses the throat and nostrils with a peculiar acridity.

Sulphate of Copper, the solution being rendered alkaline by potass, throws down a green precipitate, which becomes nearly white on adding a little hydrochloric acid, *the cyanide of copper.*

Sulphate of the Protoxide of Iron, similarly employed, gives a greenish precipitate, which becomes of a deep blue color on the addition of sulphuric acid, *the ferrocyanate of the protoxide of iron.*

Nitrate of Silver throws down a white precipitate, *the cyanide of silver,* which is soluble in nitric acid only at its boiling temperature, and which, when dried and heated in a tube, emits cyanogen gas, which burns with a rose-colored flame.

AGENT.

CARBONIC ACID.

Symptoms. — Instances are known of horses having been suffocated during fires, arising from the disengagement of this gas, with, perhaps, some of the compounds of hydrogen. Its sources otherwise are abundant. In a state of dilution it causes coma; when pure, spasm of the glottis, and death by asphyxia.

Treatment. — Removal to the air; cold affusions over the body; bloodletting; diffusible stimulants.

Morbid Appearances. — Engorgement of the vessels of the lungs with black blood. The vessels of the brain and of the heart are in a similar state. The bronchi and trachea filled with frothy mucus.

Tests. — The tests for carbonic acid gas are simple enough, but here they are inapplicable.

AGENT.

SULPHURETTED HYDROGEN.

Symptoms. — This gas, given off from cesspools and other places, has been at times the cause of death. It is rapidly absorbed by the blood, and produces coma and tetanic convulsions. Sometimes death takes place from asphyxia.

Treatment. — The same as the above; to which, perhaps, may be added the inhalation of dilute chlorine.

Morbid Appearances. — The muscles have lost their power of contractility. The blood-vessels are gorged with fluid black blood; the bronchial tubes inflamed, with increased secretion of mucus both in them and the trachea; the odor from the body is highly offensive.

Tests. — *Carbonate of Lead* on a piece of card paper, exposed to an atmosphere impregnated with this gas, is turned black by the formation of the *sulphuret of lead;* but as the body when undergoing decomposition emits the same kind of gas, this test can only be accepted as a corroborative proof.

III. — NARCOTICO-ACRID POISONS.

These cause Death, either by irritation or narcotism, and sometimes by both combined. Their influence is first local and then remote, impressing the Nervous System. They are principally derived from the Vegetable Kingdom.

AGENT.

NUX VOMICA ET STRYCHNIA.

Vomic Nut and Strychnia.

Symptoms. — The vomic nut induces a quickened and irritable pulse, highly labored respiration, snortings, tetanic spasms, loss of muscular power, injection of the mucous tissues, extreme thirst, and death from asphyxia; previous to which there is intense suffering. The action of its alkaloid, *strychnia,* is more energetic. It is shown by tremors, followed by a quickness of the pulse and labored respiration, extreme irritability, loss of power in the extremities, tetanic convulsions increasing in violence, the legs being thrust from the body, the muscles rigid, opisthotonus, profuse perspiration, insensibility, and the pulse and respiration being scarcely perceptible; the paroxysm exists for a few minutes only, and is followed by a remission of the symptoms, leaving the animal much exhausted and extremely irritable. The exacerbations, however, continue until death takes place from suffocation.

Treatment. — From the tenacity with which the powder of the nut adheres to the stomach and intestines, it is with difficulty dislodged. Its removal may be attempted by means of active purgatives, or antidotes may be thrown in; these consist of chlorine and of iodine, which form inert compounds with the active principle, strychnia; but, as the action of the alkaloid is on the spinal marrow and the brain, little good can be hoped to be obtained when a dose sufficiently large to destroy life has been given, unless active measures be immediately adopted. If the dose be not sufficiently large for this purpose, there will be a succession of paroxysms, leaving behind them much debility, which is to be counteracted by tonics and diffusible stimulants, with, perhaps, counter-irritants along the course of the spine, lest effusion should take place.

Morbid Appearances. — Mucous lining of the alimentary tube inflamed, lungs gorged with blood, and the vascular system throughout the body in a state of congestion. The spinal canal much inflamed. Effusion of bloody serum into the theca vertebralis; motor division of the spinal cord more injected than the other, and the nerves taking their origin from it inflamed. The membranes of the brain have been found inflamed, with effusion on the surface of the cerebellum, and a softening of the whole cortical portion of the brain. Rigidity of the muscles of the body. Rapid decomposition, accompanied with much fœtor.

Tests. — The powder of the nut has a greenish-gray color, an intensely bitter taste, and the odor of liquorice. Being collected, it is to be boiled in water acidulated with sulphuric acid, filtered, and the solution neutralized by carbonate of lime and evaporated to dryness. The dry mass being acted upon by successive portions of alcohol, these are to be evaporated to the consistence of syrup, when the product will be found to have an intensely bitter taste, and it becomes of a deep orange-red color with nitric acid, which color is destroyed by the protochloride of tin. Sometimes it deposits crystals of strychnia on standing. These tests will also be available for the alkaloid; to which may be added its sparing solubility in water, the alkaline reaction of its alcoholic solution, and its forming neutral and crystallizable salts with acids.

AGENT.
SEMEN CROTONI.
Croton Seed.

Symptoms. — This purgative, when incautiously administered, has produced death by inducing violent inflammation of the intestinal canal, followed by superpurgation; the alvine dejections being profuse, watery, and offensive.

Treatment. — A free use of demulcents, with astringents, as catechu, opium, and chalk. Bloodletting; opiate enemas. Hot rugs to the abdomen, counter irritants, etc.

Morbid Appearances. — Violent inflammation of the intestines, particularly the cæcum and colon, involving all the tunics, the mucous lining being easily torn. Fæces abundant and semi-fluid. Lungs in a state of congestion.

Tests. — None definite.

AGENT.
DIGITALIS PURPUREA.
Fox Glove.

Symptoms. — Languor, gastric irritation, coldness of the body and extremities, paleness of the mucous tissues, cold and clammy perspiration, quickened and feeble pulse, death.

When it accumulates in the system, after having been repeatedly given in comparatively small doses, it produces loss of appetite, nausea, languor, a quick and irregular pulse, followed by purgation, and the effects then gradually disappear.

Treatment. — Expulsion of the agent by means of a solution of aloes, combined with linseed oil. The free use of demulcents; diffusible stimulants; counter-irritants.

Morbid Appearances. — Depending upon the condition and previous state of the animal. If much debilitated, inflammation of the mucous lining of the stomach and alimentary tube may be seen to exist. At other times no trace of its influence on any of the tissues can be detected, and it is then supposed to cause death by exhaustion of the nervous energy.

Tests. — None definite.

AGENT.
VERATRUM ALBUM.
White Hellebore.

Symptoms. — Efforts to vomit, accelerated pulse, untranquil respiration, intestinal irritation, which, if followed by purging, affords relief; if not, these symptoms become more urgent, the body is covered with perspiration, saliva is secreted in increased quantities, the legs become deathly cold, inflammation of the bowels supervenes, and death.

Treatment. — A free use of demulcents. Milk has been strongly advocated; on what grounds beyond that of its being a bland fluid, I am at a loss to conjecture. Oleaginous purgatives; counter-irritants.

Morbid Appearances. — The villous coat of the stomach will be found inflamed; the intestines also in a high state of inflammation, particularly the cæcum and colon; the heart pale and flabby; and the lungs congested.

Tests. — None definite.

AGENT.

NICOTIANA TABACUM.
Tobacco.

Symptoms. — Nausea, giddiness, coma, feeble and irritable pulse. Sometimes general excitement of the system, profuse perspiration, labored respiration, pulse much quickened, partial insensibility.

Treatment. — Expulsion of the agent by purgatives; diffusible stimulants when coma exists; demulcents.

Morbid Appearances. — I am not acquainted with an instance of death having taken place, although this agent is frequently given as a vermifuge in very large quantities.

Tests. — None definite.

AGENT.

JUNIPERUS SABINA.
Savin.

Symptoms. — This, like the preceding agent, is given as a vermifuge, and sometimes incautiously. Gastric irritation is then evinced, the animal refuses food, and is languid; this is followed by diuresis, and sometimes by purging; the pulse becomes irregular and full, and the respiration hurried.

Treatment. — Expulsion of the agent from out the alimentary canal by oleaginous purgatives; demulcents.

Morbid Appearances. — Esophagus and stomach inflamed, particularly the villous portion of the latter viscus, on which patches of extravasated blood are seen to exist; the small intestines contain much mucus, and are slightly inflamed; lungs congested; larynx and trachea of a rusty yellow color; glands at the root of the tongue much enlarged.

Tests. — The partially digested vegetable matter found in the alimentary tube, which may be distinguished by its odor.

Under the head of Narcotico-Acrid Poisons, perhaps, should be placed the ATROPA BELLADONNA, *Deadly Nightshade*, which, in large doses, induces singultus, a dilatation of the pupils, feeble and irritable pulse, and a relaxed state of the bowels. Also many of the umbelliferous order of plants, as CONIUM MACULATUM, *Common Hemlock*, the influence of which is probably that of a Narcotic; CICUTA VIROSA, *Water Hemlock*, which, to some animals proves an energetic Poison; with a few of the natural family of the Ranunculaceæ, as the ACONITUM NAPELLUS, *Monkshood*, and HELLEBORIS NIGER, *Black Hellebore*, which cause death by irritation, producing gastro-enteritis, followed by delirium; likewise DELPHINIUM STAPHYSAGRIA, *Staresacre;* BRYONIA ALBA, *Wild-vine* or *Bryony*, and FELIS FŒMINA, *Female Fern;* of which latter very large quantities are required to effect any marked change in the animal system; and, indeed, it may be said of the Vegetable Poisons generally, that the Horse is enabled to resist the influence of comparatively immense doses of them, which in all probability arises from the peculiar structure of his stomach.

Wheat and Barley have been designated as poisons to this animal; and occasionally they have proved to be so, by setting up acute gastritis. A very common sequela of poisoning by Wheat is inflammation of the laminæ, the result of metastasis; and of Barley, a depilation of the skin. We are, however, in want of more correct information than at present we possess, before anything definite can be laid down under this head, as both wheat and barley, given in moderate quantities and with judgment, often prove beneficial.

I am induced to pass the agents above enumerated thus cursorily over, my object having been to give a condensed and tabular view of such substances as are known to destroy life in the horse when incautiously or maliciously administered, and to elucidate a Thesis on Poisons which I had the honor to read before the Members of the Veterinary Medical Association in 1836: at the same time, I hope that this attempt may prove of some use to the Student of Veterinary Medicine.—*Morton.*

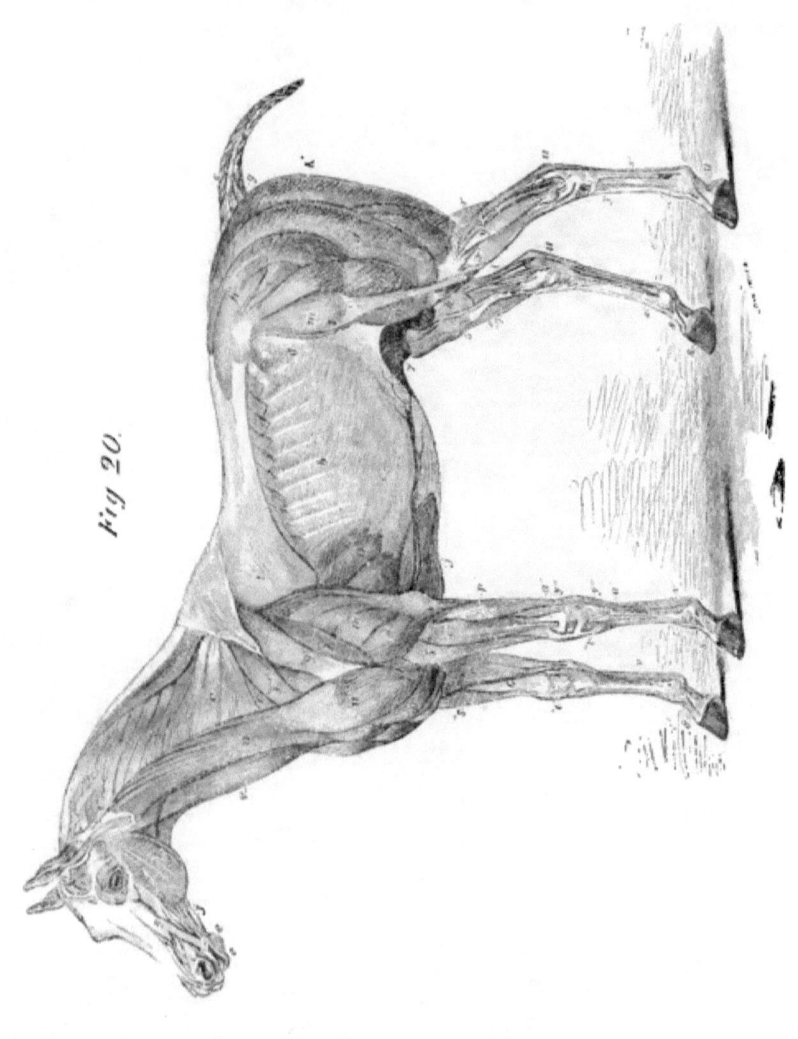

Fig 20.

EXPLANATION OF FIGURE XX.

THE HEAD.

- *a.* Orbicularis palpebrarum.
- *b.* Levator palpebrae.
- *c.* Dilator naris lateralis.
- *d.* Dilator naris anterior.
- *e, e.* Orbicularis oris, the circular muscle of the mouth; the letters are rather too low to indicate the muscle.
- *f.* Nasalis longus.
- *g.* Levator labii superioris.
- *k.* Masseter.
- *m.* Attolentes et abducens aurem.

THE NECK.

- *c''.* Trachelo subscapularis. — Scalenus.
- *s.* Splenius.
- *r, t.* Tendon of the splenius and complexus major.
- *u.* Levator humeri.
- *v.* Sterno maxillaris. The jugular vein is here shown between the two preceding muscles.

THE SHOULDER AND FORE EXTREMITIES.

- *e''.* Sterno scapularis. — Pectoralis transversus.
- *f''.* Antea spinatus.
- *g''.* Postea spinatus.
- *h.* Teres major.
- *m'', n''.* Triceps extensor brachii.
- *l.* Scapulo ulnaris.
- *o''.* Pectoralis magnus.
- *p'', u''.* Flexor metacarpi externus.
- *r.* " " internus.
- *r''.* Knee joint.
- *s'', 5.* Extensor metacarpi magnus.
- *t.* Extensor metacarpi obliquus.
- *u', v'', x.* Tendons perforans and perforatus.
- *y'', y''.* Extensor suffraginis.
- *z'', z''.* Pastern joint.
- 8. 8. Extensor tendons.
- 6. Radial vein.
- *g.* Flexor pedis.
- *k.* Humero cubital.

ABDOMINAL REGION.

- *b.* Intercostales.
- *c.* Transversalis abdominis externus.
- *d.* " " internus.
- *D.* Serratus magnus.
- *J.* Pectoralis magnus.
- 7. The sheath.
- 4. Superficial thoracic vein.

EXPLANATION OF FIGURE XX. CONTINUED.

POSTERIOR EXTREMITIES.

- *s'*. Erector coccygis.
- 9. Compressor coccygis.
- *k''. i.* The three glutei.
- *J'.* Triceps.
- *k'. k.* Biceps abductor tibialis.
- 5. Tibia.
- 6. *z. x. g.* Fleshy belly of the extensors.
- *l.* Plantaris.
- *m.* Tensor vaginæ.
- *n'.* Rectus.
- *O'.* Vastus externus.
- *r.* Gastrocnemius internus.
- *u. u. u. v.* Gastrocnemii.
- *y.* Peroneus.
- *y. z. x.* 8. Extensor tendons.
- *z.* (Off leg.) Flexors perforans and perforatus.

A

DICTIONARY OF VETERINARY SCIENCE:

CONTAINING

MANY PRACTICAL OBSERVATIONS,

OF MUCH IMPORTANCE TO

HUSBANDMEN AND HORSE OWNERS.

SELECTED FROM VARIOUS SOURCES, WITH ADDITIONS.

A DICTIONARY OF VETERINARY SCIENCE.

ABDOMEN. — That part of the animal usually denominated the belly. This cavity contains the intestines, or bowels, liver, spleen, pancreas, kidneys, etc., and is separated from the thorax, or chest, by the diaphragm.

ABORTION. — Our attention was called, a short time ago, to a mare, about eight years old, said to be laboring under colic. She had been driven very fast during the early part of the day; and about noon, when we saw her in the stable, she seemed to manifest considerable uneasiness. The surface of the body was cold, pulse small and intermittent. The genitals were considerably swollen, and a slight discharge from the vagina was observed. She had occasional uterine pains, which, however, were very feeble.

We immediately gave the following diffusible stimulant:

Powdered grains of paradise,	1 drachm.
" bethroot (*trillium purpureum*)	½ drachm.
Hot water,	1 quart.

This was administered from a bottle. In a few minutes, the parturient process commenced, and she shortly gave birth to a dead fœtus. The mare was in her fifth parturient month. She was put on a generous diet, and rapidly convalesced without any after treatment.

Remarks. — Great care and gentleness should be exercised toward mares during pregnancy. Hard work in harness, over bad roads, is likely to produce abortion; and mares that have once aborted are liable to a recurrence of the same. Light work and moderate exercise, however, are essential to their general health.

When the period of foaling draws nigh, the mare should be separated from her companions. Having foaled, she should be turned into a pasture, where there is a barn. The foal may be weaned at six months; if it should die, or be taken from the dam, humanity would suggest the propriety of a few weeks' rest, to enable the mother to recover from the effects of parturition.

Miscarriage, slipping, or slinking foal or calf, warping. — In mares, miscarriage is very generally caused by over-exertion during the latter period of gestation. It is not unfrequently brought about by accidents at grass, such as falling in a ditch or hole, and, struggling violently to extricate themselves. Kicks on the belly are by no means an uncommon cause of miscarriage; for which reason, a mare, when near her time, should be kept by herself: after foaling she will require a few weeks' rest, in order to recover from the effects of parturition; and, when first brought into work again, the services required of her should be very slight. Exposure to wet and cold will occasion miscarriage; also, high feeding and want of proper exercise. Abortion is of more frequent occurrence in sheep than in mares, and is caused by fright, overdriving, and being worried by dogs, and by being kept in cold, damp situations, and on improper food.

Cows are particularly liable to the accident of warping, or slinking the calf. The common cause of abortion is improper feeding. The filthy, stagnant water they are often compelled to drink is likewise a serious cause, not only of abortion, but also of general derangement of the animal functions. Dr. White states that "a farm in Gloucestershire had been given up three successive times in consequence of the loss the owners sustained by abortion in their cattle: at length the fourth proprietor, after suffering considerably in his live stock for the first five years, suspected that the water of his ponds, which was extremely filthy, might be the cause of the mischief; he therefore dug three wells upon his farm, and, having fenced round the ponds to prevent his cattle from drinking there, caused them to be supplied with the well-water in stone troughs erected for the purpose; and from this moment his live stock began to thrive, and the quality of the butter and cheese made on his farm was greatly improved. In order to show," says the same author, "that the accident of warping may arise from a vitiated state of the digestive organs, I will here notice a few circumstances tending to corroborate this opinion. In 1782, all the cows in possession of farmer D'Euruse, in Picardy, miscarried. The period at which they warped was about the fourth or fifth month. The accident was attributed to the excessive heat of the preceding summer; but, as the water they were in the habit of drinking was extremely bad, and they had been kept upon oat, wheat, and and rye straw, it appears to me more probable that the great quantity of straw they were obliged to eat, in order to obtain sufficient nourishment, and the injury sustained by the third stomach, in expressing the fluid parts of the masticated or ruminated mass, together with the large quantity of water they drank, while kept on this dry food, was the real cause of their miscarrying.

"A farmer at Chareton, out of a dairy of twenty-eight cows, had sixteen slip their calves at different periods of gestation. The summer had been very dry, and, during the whole of this season, they had been pastured in a muddy place, which was flooded by the Seine. Here the cows were generally up to their knees in mud and water. In 1789, all the cows in a village near Mantes miscarried. All the land in this place was so stiff as to hold water for some time; and, as a vast quantity of rain fell that year, the pastures were for a long time completely inundated, on which account the grass became bad: this shows that keeping cows on food that is deficient in nutrition, and difficult of digestion, is one of the principal causes of miscarriage." It is supposed that the sight of a slipped calf, the smell of putrid animal substance, is apt to produce warping. Some curious cases of abortion which are worthy of notice happened in the dairy of a French farmer. For thirty years his cows had been subject to abortion. His cow-house was large and well ventilated; his cows were in apparent health; they were fed like others in the village; they drank the same water; there was nothing different in the pasture; he had changed his servants many times in the course of thirty years; he pulled down the barn or cow-house, and built another, on a different plan; he even, agreeably to superstition, took away the aborted calf through the window, that the curse of future abortion might not be entailed on the cow that passed over the same threshold. To make all sure, he had broken through the wall at the end of the cow-house, and opened a new door. But still the trouble continued. Several of his cows had died in the act of abortion, and he had replaced them by others: many had been sold, and their vacancies filled up. He was advised to make a thorough change. This had never occurred to him; but at once he saw the propriety of the counsel. He sold every beast, and the pest was stayed, and never appeared in his new stock. This was owing, probably, to sympathetic influence; and the result of such influence is as fatal as the direct contagion." (See Youatt.)

The usual symptoms preceding abortion are a sudden filling of the udder, and a loose, flabby, and sometimes swollen appearance of the genitals, which discharge a little red-colored fluid. The lancet and medicine have been resorted to with very little success. Both of them are decidedly injurious; the animal should be put into some dry, sheltered place, by herself, and kept on boiled mashes and gruel for a few days.

ABSORBENTS.—Medicines which are given in view of absorbing gas or neutralizing acidity in the digestive cavity.

ABSORBENT VESSELS.—(See Lacteals, part first.)

ABSINTHIUM.—Common wormwood, used for the purpose of reducing swellings that have resulted from violence. Two ounces of wormwood are steeped in one quart of New England rum; if a limb of the animal is involved, the wormwood is then bound on with bandages, and the parts occasionally wetted with the fluid.

ACACIA.—Gum arabic, used as a demulcent and lubricant. In poisoning, it is useful to sheathe the membranes of the stomach and alimentary canal, and will defend them from the action of drastic purges.

ACACIA CATECHU.—Gum catechu. This is a powerful astringent: it is obtained from a tree that grows in Japan.

ACETABULUM.—The socket in which the head of the thigh bone is lodged.

ACETATED LIQUOR OF AMMONIA.—This has been long known by the popular term of Mindererus' spirit, and is made by pouring any quantity of acetic acid, diluted with seven times its amount of water, upon carbonate of ammonia, until all fermentation ceases, or until a neutral solution has been formed. It is useful in horse practice; it gently invigorates, is diaphoretic, and sometimes it proves mildly diuretic. It principally shows its salutary effects in the commencement of the febrile stage, or at the close of lingering febrile diseases, particularly of influenza. In the more early stages of epidemic catarrh, it may also be exhibited: the dose is from four ounces to an almost unlimited quantity. The author used this preparation with remarkable success in the treatment of influenza, which prevailed, in Massachusetts, in the fall and winter of 1855. The dose for horses and cattle is from three to four fluid ounces. It is generally given diluted with an equal quantity of water.

ACIDS.—Are distinguished by their sour taste; they readily combine with alkalies, producing effervescence. Those commonly used in veterinary practice are: sulphuric, nitric, hydrochloric, and acetic. They are all more or less corrosive, and decompose the vital tissues, by uniting with their serous, albuminous, and saline constituents. Therefore, when administered to the horse, they should be diluted with water.

ACONITA.—WOLFSBANE.—A powerful sedative; it moderates the action of the heart, and produces depression of the vital energies. It is generally used in the form of tincture. Dose, from ten to fifteen drops.

ACTION.—The gait of a horse; which depends on his powers and the mode of training.

ACTION OF MEDICINES.*—Every medicine is endowed with certain inherent characteristic actions, which distinguish it as decidedly as its physical and chemical properties. Thus, some medicines act on the bowels, causing purgation; others on the kidneys, stimulating the secretion of urine; and others on the brain and nervous system, causing insensibility; in fact, there is no part or organ of the body, except the spleen and pancreas, which is not influenced, and that often in several different ways, by some medicinal agent. It is impossible, however, to explain why a medicine should act in one way rather than in another; why, for example, aloes is purgative, and not diuretic, narcotic, or anæsthetic; or why chloroform is anæsthetic, and not vesicant, diuretic, or purgative. The student must therefore endeavor to conceive of these actions, or dynamical effects of medicines, in the same manner as he does of their more familiar properties of color, odor, taste, or density.

Some medicines, as demulcents, caustics, and astringents, have merely a local or topical action—soothing, irritating, corroding, or altering the animal tissues, but not extending their influence beyond the part to which they are first applied. Others, either with or without such a local effect, have a remote or indirect action on organs at a distance from the part with which they are first brought in contact. Medicines which act thus remotely or indirectly are thought to produce their effects in either or both of the two following ways: (a) They are absorbed into the circulation, and carried by the blood to remote organs; or (b), The

* Finlay Dun.

impression, which they produce on the parts with which they are first brought in contact, is transmitted along the nerves to other parts. The latter mode of operation is sometimes called action by sympathy.

(*a.*) The great majority of medicines appear to act in the former of these two ways, being taken up by the blood-vessels from the surface of the mucous membranes, skin, or other part to which they have been applied. Thus, most medicines given by the mouth, after having, if solid, undergone solution in the acid gastric juice or alkaline bile, pass, by a process of endosmose, into the capillary veins which ramify on the surface of the stomach and intestines, enter the general circulation by the mesenteric and portal vessels, and are thus carried to all parts of the body, altering, it may be, the nutritive processes of various organs and tissues, and at length expelled through some of the excretory channels, as the skin, kidneys, or bowels. The rapidity with which most substances are thus absorbed, and make the round of the circulation, is almost incredible. Professor Hering, of the Veterinary College, Stuttgardt, found that yellow prussiate of potash injected into one of the jugular veins of a horse appeared in the other in twenty-five seconds, and was exhaled from the mucous and serous membranes in a few minutes; and also that chloride of barium injected into the jugular vein of a dog reached the carotid artery in seven seconds. Dr. Blake observed that chloride of barium and nitrate of barytes traversed the whole circulation of a dog in nine seconds, and that of a horse in twenty seconds; and a similar rapidity of distribution doubtless obtains with substances which cannot easily be detected in the blood.

(*b.*) The other hypothesis, regarding the action of medicines, is that they owe the development of their effects to the production of some nervous impression on the part to which they are first applied, and its subsequent transmission to remote organs by means of the nervous system. A recent modification of this theory, advanced by Messrs. Morgan and Addison, assumes that the nervous impression is produced, not upon the part with which the medicine is first brought in contact, but on the interior of the blood-vessels after partial absorption. Part of the evidence in support of these hypotheses is derived from the fact that some poisons operate with such extreme rapidity as to render it doubtful whether there could be time for their being absorbed and making the round of the circulation. Thus: anhydrous prussic acid, conia the alkaloid of hemlock, and aconita the alkaloid of aconite, when injected into the veins, applied to the cellular tissue, or given by the mouth, produce almost instantaneous effects, and death in a few seconds. It appears, however, that the strongest evidence in favor of the theories under consideration consists in the effects of local injuries in producing constitutional disturbance. For example, a blow on the region of the stomach sometimes causes fatal swooning; distention of the stomach often produces hiccough; the presence of worms in the intestines sometimes induces epilepsy; and a local injury frequently causes fever and constitutional disturbance of all the more important organs of the body. In such cases the connection between cause and effect obviously depends on the transmission of nervous impressions only. And if topical causes are thus productive of remote effects, it is surely fair to infer that medicines and poisons may operate in a similar manner.

In fine, although it would appear that most medicines are absorbed and actually conveyed to the parts on which they act, and that such absorption and actual contact are essential to their action, yet it is highly probable (though not yet positively ascertained) that some substances, especially the more active poisons, astringents, and emetics, owe their effects to the production of a nervous impression, and its propagation to remote organs. Nor is it at all improbable that, under different modifying influences, certain substances will operate sometimes in one and sometimes in the other of these two ways.

ACTUAL CAUTERY.—Red-hot iron. (See FIRING.)

ACUTE.—A term applied to those diseases which are sudden in their attack and violence, accompanied with great pain.

ACUTE INDIGESTION IN CATTLE, CALLED HOVE, OR BLOWN.—When cattle have become fatigued by driving or by long fasting, and suddenly find themselves with plenty of food before them, particularly such as requires little mastication, as chaff, bran, grains, etc., etc.; and also at all times when they meet with food they have long been deprived of, as various artificial grasses, particularly red clover, they are apt to eat greedily, and omit to stop for the purposes of rumination; by which means the rumen or paunch becomes so distended as to be incapable of expelling its contents. From this, fermentation begins to take place, and a large quantity of gas generates which increases the distention, until the stomach, by its pressure on the diaphragm, suffocates the animal.

The *symptoms* are uneasiness and distress, with quickened respiration; sometimes there is a degree of phrensy present. When it is occasioned by green food, the evolution of gas is enormous, and the tympanitis gives a drum-like distention to the belly; but when dry food, as chaff, bran, etc., etc., has been taken, the impacted matter does not distend so quickly, and the symptoms are less acute; they resemble those of constipation; and sounding the side gives back a response as though a solid matter were hit against. It is thought to be more likely to occur in warm and wet weather than in any other; and, if such be the case, it must arise from the state of the vegetable matter and the surrounding warmth both being favorable to fermentation.

The *treatment* will consist in attempting to lessen the distention by *evacuating* the distending gas, or otherwise trying to neutralize it. Purgatives have little or no effect. The evacuation of the gases is effected by the introduction of a probang, which is passed down the esophagus; or it is brought about by puncturing the side, when the distention is urgent, or the want of assistance renders it imperative to evacuate the gas immediately, to prevent suffocation; a puncture is therefore at once made into it, which, among graziers, is called *paunching*. When nothing better is at hand, this may be performed with a lancet, or even a pen-knife; the wound made being kept open by the introduction of a piece of hollow elder or common wood; the place of puncture being midway between the ileum or haunch-bone and the last rib, a span below the transverse processes of the lumbar vertebræ on the *left* side, to which the first stomach or paunch inclines. A cattle trocar will, however, permit the gas to escape with certainty and speed, and should always be used to make the puncture, in preference to any other instrument. As soon as the air is perfectly evacuated, and the paunch is observed to resume its office, the *trocar* may be removed; the wound being carefully closed by a pitch plaster, or other adhesive matter. It is necessary to observe, that this operation is so simple and safe, that, whenever a medical assistant cannot be obtained, no person should hesitate a moment about doing it himself. The domestic remedies for lessening the distention, by condensing the gas, have been various; as oil of turpentine, and particularly ammonia, a strong solution of which in water has been found serviceable. The alkalies generally have long been used with variable success. Vinegar, in the *Quarterly Journal of Agriculture*, is strongly recommended; but, as it is observed that the elastic fluids developed are not always alike, so the effects resulting from the most reputed agents have too often failed. Mr. Youatt recommends the introduction of chlorinated lime, given in doses of from ʒij to ʒiv suspended in water.

But it is to a foreign veterinarian we are indebted for the best agent for neutralizing the gases given off when the rumen is distended. M. Charlet has recommended the chloride of potash, which substance has a great affinity for the compounds of hydrogen that usually form the major portion of those which exist in the stomach. This substance is to be given in doses of an ounce to a horse, half an ounce to a cow, and three drachms to a sheep. Occasionally, however, from the contents of the stomach being in a state of fermentation, no gas will escape, upon the probang or trochar being introduced. The chloride of potash is then to be poured down the probang or trochar, which ever may be used; and this substance must not be mixed with either mucilage or aromatic bitters, but sulphuric ether and cold water may be joined to it without injuring its effects.

ÆTHER NITROSUS. — Sweet spirit of nitre. Sweet spirit of nitre is stimulant, anti-spasmodic, diuretic, and diaphoretic. The dose is from one to two ounces, diluted with water.

AGE. — The age of a horse may be known by marks in the front teeth and tusks of the under jaw, until he is about eight years old, after which period it is a matter of guess-work; yet those who are expert can tell very near the exact age. There are many circumstances which tend to show whether a horse be old or not. The number of a horse's teeth is forty, — twenty-four grinders, and sixteen others, — by some of which his age may be known up to a certain period. Mares have only thirty-six teeth, as in them the tushes are usually wanting. A few days after birth, the colt puts forth two small front teeth in the upper and under jaws, and soon after two more; these are called nippers. The next four shortly afterwards make their appearance. The four corner teeth — as they are termed — come a few months after the last named. These twelve teeth, in the front of the mouth, are small and white, and continue without much alteration until the colt is about two years and a half old, when he begins to shed them. The two teeth that first make their appearance are the first that are lost, and are replaced by two others, called horse's teeth, considerably stronger and larger than those that have made way for them. Between the third and fourth year the two teeth next the first fall out, and are in like manner replaced by horse's teeth. Between the fourth and fifth year, the corner teeth are changed; the tushes make their appearance. About the fifth year, the horse is said to have a full mouth. After this period, up to the eighth year, the age of a horse can, with some degree of certainty, be known by the cavities in the teeth, which at first are deep, but are gradually, by the process of mastication, worn down, and about the eighth year disappear. After the fifth year, the above criterion of age may be corroborated by the grooves in the tushes of the male, which are inside; they are two in number. At six, one of these cavities, viz., the one next the grinder, disappears; at seven, the other is considerably diminished; and at eight is almost, but not always, entirely gone. After this period, the tushes become more blunt and round. The marks in the upper teeth are by some considered indicative of the horse's age; those in the two front teeth disappearing at eight, in the two next at ten, and in the corner teeth at twelve. The marks in the lower teeth will disappear about the eighth year.

As a horse grows old, he generally turns more or less gray; the cavities above the eyes become deeper; the under lip falls; the gums shrink away from the teeth, giving them the appearance of a greater length; the back becomes hollow, or curved.

AGE OF NEAT CATTLE is known by their horns. At the age of about two years, they shed their first fore teeth, which are replaced by others, larger and more prominent; about five, the early teeth are all replaced by the permanent ones. As the animal advances in years, these teeth wear down, the enamel disappears, and they assume a black or brown appearance. When three years old, a change takes place in the structure of the horns; after which period these appendages, like the permanent teeth, preserve the same character. After the third year, the horns continue to grow as long as the animal lives, and the age is indicated by the rings, or prominences, which are easily distinguished on the horn, and by which the age of the creature may be nearly ascertained, by adding three years to the number of rings.

AIRING, in the management of horses, implies exercising them in open air.

ALBUMEN.— That part of the white of an egg which coagulates into a solid mass when boiled; it abounds in the bones, muscles, cartilage, hoof, hair, etc.

ALCOHOL.— Rectified spirit. This is extensively used in medicine for making tinctures. With an equal quantity of water it is termed New England rum. Alcohol is a powerful irritant and caustic poison, to whatever part of the horse it is applied. If applied externally, it causes swelling, pain, and irritation; if given internally, it absorbs from the living parts the serous or watery portion, and condenses the fibrous structure. Alcohol, diluted in any form, acts on the horse as a diuretic, causing the kidneys to secrete a large amount of urine, in consequence of which they become overworked, and finally diseased. It is used as a diffusible stimulant. The best substitute is warm ginger tea.

ALIMENT.— That which nourishes the system.

ALIMENTARY CANAL.— The interior of the stomach and intestines.

ALKALIES.— There are different sorts: soda, potash, and ammonia, are alkalies.

ALOES.— Obtained from the aloe plant. The aloes now in use as a cathartic for horses, cattle, and sheep, are the Barbadoes. Pure Barbadoes aloes are of a dark brown color, present a rough appearance when broken, and have a rather pleasant aroma.

ALTERATIVES.— A class of medicines that act gradually and permanently upon the horse, by increasing the tone and vigor of the secreting, excreting, and absorbing system, without diminishing or destroying their power.

ALTHEA.— Marsh mallows. This plant is generally used in the formation of emollient drinks, as it contains a large amount of mucilage.

ALUM.— A mineral astringent, used to destroy proud flesh. "Alum is a powerful astringent, whether administered internally or applied externally. It may be given to the horse in does of from ʒij to ʒiv, and its employment has been attended with some benefit in obstinate cases of diabetes, also in diarrhœa, the primæ viæ having been previously emptied by means of laxatives. It has likewise been found useful in dysentery and lead colic. For either of these diseases it may be advantageously conjoined with opium and aromatics." (*Morton's Pharmacy.*)

AMAUROSIS, OR GUTTA SERENA.*— This disease, known by the term *glass eyes*, from the peculiar *glassy* appearance the organs assume, is generally considered as dependent on a paralytic state of the optic nerves, or of their expansions, the retinæ. By others, it is, however, thought to arise from the effects of inflammation, by which coagulable lymph is placed over the optic nerve, rendering the retina inaccessible to the stimulus of light; this can hardly be an occasional, and is certainly not the usual, cause. The disease, however, is likely to arise from any irritation of the brain; thus, it is found to follow staggers and the loss of large quantities of blood; which last-mentioned cause especially affects the nervous system. The veterinarian should make himself familiar with the appearances of this complaint, otherwise he may lie open to serious imposition. In amaurosis, a horse presents indications of blindness in his *manner*, though but little in his *eyes;* he seems cautious in stepping; lifts his legs high, and moves his ears quickly, as though endeavoring to make up by sound the intelligence lost by the deprivation of sight: but, above all, a hand moved close to the eye occasions no winking, unless held near enough for the motion to influence the air around, which an artful person might manage with ease. When this kind of eye is examined closely, the pupil will be found of one *invariable* see

* Blaine.

and unvarying hue; it will not enlarge and diminish as in a healthy horse, when removed farther from, or nearer to, the light; for the retina, ceasing to be influenced by the luminous ray, no longer controls the movements of the iris. It is, therefore, from the peculiarities in the manner of the horse, the invariable size of the pupil, and a *greenish glassy cast* in such eyes, that these cases may be distinguished. As it has hitherto proved incurable, we shall waste no time on its treatment.

AMMONIA. — This is a volatile alkali; is rapidly absorbed by water; and, by union with acids, forms several salts. The compounds of ammonia employed medicinally are: hydrochlorate of ammonia, sesqui-carbonate, and solution of the acetate of ammonia. The aromatic spirit of ammonia is a valuable stimulant and anti-spasmodic in colic or hoven. For the preparation of the latter, Mr. Morton gives the following formula:

Take of spirit of ammonia, 8 fluid ounces; volatile oil of lemons, 1 fluid drachm; volatile oil of rosemary, 1½ fluid drachm. Dissolve the oil in the spirit by agitation.

ANASARCA. — That form of dropsy that affects the whole, or nearly the whole system, or, in other words, an effusion of serum into the meshes of the cellular tissue.

ANASARCA, ŒDEMA, AND WATER FARCY.* — We need make no distinction between these terms, particularly the two first. As generally accepted, œdema carries probably rather a more local definition with it; thus, we say an œdematous swelling; but anasarca is more frequently used to designate an extensive dropsy of the cellular membrane. Both, however, have the same origin, and are accompanied by the same symptoms. It differs from ascites principally in its external seat, which is sometimes partial and sometimes general. It also appears under different forms, as it has different origins; and its terminations are also under the influence of these circumstances. A debility of the absorbent system is usually observed in the spring and autumn. There is, however, some general atony of the whole system, and the œdema disappears as the constitution establishes itself. At other times anasarca and œdema appear as accompaniments or sequelæ to acute diseases that have disturbed the functions generally, in the which case the absorbents become irritated; or to the œdema is added tumefied lymphatics.

ANÆSTHETICS. — Agents which produce insensibility to external impressions and to pain. The author uses, for inhalation, three parts of sulphuric ether to one of chloroform. In allusion to the use of anæsthetics, Mr. Morton writes: "Anæsthetics are less used in surgical and other painful operations in the lower animals than in man, on account of the larger quantities required, the difficulty of administration, and the undue prolongation of the preliminary stage of excitement. They have been used in parturition, and afford, as in the human subject, immunity from pain, but without apparent interference with the force or frequency of the involuntary contractions of the uterus. They have further been used for relieving the irritability and pain of such diseases as peritonitis, pleurisy, and pneumonia; for removing the spasms of tetanus, colic, and asthma; and for alleviating, by local application, the irritability of severe wounds. For all such purposes their use might, with advantage, be much extended."

ANASTOMOSIS. — The communication of blood-vessels with each other, or their opening one into the other, by which means, when the passage of blood through an artery or vein is prevented by ligature, compression, or any other cause, the circulation is still kept up by means of the anastomosing vessels.

ANATOMY. — The science that teaches the structure of the animal economy.

ANALYSIS. — The resolution of compound bodies into their original or constituent principles.

ANCHYLOSIS. — The loss of motion in a joint. There are two kinds, called com-

* Blaine.

plete and incomplete. In the former, the joint has grown together so as to be immovable; in the latter, some motion remains, and the rigidity is owing to the contraction and thickening of the ligaments. Anchylosis in the horse is not unfrequently a consequence of wounds or bruises; the latter, causing an absorption of the fluids that nourish the joint, anchylosis is the result. In bad spavins and ringbones, there is frequently anchylosis of the hock and pastern joints. The author's attention has lately been called to a case of ringbone that had been operated upon by some person totally unacquainted with the nature of the disease. The operation was performed in the most cruel and barbarous manner. The operator having never studied the anatomy of the parts, it could not be otherwise expected. On an examination of the animal, ossific or bony deposits were found inside the hind legs, in the form of a spavin; deposits also existed on the canon bones, and on the pasterns, thus proving that the disease was incurable; the general health was impaired, the knees sprung, and the animal was pronounced by the owner to be worthless; yet this specimen of inhumanity, the self-styled "doctor," had the audacity to state that he could perform a cure for the trifling sum of five dollars. The fact of his attempting to cure a constitutional disease by local means, under such unfavorable circumstances, shows that he was an ignoramus; and the barbarous manner in which he performed the operation, shows that he was destitute of every particle of humanity. The author has digressed merely for the purpose of warning owners of domestic animals against trusting them, when diseased, in the hands of those who are unacquainted with their mechanism.

ANEURISM. — A tumor filled with blood, communicating with an artery. It usually occurs from rupture of one of the coats of the artery, and dilitation of the cellular coat: it is then denominated true aneurism. When an artery is wounded, and the blood escapes into the surrounding tissues, it is called false aneurism.

The general mode of curing aneurism is by tying a ligature around the artery; the coats of the artery become united, and part of the artery obliterated; the circulation is carried on by anastomosing vessels. (See ANASTOMOSIS.) Some aneurisms have been known to undergo certain natural changes, by which they have been spontaneously cured, thus proving that the vital power is more efficient "than an evil system of medication."

ANISE SEED. — A mild carminative. It is much used in veterinary practice, and is one of the ingredients in cordial balls.

ANODYNES. — Medicines that relieve pain, procure sleep, and lessen the irritability of the nervous system.

ANTACIDS. — ALKALIES.

ANTHELMINTICS. — Medicines that are said to destroy worms, and are supposed to cause their expulsion from the animal. Many of the remedies recommended by some writers would be more likely to kill the horse, instead of the former. The proper method of preventing the generation of worms in the alimentary canal, is to pay attention to feeding, watering, etc., and give cathartics.

ANTIMONY. — A mineral poison. It has been extensively used in veterinary practice. There are numerous preparations of antimony, but they are all more or less objectionable. Large quantities of this mineral have been used on horses; yet, in some cases, where there is vital power enough in the animal to dispossess it from the system, no immediately unfavorable results were observed. Yet it is an agent of such diversified therapeutical powers, that the wisest of the faculty have never ventured to prescribe and fix limits to its action. (See TOXICOLOGICAL CHART.)

Mr. Finlay Dun, of the Edinburgh college, has lately made a series of experiments with tartar emetic, on horses, and he speaks very highly of it as antiphlogistic. The dose for a horse is from one to four drachms, either in bolus or solution, repeated as occasion may require.

ANTIDOTES. — See TOXICOLOGICAL CHART.

ANTISEPTICS. — Medicines that correct and prevent putridity. The best and most efficient are charcoal, Peruvian bark, acetic acid, and bayberry bark.

ANTI-SPASMODICS. — Medicines that are employed in spasmodic and convulsive disorders. The most efficient are assafœtida, pennyroyal, or any of the mints. The most powerful in spasm, or lockjaw, are lobelia, warmth and moisture, castor, musk, ginseng, and Indian hemp, or milk weed.

APOPLEXY. — A lesion of some of the vessels of the brain.

ARM. — A term applied to the upper part of the fore leg.

AROMATICS. — Medicines that have a warm, pungent taste, and fragrant smell; of this kind are cardamom seeds, cloves, and nutmegs, sweet flag, etc.

ARSENIC. — A destructive mineral poison. It has been used, in many diseases of the horse, without the slightest benefit. Dr. White states, "So various are its effects, that he has known a very small quantity to terminate fatally."

ARTERIOTOMY. — When blood is taken from an artery, the process is called arteriotomy. The proper place for puncturing the temporal artery, is at the precise spot where this vessel leaves the parotid gland to curve upwards and forwards around the jaw, which is just below its condyle. The operation should be performed with a lancet.

ARTERIO-PHLEBOTOMY is sometimes resorted to for the abstracting blood from the roof of the mouth and the toe of the foot; in such cases, however, a want of knowledge, as regards the anatomy of the parts, may occasion a serious hæmorrhage.

ASCITES. — Dropsy of the abdomen.

ASTHMA. — Supposed to originate in the muscles of respiration. (See COUGH, ROARING, etc.)

ASTRINGENTS. — Medicines that contract and condense muscular fibre. The principal are kino, catechu, oak bark, nutgalls, and bayberry bark.

ATMOSPHERE. — The name given to an elastic invisible fluid which surrounds the globe; it is composed of oxygen, nitrogen, and a small portion of carbonic acid gas. In stables that are not ventilated, the vapor arising from the dung and urine combine with it, and render it unfit for respiration.

ATLAS. — The first vertebra, or bone of the neck.

ATROPHY. — A wasting of the body.

AURICLES. — The two small cavities of the heart.

BACKGALLED. — When accidents of this kind occur, the saddle or harness should be padded, or chambered, so as to remove pressure from the part; sometimes they are difficult to heal, owing to the presence of morbific matter in the system.

BACKRAKING. — This is a name given by farriers to the operation of introducing the hand into the fundament, and emptying the rectum of its contents. The use of injections will, ere long, supersede this beastly practice. The most suitable injection to soften the fæces is warm soapsuds.

BACK SINEWS. — The flexor tendons of the fore and hind legs are so named. They are frequently strained, or otherwise injured, by over exertion or accidents.

BALL. — Bolus, or large pill. The mode of giving a ball is by drawing out the tongue to the right side, and holding it in the left hand, while an assistant stands on the left side and holds the mouth open. The ball is to be held by the finger and thumb of the right hand, drawn into as small a compass as possible, and passed as far as the horse's throat. This must be done by a quick motion of the hand, which should be kept toward the roof of the mouth, as there is more room for it in that direction.

BALSAM. — A name applied to several resinous substances, such as balsam of tolu, Peruvian balsam, balsam copaiba, etc., Canada or fir balsam; the medicinal properties are stimulant and diuretic.

BALSAM COPAIBA, or CAPIVI, is used for chronic cough; the dose is about one ounce.

BALSAM OF SULPHUR. — A preparation made by boiling sulphur and olive oil together, until united in the form of a dark-colored tenacious mass. This has been much esteemed by old farriers in obstinate coughs. When mixed with a small proportion of oil of anise-seed, it has been thought more efficacious, and is then named anisated balsam of sulphur.

BANDAGE.— Strips of linen, cotton, or flannel, about three or four inches wide. They are serviceable in habitual swellings of the legs, or weakness of the fetlock joint. They are likewise used for the purpose of keeping on dressings, or assisting in uniting parts that are cut or lacerated; they assist by pressure in expelling matter, or preventing the descent of ruptures, and as compresses for restraining bleeding or hemorrhage. The mode of applying the bandage to the leg is as follows: the material, after being cut the proper width, must be rolled up, and the bandage fixed by taking two or three turns in the same place; after which, the roller may be carried round spirally, taking care that every turn of the bandage overlaps about two-thirds of the preceding one. When the inequality of the parts cause the margin to slack, it must be reversed, or folded over; that is, its upper margin must become the lower, etc. A bandage should be moderately tight, so as to support the parts without intercepting the circulation, and should be so applied as to press equally on every part. In bandaging a horse's leg, the roller should be applied from the upper part of the hoof to the knee; in every case it is advisable to bandage from joint to joint, thus leaving the joint at liberty. When it is found necessary to bandage a joint, the bandage should be put on in the form of a figure 8.

BARB.— A general name for horses imported from Barbary. The barb, one of the most celebrated of the African races, is to be met with in Barbary, Tripoli, and Morocco; he seldom exceeds more than fourteen hands and a half in height. The barb requires more excitement to call out his powers than the Arabian; but, when sufficiently excited, his qualities of speed and endurance render him a powerful antagonist to the Arabian.

BARK.— This name is generally applied to several different species of Peruvian bark, the yellow and the red. The active principle of the yellow bark is an alkaloid principle, called quinæ, combined with a peculiar acid, called kinic, or cinchonic, in the state of an acid salt; besides these, it contains an oily and a yellow coloring matter, tannin, kinate of lime, and woody fibre. Their value in treating diseases of the horse consists in their tonic and astringent properties. It should be given to the horse in the form of infusion; one ounce of powdered bark to a quart of boiling water. It is also useful to restore indolent ulcers to a healthy state. The best tonic for a horse is hydrastis Canadensis (golden seal).

BAR SHOE.— A particular kind of shoe, which is sometimes used to protect the frog from injury; also in corns.

BARS OF THE FOOT.— (See FOOT, part first.)

BARS OF THE MOUTH.— Transverse ridges on the roof of the mouth; they are most conspicuous, or full, in a young horse. When swollen, or fuller than usual, the horse is said to have the lampas.

BASILICON.— A digestive ointment, composed of equal parts of olive oil, yellow beeswax, and common resin (or rosin). These are to be melted over a slow fire, and stirred until the mixture is quite cool.

BAY.— A bay color, in horses, is so named from its resemblance to dried bay leaves.

BICEPS.— The biceps is a double-headed muscle, which serves to bend a limb.

BILE, or GALL.— A bitter, greenish fluid, secreted by the liver for the purpose of assisting digestion. In the horse there is no gall bladder, or receptacle for the bile; it passes directly into the duodenum, or first part of the small intestines, a few inches from the stomach.

BILIOUS.— Diseases are called bilious when they depend on a morbid state of the liver.

BITS.— There are various kinds of bits in use; among them are the snaffle and curb. A snaffle may be either plain, or twisted, but the latter is apt to make the mouth callous; it consists of two pieces, having a sort of hinge joint in the centre. When used for the purpose of breaking young colts, it should be made large, so as not to hurt the mouth. The form of the curb bit resembles somewhat the letter H. The bridle is fastened to the side pieces, which act as levers of different powers, according to the distance from the cross-bar, to which the bridle is attached. The humane man will never inflict unnecessary severity on the horse, and will avoid continual strain on the reins or bridle, which, aside from the torture they inflict, tend to render the horse's mouth callous. The best form of bit, and the most simple, is the stiff, arched bit. The author has seen a very fine specimen of this article, manufactured by Messrs. Hannaford & Ilsley, of this city. The centre piece is large and curved; the cheeks are movable, and their upper ends curved outwards, which prevents their injuring the cheek bones. It is very important that a horse should be properly bitted; many docile horses are rendered stubborn and unmanageable, by having a bit that is too narrow. Many young horses are injured while they are teething, and the mouth is tender, by bearing too hard on the rein. The author would suggest a trial of an India rubber centre piece, in such cases.

BITE OF ANY RABID ANIMAL. — In most works on veterinary science, the writers recommend excision, or cutting out the bitten part, and afterwards cauterizing with the firing iron; but this method is very unsatisfactory, and only puts the animal to unnecessary torment. The morbid matter from a rabid animal is generally taken up by the absorbents, sometimes in a few seconds, and the operation of cauterizing would then be of no avail. The treatment we recommend is, to dose the animal with a tea of lobelia; half a pound of the herb and seed may be steeped in two quarts of scalding water, and given in doses of half a pint, at intervals of an hour. A large poultice of the same should be bound on the bitten part, and kept in contact with the parts by bandages, and the poultice renewed every six hours, until all signs of poisoning disappear. The animal should be kept on scalded shorts, in moderate quantities.

BLADDER. — The bladder is a musculo-membranous bag, situated, when empty, in the cavity of the pelvis. Its use is to contain the urine, which flows into it through the ureters, from the kidneys. It is divided into three parts, viz., the fundus or bottom, the body, and the neck. When full, the fundus of the bladder protrudes out of the pelvis, into the abdominal cavity; it then receives a covering from the peritoneum. Its other coats are an internal mucous membrane, and an external muscular coat, formed of two distinct sets of fibres; the one longitudinal, and the other circular. The former are thickest about the fundus, the latter about the neck or cervix, — which, by this arrangement, is always kept closed, except during the time of voiding the urine. On opening horses that have died from accident, we sometimes find the bladder empty, and its muscular fibres so condensed that it appears like a solid mass of small dimensions; such is the contractile power of its muscular coat, by which, with some assistance from the abdominal muscles and diaphragm, the urine is expelled. The author has opened several horses that have died from lockjaw, and found the bladder distended to its utmost capacity, containing about a gallon and a half of dark-colored fluid, resembling coffee-grounds. In one case, the muscular fibres about the neck of the bladder were lacerated by the over-distention and spasm of the neck of that organ. When horses are accustomed to drink too much water, without being allowed to stale often enough at work, the bladder becomes over-distended, and often paralysis, weakness, or local debility sets in, and the neck of the bladder becomes at length so relaxed as to be unable to offer sufficient resistance to the muscles that propel the urine into the urethra, so that it is constantly dribbling off as fast as it is secreted. This is termed incontinence of urine.

Sometimes the irritability of the bladder, in the latter case, depends on the acrimony of the urine; and, whenever this is the case, attention to feeding, watering, etc., will remove it. Diseases of the kidneys and bladder are accompanied with tenderness over the loins, and a remarkable stiffness of the hind legs. Whenever the bladder is distended with urine, recourse should be had to the catheter.

Palsy, or paralysis, of the bladder, is sometimes dependent on functional derangements, as stomach staggers, or injuries to the brain and spinal marrow.

BLASTING. — When cattle or sheep are first turned into luxuriant pasture, after being poorly fed, they frequently gorge themselves with food, which, fermenting in the rumen, or paunch, so distends it with gas that the animal is often in danger of suffocation. The symptoms are most distressing; and, unless relief be speedily afforded, death very commonly ensues. If the symptoms are very alarming, a flexible tube may be passed down the gullet: this will generally allow the gas to escape, and afford temporary relief, until more efficient means are resorted to; these will consist in arousing the stomach and digestive organs to action, by stimulants and carminatives, and counteracting the tendency to putrescence by doses of charcoal or soda. Some practitioners recommend puncturing the rumen, or paunch; but there is always some danger attending it, and, at best, it is only palliative, and the process of fermentation will proceed; the gas may escape, but the materials that furnished it still remain. Youatt states: "A cow had eaten a large quantity of food, and was hoven. A neighbor, who was supposed to know a great deal about cattle, made an incision into the paunch; the gas escaped, a great portion of the food was removed with the hand, and the animal appeared to be considerably relieved, but rumination did not return: on the following day the animal was dull; she refused her food, but was eager to drink. She became worse and worse, and on the sixth day she died;" thus proving that the remedy was worse than the disease.

When animals are blasted in a moderate degree, the carminative drink, and decoction of lobelia, will prove effectual. In all cases of hove, it will be advisable to give injections of warm water, to which add a handful of salt, and the same quantity of charcoal. As a means of preventing the blast, it may be remarked, that animals should never be turned into any nutritive pasture while the dew is on the ground, or after rain.

BLEEDING. — The practice of abstracting blood has received the seal of antiquity, yet that is no argument in favor of its usefulness; and, in view of improving in the future, the author here introduces an article on the subject, by Professor Buchanan:

"We affirm that bleeding is a barbarous and unscientific remedy, and deny that it is ever necessary. In this matter we take our stand upon the facts recognized by the highest authorities in medical literature. We refer to the most recent and accurate researches in chemistry and pathology; to the experimental investigations of Andral, Magendie, Louis, Simon, and many others, which have settled, beyond all doubt, and placed among the permanent facts of medical science, to be received by all medical schools of whatever therapeutic faith, the phenomena of the blood, when its composition has been affected by hemorrhage, by bleeding, and by various other agencies.

"It is indisputably established that bleeding produces a special change in the composition of the blood. The change which it produces is not a removal of any *effete* or morbid materials, — not a removal of any element which tends to create or aggravate disease, — but a removal of the most necessary and healthy portion, upon the presence of which we depend for the maintenance of health and vigor. Bleeding inevitably reduces the red or globulous portion of the blood, because it removes or destroys a certain amount of the red globules, and the loss which it produces is readily supplied by absorption of water and of comparatively crude materials, while the highly-organized globules are regenerated with great slowness and difficulty.

"It is a well-established fact, that the red globules of the blood are essential to life, and that their abundance or scarcity is a criterion of the vital force and activity of the constitution. As the proportion of the red globules increases, the general vital power rises, and the activity or energy of all the organs increases; while a diminution of their ratio enfeebles or disorders the various organs, and predisposes to nervous and tuberculous disorders, and to the whole range of adynamic and cachectic diseases. If the ratio is diminished as much as one-seventh, general debility is the consequence, predisposing to disease and diminishing the power of recovery; if as much as one-fourth or more, this reduction of vital power is incompatible with health, and inevitably results in some form of disorder.

"Is it not, then, exquisitely absurd to adopt, as a remedy in disease, a measure which, even in the most vigorous health, tends directly, with rigorous precision, to destroy the vital powers and *bring on disease?* Yet this measure has been, and still is, sustained by many medical men, although clinical experience, as well as chemical science, has shown its injurious effects, and thousands in America and Europe have been, and are now, demonstrating that all forms of disease may be better treated without bloodletting than with it.

"We affirm that, in disease, the pathogenetic elements of the blood should be removed, instead of its healthful and necessary constituents. Nature has provided for the removal of all noxious materials, by numerous appropriate outlets, which discharge every thing that is injurious to human health. It is the duty of the physician to aid nature by such medicines and means as will rouse the secretions and excretions, and thus insure the restoration of the blood to a perfectly healthy condition. When, for want of knowledge how to accomplish this, he destroys with unnatural violence a large portion of the vital blood itself, which is as necessary to the body as its solid tissues, he acts with as much scientific precision as the savage, who would treat a case of convulsions, not by removing its causes, but by cutting out a portion of the convulsed muscles."

It will be very difficult, however, to convince some of the "*older heads*," and the world in general, that bleeding can be dispensed with; therefore the veterinarian must be prepared to please his employer, and do just as his superiors have done, — or else "*loose caste*" and practice.

BLEMISHES. — They consist of broken knees, loss of hair, cracked heels, false quarters, splents, windgalls, spavins, etc.

BLIND, MOON. — A disease of the horse's eyes, which is supposed to be the forerunner of cataract, and often ends in total blindness.

BLISTER FLY. — Cantharides, or Spanish fly. The object in applying a blister is to promote absorptions and to combat deep-seated inflammations.

BLOODROOT. — Sanguinaria Canadensis, used to prevent the growth of fungus, or proud flesh; a substitute for caustic.

BLOOD SPAVIN. — Enlarged bursæ.

BOTS. — Short reddish worms, which are often found attached to the horse's stomach. Mr. Clark says "that bots are not, properly speaking, worms, but the larvæ of the gad-fly, which deposits its eggs on the horse's coat in such a manner as that they shall be received into his stomach, and then become bots. When the female fly has become impregnated, and the eggs are sufficiently matured, she seeks among the horses a subject for her purpose, and, approaching it on the wing, she holds her body nearly upright in the air, and her tail, which is lengthened for the purpose, carried inwards and upwards. In this way she approaches the part where she designs to deposit the eggs; and, suspending herself for a few seconds before it, suddenly darts upon it, and leaves the egg adhering to the hair by means of a glutinous liquor secreted with it. She then leaves the horse at a small distance, and prepares the second egg; and, poising herself before the part, deposits it in the same way; the liquor dries, and the egg becomes firmly glued to the hair. This is repeated by various flies, till four or five hundred eggs are sometimes deposited on one horse. They are usually deposited on the legs, side, and back of the shoulder, — those parts most exposed to be licked by the animal: in licking, the eggs adhere to the tongue, and are carried into the horse's stomach in the act of swallowing. The bots attach themselves to the horse's stomach, and are sometimes, though less frequently, found in the first intestine. The number varies considerably; sometimes they are not half a dozen, at others they exceed a hundred. They are fixed by the small end to the inner coat of the stomach, to which they attach themselves by means of two hooks. The slowness of their growth, and the purity of their food, which is supposed to be the chyle, must occasion what they receive in a given time to be proportionably small; from which, perhaps, arises the extreme difficulty of destroying them by any medicine or poison thrown into the stomach." A large amount of opium, tobacco, and corrosive sublimate, sufficient to destroy the horse, have from time to time been given; and, on opening the stomach, these animals have been found uninjured. "The presence of bots in the horse's stomach is not easily ascertained, as it is certain that great numbers have been found after death in the stomach, without appearing to have produced any kind of inconvenience to the animal while alive. It does not appear that any effectual remedy has yet been discovered for bots." Mr. Blaine says, "that he has kept them alive for some days in olive oil, and in oil of turpentine, and that even the nitrous and sulphuric acids do not immediately destroy them. At a certain season of the year, they detach themselves from the stomach, and pass off with the excrement." A run at grass is the most effectual remedy.

COMPOUND FOR BOTS. — Persons desirous of treating a horse for bots, can use the following:

Powdered poplar bark, ... 4 ounces.
" mandrake, ... 2 ounces.
" balmony (snakehead), . . 4 ounces.
" wormseed, ... 2 ounces.
" golden seal, ... 1 ounce.
" slippery elm, ... 4 ounces.

Mix. Divide into sixteen powders, and give one, night and morning, in the food.

Regimen. — The animal should be kept on a generous diet; green food or succulent and agreeable vegetables will effect a change and assist to detach the bots. If, however, such articles cannot be procured, let the horse have a mess of scalded shorts every night.

BOW-LEGGED. — Defective conformation of the legs.

BOX, LOOSE. — A loose box, as it is generally called, is a place wherein a horse is turned without being fastened to the manger or rack: such a place is useful to turn a horse into when he is sick, or when the mare is about foaling.

BRAIN. — The connection that exists between the brain and stomach, by means of the eighth pair of nerves, or par vagum, is the cause of this important organ being often disturbed in its function. Thus it is that, when the stomach is loaded with food, its function becomes deranged, and the brain is affected sympathetically. A diseased action is then set up, and all the functions become more or less deranged. A horse in this case will become dull and languid, and sometimes labor under symptoms of apoplexy. In consequence of this nervous communication between the stomach and brain, the latter organ is sometimes affected by the irritation of bots in the stomach. The best way to prevent apoplexy, staggers, etc., is, by attention to diet, exercise, etc.

Dropsy of the brain does not often occur to horses or cows; but sheep appear to be more liable to the disease than other quadrupeds. The symptoms of the disorder in horses are variable. "In one case there was a considerable degree of dulness and heaviness about the head; the pulse was not much affected, but there was loss of appetite. The animal appeared as if suffering much pain in the head, generally keeping it lower than the manger. These symptoms were followed by delirium, convulsions, and death. In another case, when probably the water had accumulated very gradually in the ventricles of the brain, the horse appeared to be free from pain, except when the circulation was hurried by brisk motion, when he would fall down in violent spasms, the fit seldom lasting but a few minutes. This horse, being of scarcely any value, was destroyed, and, upon opening the brain, about six ounces of water escaped." Sir George Mackenzie has described two kinds of this disease which sometimes happen to sheep: "The first consists of an accumulation of water in the ventricles of the brain; the other — which is most common — arises from animalculæ, called hydatids. In this case, the water is contained in cysts, or bags, unconnected with the substance of the brain, on which it acts fatally by pressure. Very soon after water has begun to collect, either in the ventricles or cysts, the animal shows evident and decisive symptoms of the disease. He starts, looks giddy and confused, as if at a loss what to do; retires from the flock, and sometimes exhibits a very affecting spectacle of misery."

BREAKING. — The breaking of young horses is a matter of great importance, and should never be intrusted to any one of a cruel or harsh disposition, as, under such a master, the very best-tempered horse may be rendered vicious. They are often broken when much too young; they are often found racing at three, and in constant work before they are four years old. This is one of the causes of contracted feet and lameness, that are continually presenting themselves to our notice. Farmers in general put their colts to work too young; and, although exercise may improve their growth and constitution, yet this advantage is more than counterbalanced by their being shod at a period when their feet are tender.

BRIDLE HAND. — The left is called the bridle hand, in contradistinction to the right, which is termed the whip hand.

BROKEN WIND. — The origin of broken wind is supposed to be a morbid secretion from the membrane lining the windpipe, bronchial tubes, and ramifications; the air-cells are somewhat ruptured, and the air is entangled in the cellular substance, or common connecting membrane. The bulk of the lungs is greatly increased, while their capacity for containing air is diminished. It is stated in Rees' Cyclopædia, under the head of broken wind, "that, after opening more than ten broken-winded horses, their lungs were uniformly found emphysematous. (See EMPHYSEMA.) This complaint is generally considered incurable; but it may often be alleviated by constant attention to diet." The animal should be fed on shorts, and green food if it can be procured, and boiled carrots. When used, his exercise should at first be moderate, and he should never be exercised immediately after feeding. If the horse shows any disposition to eat the litter, a muzzle must be provided.

According to Mr. Richard Lawrence, "the most common appearance of the lungs in broken-winded horses is a general thickening of their substance, by which their elasticity is in a great measure destroyed, and their weight specifically increased. At the same time, their capacity for receiving air is diminished." Dr. White writes, "that he has examined the lungs of broken-winded horses without observing this general thickening of their substance; on the contrary, they have appeared superficially lighter and larger than in their natural state. Two horses were purchased for the purpose of making experiments, and so badly broken-winded as to be useless. In the first, the lungs were unusually large, and there was a considerable quantity of air in the cellular membrane; but it was not ascertained whether the air had escaped from the air-cells, or had been generated within the common cellular membrane. The other horse was kept about a month in a field where there was no water and very little grass. When taken up, he appeared perfectly free from the disorder. He was shot; and, upon examining the lungs, they had not the slightest appearance of disease." This proved the superiority of nature's remedies over those of man. The same author relates that he purchased a broken-winded horse that was incapable of working. By allowing him only a small quantity of hay, sprinkled with water, giving mashes, mixed with a small quantity of oats, and only a small quantity of water, taking care at the same time that he had regular and moderate exercise, his wind became gradually better, and he afterwards was perfectly free from the complaint.

The author has examined the lungs of two horses which were said to be afflicted, for some time previous to death, with broken wind, without detecting a loss of continuity in their structure; neither was their specific gravity diminished.

BRONCHIA. — (See WINDPIPE.)

BRONCHOTOMY. — The operation of opening the windpipe for the purpose of producing artificial respiration, or to remove any substances that may have lodged in the upper part of the larynx.

BURNS are best treated by a mixture of equal portions of lime-water and linseed oil, the parts being frequently anointed with the mixture.

BURSÆ MUCOSÆ. MUCOUS BAGS, or SACS. — These are described as membranous sacs, containing a fluid similar to synovia, or joint oil, and interposed between tendons and the parts on which they move. In violent exertions these vascular membranes, which secrete and confine the synovia, are injured; hence we have windgalls, bog-spavin, etc.

BUTTERIS. — An instrument used by horse-shoers for paring the horse's hoofs.

CÆCUM. — The blind gut. So named because it is open at one end only. In the horse this part of the intestines is remarkably large.

CALF, DISEASES OF. — Many of the diseases of calves originate in a disordered state of the stomach, either from taking too much milk at a time, or from the milk not being sufficiently fresh, or being taken from a cow whose health is impaired. Whenever the stomach is disordered, either by the quantity or quality of the milk, it causes a variety of disorders, such as scouring, want of appetite, costiveness, colic, yellows, convulsions, etc.

CALKINS. — A name given to the prominences on horses' shoes, which are turned downward for the purpose of preventing their slipping.

CALVING. — At the end of nine lunar months the period of the cow's gestation is complete; but the parturition does not exactly take place at that time, — it is sometimes earlier, at others later. "One hundred and sixteen cows had their time of calving registered: fourteen of them calved from the two hundred and forty-first day to the two hundred and sixty-sixth day, — that is, eight months and one day to eight months and twenty-six days; three on the two hundred and seventieth day; fifty-six from the two hundred and seventieth to the two hundred and eightieth day; eighteen from the two hundred and eightieth to the two hundred and ninetieth day; twenty on the three hundredth day; five on the three hundred and eighth day; consequently there were sixty-seven days between the two extremes. Immediately before calving, the animal appears uneasy; the tail is elevated; she shifts about from place to place, and is frequently lying down and getting up again. The labor pains then come on, and, by the expulsive power of the womb, the fœtus, with the membranes enveloping it, is pushed forward. At first the membranes appear beyond the vagina or shape, in the form of a bladder of water: this soon bursts, the water is discharged, the head and fore feet of the calf are protruded (in natural labor) beyond the shape. The body next appears, and the delivery is complete. In a little time afterwards, some trifling pains take place, which separates the afterbirth, or cleansings; and these being expelled, the process is finished.

When the membrane breaks, and the

fluid escapes early in calving, and before the mouth of the uterus is sufficiently expanded, the process is often slow, and it is a considerable time before any part of the calf makes its appearance. The practice of hurrying the process by introducing the hand, or driving the animal about when symptoms of calving appear, is very improper. It has been known in many instances to cause the animal's death. It sometimes appears that a wrong presentation takes place, and renders the calving impracticable without assistance. In such cases it is necessary to introduce the hand in order to ascertain the position of the calf, and change it when it is found unfavorable. When, for example, the head presents without the fore legs, which are bent under the breast, it cannot, in this position, be well drawn away without danger. In this case the calf should be gently pushed back in the uterus, placing the cow in the most favorable position, and taking the opportunity for so doing while there are no pains nor straining. When the calf is pushed back, the fore legs are to be carefully drawn downward, in a line with the head, and brought out into the vagina. The author has known several cases, where parturition was seemingly difficult, of a resort to force in extracting the calf; but it should be recollected that nature is never to be interfered with in the process of delivery, or in any of her operations, unless it is clearly ascertained that assistance is necessary. When much force is used in drawing the calf, and especially if the animal be rather fat, a disease of the womb is apt to follow, puerperal fever sets in, which often proves fatal. Great mischief is also done by endeavoring to extract the calf without regard to its position in the uterus: it is sometimes so placed that delivery is not practicable until the position of the calf is shifted. When much force is used in drawing the calf, it sometimes happens that the womb falls out or is inverted, and great care is required in putting it back, so that it may remain in its situation. In doing so, there is an advantage derived from placing the cow in such a position that the hind parts may be higher than the fore. If any dust or straw remain about the womb, they should be carefully removed before the womb is put back. A linen cloth is then to be put under the womb, which is to be held by two assistants. The cow should be made to rise, that being the most favorable position, and the operator is then to grasp the body of the womb with both hands, and gently return it. When so returned, one hand is to be immediately withdrawn, while the other remains to prevent that part from falling down again. The hand at liberty is then to grasp another portion of the womb, which is to be pushed into the body like the former, and retained with one hand. This is to be repeated until the whole of the womb is put back; if the womb does not contract, friction with a brush around the belly and back may excite the muscles to contraction: should this fail, the animal may have an astringent and aromatic drink, made by infusing three ounces of ground poplar bark in about three pints of hot water; when cool, administer with a horn or bottle, taking care, while pouring down the œsophagus, to let it fall gently and gradually; by that means it will pass over the pillars of the œsophagean canal, and on to the third stomach; otherwise it would fall into the rumen, and defeat the object in view.

CAMPHOR.— A narcotic vegetable concretion. This medicine, says Dr. White, "is employed both internally and externally. It is given inwardly as an anti-spasmodic, as in lockjaw, when it is commonly joined with opium; and as a febrifuge, or fever medicine, joined with nitre and antimonial powder." Mr. Morton writes: Camphor has been occasionally given in tympanitis, and it has been supposed to act by rousing the vital energies. In a state of fine powder it is sometimes sprinkled over a linseed-meal poultice, when it has been found to allay irritation; although, as a sedative, thus applied, it is not equal to the extract of the deadly nightshade.

On account of its sedative influence, it may be advantageously combined with opium or digitalis for chronic coughs. Given for any length of time, it pervades the system, and is excreted by the lungs and kidneys.

CANTHARIDES. — *Spanish Fly.*[*] — Several preparations of cantharides are now in use; for example,

Vinegar of Cantharides.
Take of Cathurides in powder, . 1 part.
" Diluted acetic acid, . . 8 parts.
Macerate for fourteen days, and filter.

Oil of Cantharides.
Cantharides in powder, . . 1 part.
Olive oil, ... 8 parts.
Digest in a water bath for two hours, and filter for use.

Ointment of Cantharides.
Take Cantharides in powder, . 1 part.
" " Hogs lard, . . 6 parts.
Digest in a water bath and filter through paper.

CAPPED HOCK. — A swelling on the point of the hock, generally occasioned by blows; they seldom cause lameness; but, as they are a considerable blemish, an attempt should be made to reduce them by counter-irritants; friction is also useful.

CAPSICUM. — In its pure state it contains tonic and stimulant properties. It increases the physiological or healthy action of the system.

CAPSULAR LIGAMENT. — The ligament by which two bones are joined together. It forms a complete sac round them, and serves to confine the synovia, or joint oil.

CARAWAY SEEDS. — These are cordial and carminative. The dose is from one to two ounces.

CARBON. — Pure charcoal, unmixed with any foreign body. It is antiseptic and absorbent; useful as a poultice for putrid sores.

CARDITIS. — Inflammation of the heart.

CARIES. — Ulceration or rottenness of a bone.

CARMINATIVES. — Medicines that correct flatulency, or expel wind; the principal of these are the caraway and fennel seeds.

CAROTID ARTERY. — A large artery, that runs on each side of the neck, near the windpipe. The jugular vein runs immediately over the artery. Yet at the upper part of the neck they are at such a distance that there is no danger of wounding the latter in bleeding.

CARTILAGE. — Gristle. A smooth, elastic substance attached to bones. Cartilages are situated in parts where elasticity is required; they render the parts connected with them capable of slight changes of form, and instant recovery, to accommodate themselves to accidents and circumstances, without serious injury to themselves. There are also inter-articular cartilages; that is, flat, smooth cartilages, between the ends of two bones. These, being covered with synovia, or joint oil, serve to facilitate the motion of the joint.

CASTOR. — A peculiar matter found in sacs, near the rectum of the beaver. It is used as an anti-spasmodic, in doses of two drachms (for a horse), mixed in thin gruel.

CASTRATION. — An operation often performed on horses, and other domestic animals. The best method of performing it is by means of the clams, and ligature.

CATAPLASM, or POULTICE. — This application, when designed to promote suppuration, or formation of matter, is best made by mixing together equal parts of slippery elm and flax-seed, pouring a sufficient quantity of boiling water on the mixture, to make it of the consistence of mush, and binding it on the part; the bandage should not be so tight as to interfere with the return of blood by the veins. A poultice should always be renewed every twelve hours.

CATARACT. — A disease of the horse's eye. A cataract may be partial or total. The partial cataract is known by specks in the pupil, which interrupt vision in proportion to their size, and according to their situation. In the total or complete cataract, the whole of the pupil becomes of a white or pearl color. A horse's sight is least injured by partial cataract, when the speck is most remote from the centre of the pupil, and near to the upper margin. When a complete cataract takes place in one eye, the strength of the other becomes estab-

---
* Morton's Manual.

lished, so that the horse soon accommodates himself to the loss.

CATARRH, or COLD. — This is, perhaps, a disorder more common in horses than any other. The author attributes some colds (in this city) to the bountiful use of Cochituate water. When the horse has just arrived from a journey, or is in a state of perspiration, the showering process, so much in fashion just now, is decidedly injurious. If the legs of the animal are sluiced with water, and he is afterwards suffered to stand where a current of air blows on him, he is likely to take cold. Horses accustomed to warm clothing and warm stables are, of course, most liable to cold. The symptoms are cough, dulness, want of appetite, discharge from the nostrils, frequently accompanied by sore throat and difficulty of swallowing.

CATARRH, EPIDEMIC. — The epidemic catarrh is so named from its spreading over a country as a general disorder, often for a considerable time. When the disease is so prevalent, it is supposed to depend on a certain state of the atmosphere.

CATHETER. — A gum elastic tube, for the purpose of drawing off the urine. The one used for the horse is about four feet in length.

CAUSTICS. — Preparations that destroy the part to which they are applied.

CELLULAR MEMBRANE. — The substance by which various parts of the body are united to each other. The cells of which this structure is composed communicate with each other; which is proved by making a small opening in the skin of an animal, introducing a blow-pipe, and blowing through it, by which the adjacent skin will puff up; if sufficient power were employed, the air may be thus forced all over the body.

CEREBELLUM. — The small brain. It is situated immediately behind the cerebrum, or large brain, and upon the origin of the spinal marrow.

CHEST FOUNDER. — (See FOUNDER.)

CHRONIC. — A term used to denote a disease of long standing, unaccompanied by fever or inflammation.

CHYLE. — A milky fluid, formed by the action of the gastric, pancreatic, and bilious fluids. Chyle is absorbed and carried by the lacteals to the thoracic duct; but, previous to its arrival there, it passes through the mesenteric glands, where, probably, it undergoes some change.

CICATRIX. — The mark that remains after a sore, wound, or ulcer has been healed.

CIRCULATION OF THE BLOOD. — (See HEART.)

CLIPPING. — Cutting the long, rough hair of a horse. It is chiefly done to improve the appearance of the horse. The author doubts its utility. (See article HAIR, part first.)

CLOTHING. — A pernicious custom is often adopted of keeping horses clothed in the stable; making no difference in the warmth of the clothes, whatever the season of the year or the state of the weather may be. (This custom is not so prevalent here as in England.) In a good stable, it is probable that even in winter it might be dispensed with; and a horse will then be much less liable to take cold, when he happens to stand in a cold wind or rain. When a horse is moulting, or shedding his coat, light clothing might be useful; and, at such periods, showering, or standing out in the rain, would be very injurious. In summer, the horse should have a net thrown over him to protect him from the flies.

CLYSTERS, OR GLYSTERS. — A liquid preparation, forced into the rectum by means of a syringe.

COFFIN BONE, or OS PEDIS. — The bone which is inclosed by the hoof.

COFFIN JOINT. — (See HOOF.)

COLIC. — A very common disease in horses. It begins with an appearance of uneasiness; he paws his litter; sometimes makes ineffectual attempts to stale; stamps with his feet; gathers up his legs, and lies down heavily; groans, and looks round to his flank; lies down heavily again, as before, and rolls on his back. The body sometimes swells. If relief is not promptly afforded, all the above symptoms gradually increase; the pulse becomes quick, the breathing disturbed, and the pain is so great that a violent perspiration breaks out, and the horse becomes almost delirious, throwing himself about the stall, so that it is dangerous to come near him.

CONDITION.— This term is used to imply a horse being in perfect health.

CONJUNCTIVA.— The external coat or membrane of the eye. (See EYE, part first.)

CONSUMPTION.— In consumption there is a gradual loss of flesh and strength, while the appetite is seldom impaired in the early stages. It is sometimes accompanied by a discharge from one or both nostrils, and a swelling of the glands under the jaw; such cases are often mistaken for glanders. Consumption does not often take place suddenly, but is very insidious in its attack; and it often happens, that the complaint is not much noticed till tubercles are formed in the lungs, and the mesenteric glands are diseased. When a horse is observed to lose flesh, his coat staring, his skin feeling as it fast to the ribs, he should be warmly clothed, and fed on scalded shorts, oats, and boiled carrots; by proper attention to stable management, he may gain flesh and strength, his coat will become smoother, and his skin looser. Should it now be the season of the year when good grass can be procured, this will perfect the recovery. The best medicines are cod liver oil and phosphate of lime.

CONTAGION.— The mode in which a disease is communicated from one animal to another. It is derived from the word *contact*, or touch, and is used in contradistinction to infection, which implies the communication of disease by unwholesome miasmata, sometimes spreading to a very considerable distance.

CONVALESCENCE.— A state of recovery from illness, or an approach to a state of health.

CONVULSIONS.— Under this name, Gibson has classed lockjaw and staggers. Modern writers treat of these diseases under their respective heads. Calves are subject to convulsive diseases, from indigestion, and the consequent formation of acid in the stomach. It is often occasioned by some bad quality in the milk they drink, when fed by hand. Taking too much milk will often bring on the disorder. Carminatives and tonics generally afford relief; after which, it is necessary to be more attentive to the future mode of feeding, giving a little gruel occasionally.

CORDIALS.— Medicines are thus termed that possess warm and stimulating properties, such as ginger, caraway seeds, anise seeds, etc.

CORNEA.— The outer transparent part of the eye.

CORNER TEETH.— The outermost of the front teeth are thus named.

CORNS.— Corns generally appear at the inner angles of the fore feet, from injuries, etc.

CORONET BONE.— Os corona. The second of the consolidated phalanges of the horse's foot.

CORONET.— The upper part of the hoof, where the horn terminates.

CORROSIVE SUBLIMATE.— Among the poisons that are given, with a view of curing disease, corrosive sublimate seems to stand foremost in the destruction of vitality, and the production of incurable diseases. Dr. White remarks: "It is necessary to observe carefully its effects; for, whenever it takes off the appetite, or causes uneasiness of the stomach or bowels, it should be immediately discontinued. A solution of corrosive sublimate in water has been employed as a lotion in mange, but is generally considered dangerous; a fatal disorder of the bowels having in several instances followed its use. Five cows, that were bathed with a solution of corrosive sublimate in tobacco water, died soon after.

COUGH.— A cough is sometimes the first symptom of a cold, or catarrh; but there is another kind of cough, which accompanies indigestion. Horses that eat too much hay, and drink a large quantity of water, often have chronic cough. This can be removed by proper attention to feeding.

COWS, DISEASE OF.— The disorders of cows are not so numerous as those of the horse; they are often brought on by feeding on improper food, or by being kept on low, marshy grounds. Cattle that are brought from a warm to a colder climate, and such as are naturally of weak constitution, are most liable to disease.

CRAMP. — A spasmodic affection of the muscles, either of a particular part, or of the whole body. In lockjaw, for example, the muscles of the jaw are at first chiefly affected; but, gradually, unless relief is afforded, the spasm, or cramp, generally extends to the neck, limbs, and at length to all parts of the body.

CRASSAMENTUM. — Red globules, or coloring matter, of the blood, mixed with coagulable lymph.

CREMASTER. — A muscle which surrounds the spermatic cord, as it passes out of the belly into the scrotum. Its use is to suspend and draw up the testicle.

CRIB BITING. — A disagreeable and injurious habit, which some horses acquire; it consists of laying hold of the manger with their teeth. It generally proceeds from indigestion.

CROPPING THE EARS. — The ear may be inclosed between the two parts of a carpenter's rule, which can be adjusted and held so as to give the ear any shape that may be required. All that part outside the rule is then cut off with one stroke of a sharp knife, and then bathed with tincture of myrrh.

CRUPPER. — A strap affixed to the saddle, with a loop at the end, for the purpose of admitting the horse's tail.

CUD. — The food contained in the first stomach, or rumen, of a ruminating animal, which is returned to the mouth to be chewed at the animal's leisure.

CUMIN SEEDS. — A carminative, or cordial.

CURB. — A swelling of the horse's hock, generally caused by blows or strain.

CUTANEOUS DISEASES. — Diseases whose seat is in the skin, as the mange, for example. They are generally dependent on a vitiated state of the secretions, and a disordered state of the bowels.

CUTICLE, or SCARF SKIN. — A thin, insensible membrane, which covers and defends the true skin. It is this which forms the bladder raised by blistering.

CUTIS. — The skin, or hide, which lies under the cuticle. Besides the cuticle and skin, horses and other large animals have a muscular expansion, which lies immediately under the latter, called the fleshy panicle, by which the skin is moved, so as to shake off dust or flies, or anything that hangs loose upon the hair.

CUTTING. — A horse is said to cut, when he strikes the inner and lower part of the fetlock joint, in travelling. The usual mode of correcting this, is to make the outer side of the shoe higher than the inside.

DEBILITY. — Debility may be permanent or temporary. In the first, the constitution is naturally weak, or has been rendered so by improper treatment, or sickness; the second generally arises from over-exertion, and, if the exciting cause be frequently repeated, terminates very commonly in a total decay of the constitution. Rest and kind treatment are the best cure for weakness induced by fatigue. The greatest attention should be paid to the degree of work that a horse is capable of enduring, as what may be salutary for him at one period may greatly exceed his strength at another; and this generally depends on the mode of stable management. The common practice of working horses too early frequently results in debility.

DECOCTION. — The process of extracting the virtues of a substance by boiling it in water. The liquid so prepared is termed decoction. Almost all the medicinal properties of plants may be extracted by pouring boiling water over them. In boiling they lose their volatile properties.

DEGLUTITION. — The act of swallowing. The power of swallowing is often impeded in the horse by sore throat, distemper, etc. This impediment is only of a temporary nature; but there is another, which is of a more serious kind, and interferes with mastication as well as with swallowing. The grinding teeth of horses often wear down in such a manner, that the outside edge of the upper grinders irritates or wounds the cheek, and the inside of the lower grinders acts similarly upon the tongue, or the skin connected with it. Whenever a horse is observed to void unbroken oats with his dung, the teeth and cheeks should be examined. It will often be found necessary to rasp the outside edges of the upper grinders, and sometimes the inner edges of the lower ones.

DEMULCENTS.— Medicines of a mucilaginous kind, which sheath the mucous membranes when they are tender and irritable, and defend them from the action of what would otherwise injure them. Of this kind are marsh mallows, linseed tea, solution of gum arabic.

DENTITION.— The act of changing the teeth, which is going on from the second to the fifth year. During this period, the horse's mouth is apt to become tender, which renders it necessary to keep him for a short time on scalded shorts, or boiled carrots.

DIABETES.— An excessive discharge of urine, accompanied by thirst and debility. There are three outlets for the fluids of the body,— the surface, the lungs, and the urinary passage. When either is deficient in action, one or both of the others must make up that deficiency; so, excess in one produces deficiency in the others; hence, in diabetes we often find a dry skin and staring coat; and in excessive perspiration, the urine is scanty, whatever be the organs affected or whether the one or the other be excessive or diminished. The indications are, to equalize the action of these opposing or sympathizing surfaces, by restoring the diminished secretions, and cleansing and toning the organ whose action is excessive. All direct efforts to produce specific effects, without regard to a balance of action through the whole animal, do more harm than good.

DIAPHORETICS.— Medicines that promote insensible perspiration, or excite moderate sweating. Of this class are lobelia and emetics, given in infusion.

DIAPHRAGM, MIDRIFF, or SKIRT.— A muscular and tendinous expansion, which divides the cavity of the chest from the abdomen, or belly.

DIARRHŒA, or PURGING.— In Professor Percivall's lectures on diarrhœa, he states that, "for the majority of cases brought to us, we are indebted to the groom, the farrier, and stable-keeper, who used to kill many horses by literally purging them to death. Thirty years ago, an ounce and a half or two ounces of aloes, occasionally combined with one or two drachms of calomel, composed the common purge; and even now, among these people, nine, ten, and eleven drachms are by no means unusual doses. Young horses, on their first arrival in the metropolis, are all physicked; they have given to them, indiscriminately, doses of aloes, every one of which would be sufficient to purge two of them; the result is, that the light-carcassed, irritable subject is carried off at once by superpurgation, while another, or two, may linger in misery and pain from a dysentery that will end in gangrene and death, or be rendered more speedily fatal by the doses of opium, or some other powerful astringent, which are so perniciously resorted to on these occasions. There is another not uncommon cause of this disease, and that is, continuous and excessive exertion. After having been ridden for many hours, a horse will often express irritation in the bowels, by frequently voiding his excrement, which will be found to be enveloped in a slimy or mucous matter, that is called by some molten grease."

DILUTENTS.— Those substances that increase the fluidity of the animal economy. Water may be justly considered as the only dilutent.

DIRECTOR.— A grooved instrument, made for the purpose of conducting the knife in opening sinuses and in several other operations of surgery.

DISLOCATION.— A displacement of a bone from its socket. A dislocation of the fetlock joint may be replaced, and kept in its position by bandages; the horse should not take any exercise until it is completely healed. A dislocation of the stifle, or patella, must be reduced by bringing the horse's leg under the belly, and then depressing the outer angle of the patella, or stifle bone, with the hand, which gives the muscles the power to draw the bone into its place. Generally speaking, dislocations are rare.

DISTEMPER. — This name is applied to diseases that prevail at particular periods, and spread to a considerable distance. (See EPIDEMIC.)

DOCKING. — Cutting off part of the tail. If this is ever necessary (and the author doubts it), then the operation should be performed before the animal is two years old.

DRASTIC. — A term applied to purgatives that operate powerfully.

DRENCHES, or DRINKS. — When it is necessary that any medicine should operate speedily, this is the best form in which it can be given. A bottle with a short neck is the best drenching instrument. In giving a drench, the tongue should be at liberty, the head moderately elevated; the drench is then poured down moderately. The head is to be kept in an elevated position until the drench is swallowed. If the animal happens to cough while the drench is in his throat, the head should be immediately let down.

DRESSING. — A term employed to designate medical applications to a wound, or ulcer, and the operation of cleaning a horse.

DROPSY. — This disease consists in a collection of serous or watery fluid, either in cavities, as the chest, belly, or ventricles of the brain, or in the cellular membrane under the skin. Dropsy is more a symptom of disease than a disease itself; but sometimes, on account of the violence and danger of the symptom, it is often treated as a disease. The proximate cause is a check to perspiration; the remote cause is bleeding, or any thing that can debilitate the general system.

DROPSY OF THE CHEST. — This is sometimes a consequence of disease of the lungs; and, when it happens, those important organs generally are so far disorganized, or injured, that there is very little chance of the animal's recovery.

DROPSY OF THE BELLY, or ASCITES. — The causes are the same as above; the only difference is, that, from circumstances predisposing, the fluid is determined on the peritoneum (see PERITONEUM) instead of the pleura.

DUCT. — A membranous tube, or canal, through which certain fluids are conveyed. Thus the lachrymal duct conveys tears from the eyes to the nose.

DUNG. — By examining a horse's dung, we are enabled to judge of the state of his health. When the dung is hard, and in small knobs, and covered with slime, laxative medicines are beneficial; and when it is passed in too great quantities, it commonly arises from too liberal allowance of food. If oats are voided whole, it will generally be found to be caused either by a defect of the teeth, or by a too voracious appetite, occasioning the food to be swallowed without mastication; in which case the animal should be fed on shorts, or scalded food.

DUODENUM. — The first intestine that comes from the stomach. (See INTESTINES.)

DURA MATER. — A strong membrane that invests the brain and divides it into two lobes. It likewise separates the large brain (cerebrum) from the small, or cerebellum.

EAR. — The horse's ear is merely an organ for collecting sound; consequently he has complete power over the muscles attached to them, and can turn them in different directions.

EFFLUVIA. — Invisible vapors that arise from bodies.

EFFUSION. — The oozing out of serum, or coagulable lymph, from the blood-vessels.

EMBROCATION. — A liquid preparation for rubbing upon the skin, and generally used for strains, bruises, and enlarged glands.

EMBRYOTOMY.* — When, from weakness, a very narrow pelvic opening on the fore part of the mother, or monstrosity on the part of

* Blaine.

the foal, no efforts can bring the fœtal mass away entire, it must be dismembered. A knife made for the purpose, having the blade concealed, with the haft lying within the hollow of the hand, is to be taken up into the vagina. We are told that, occasionally, hydrocephalus in the colt prevents the head from passing. Such a case will detect itself by the volume that will be felt on examination, and which will be easily lessened by plunging the point of the knife in the forehead, and evacuating the contents by pressing the skull in; when, laying hold of the muzzle, the head may be brought through the pelvic opening. But it is usually the natural size of the head which forms the obstruction; in which case the head itself must be removed. When the head has been dissected off and brought away, it will be necessary probably to contract the volume of the chest; which will not be difficult, by cutting the cartilaginous portions of the ribs, detaching the thoracic viscera, and then crushing, or rather moulding, the empty thorax together; after which the rest of the body will offer little obstruction. When the head cannot be got at, the limbs must, one by one, be detached; after which the body, and at last the head, may be drawn out either entire, or lessened considerably.

EMETIC, TARTAR. — Tartarized antimony. A corrosive metallic poison. Dr. White relates: "From examining the stomachs of horses that have taken this mineral, I am satisfied that irreparable mischief may be done with it, and certain it is that a vast deal of unnecessary pain has been thus inflicted."

EMPHYSEMA. — Swellings which contain air. Such swellings are known by a kind of crackling noise, or sensation, when they are pressed with the finger.

EMULGENT ARTERIES. — The arteries which convey blood to the kidneys.

ENCANTHIS. — A disease of the inner corner of the eye.

ENCYSTED. — A term applied to tumors which consist of a solid or liquid substance, contained in a sac, or cyst.

ENTERITIS. — Inflammation of the bowels. This is a very serious form of disease, and is the result of plethora, or the sudden application of cold to the surface. It is sometimes owing to an overloaded state of the stomach and bowels. Obstinate spasm will also produce it.

EPIDEMIC. — Diseases which spread over a whole country, at certain seasons, are thus named. If many suffer in the same manner, it is called epizootic. There are very few diseases which assume, in its latter or earlier stages, such a variety of forms; perhaps depending on the location, and the peculiar state of the constitution. Youatt writes: "In 1711, an epidemic commenced, which, although it sometimes suspended its ravages, would visit new districts; it also appeared in a certain district, and confined itself to that location. In 1747, it appeared, and would seem as if there was a strange caprice about it. It would select its victims, the best of the herd, around a certain district, and confine itself to that location for a short time; then disappear for several months, return, and pounce upon this privileged spot. In some districts, it would attack the mouth and throat; commencing with a loss of appetite and difficult respiration, terminating with a discharge of blood from the anus. Sometimes the animals will eat and work until they suddenly expire; others will linger in dreadful agony." It appears that this malady is not infectious; for the same author writes: "Cattle were in the same barn as those infected; they ate of the same fodder that the distempered beasts had slavered upon, drank after them, and constantly received their breath and odor, without being the least affected. In 1756, it assumed a different form; some cattle were taken all at once with violent trembling of every limb, and blood ran from the nose, and bloody slime from the mouth, and the animals died in a few hours."

EPIGLOTTIS. — The cartilage which covers the larynx, or top of the windpipe, at the time food or water is passing into the gullet.

EPILEPSY. — The falling sickness; fits. Horses, cats, and dogs are subject to fits, which often depend upon an accumulation of water in the ventricles of the brain, or upon the irritation of worms in the stomach or bowels. During the present year, the author's attention was called to a horse, (the property of Mr. Downs, of this city;) the horse was lying down, and at times appeared insensible; convulsive struggling would take place occasionally. The muscles of the eye were affected by spasm, and distorted; the duration of the fit varied. As the disease progressed, the hind extremities were paralyzed, and the horse would struggle violently at intervals of fifteen minutes. On an examination, after death, nearly a peck-measure-full of the long round worm was found in the small intestines. The author examined the brain of a horse that was said to die in a fit, and found about five ounces of water in the ventricles of the brain. These fits in horses do not exactly resemble those occurring in man.

EPISTAXIS.— Bleeding at the nose. This sometimes occurs in glanders, and denotes a considerable ulceration within the nostrils. When it happens to a horse in health, it shows an unequal circulation of the blood.

EPSOM SALTS.— Sulphate of magnesia. A neutral salt, often employed as an aperient for cattle; but it is very uncertain in horses, and is apt to gripe them.

ESCHAR.— A slough formed by the application of caustic.

EXCRESCENCE.— Any preternatural formation on any part of the body, as warts, wens, etc.

EXOSTOSIS.— An osseous tumor originating from a bone; such as splent, spavin, ringbone, etc. Perhaps no animal is more subject to this disease than the horse; and in no department of the veterinary science is there a greater need of reform than in the treatment of the disease now under consideration. Almost every man who knows anything about a horse can detect a spavin, etc.; but not one in a hundred can tell anything about the true nature of the malady, or the indications to be fulfilled in the treatment; and in consequence of a lack of knowledge on this subject many a poor animal has suffered immensely, who, if he was not deprived of the power of speech, would make the ears of his oppressors tingle with a tale of man's barbarity and inconsistency.

The bony structure, being composed of vital solids, although studded with crystallizations of saline carbonates and phosphates, is liable, like other parts of the structure, to take on preternatural or morbid action, and may result from or accompany constitutional idiosyncrasies, resulting from hereditary taints on the side of the dam or stallion. The most frequent causes of splent, spavin, etc., are undue acts of exertion on hard pavements, and the imposition of weight disproportioned to the strength of the animal: young horses are particularly liable to exostosis when severely worked or over-burdened. Any sudden or extraordinary efforts in backing or suddenly pulling up at full speed, racing before the horse shall have arrived at maturity, while the joints are yet in a state of imperfection, very frequently lay the foundation of exostosis. The parts being sprained and taxed beyond endurance, disease is excited in the ligamentous substance, and extends itself to the periosteum and bones; the ligaments often become ossified, and are rendered fixtures; the periosteum, being raised by bony accumulations, presents itself in the situation of splents, spavin, or ringbone.

Sir A. Cooper divides exostosis, in reference to its seat, into two kinds, periosteal and medullary; and again, as to its nature, into cartilaginous and fungous. "But," says Mr. Percivall, "it is to that kind only which is situated between the shell of the bone and the periosteum covering it, that we have to attend in veterinary practice. On dissection we find the periosteum thicker than usual, with cartilage beneath it, and ossific matter within the cartilage, extending from the shell of the bone nearly to the internal surface of the periosteum, still leaving on the surface of the swelling a thin portion of cartilage unossified."

When the accretion of these swellings ceases, and the disease has been of long standing, they are found to consist on their exterior surface, of a hell of osseous matter similar to that of the original bone; consequently, when an exostosis has been formed in the manner here described, the shell of the original bone becomes absorbed, and cancelli are desposited in its place.

"In the mean time, the outer surface of the exostosis acquires a shell resembling that of the bone itself. When the exostosis has been steeped in an acid, and by this means deprived of its phosphate of lime, the cartilaginous structure remains of the same form and magnitude as the diseased deposits; and, as far as I have been able to discover, it is effused precisely in the same manner as healthy bone.

"An exostosis, abstractedly considered, does not appear to occasion much inconvenience to the animal, except in the early stages. A ringbone, confined to the pastern bones, is of little consideration; but, should it show itself at or near the joint, it seldom fails to produce lameness, which is often of a permanent nature. Lameness, therefore, is not an invariable symptom of exostosis; for most splents, and many ringbones, and even spavins, exist without lameness. When this disease invades ligamentous structure, however, lameness generally accompanies it, — an effect we would refer to the excessive tenderness of the part. Should the tumor interfere, either from its bulk or situation, with the motions of joints, muscles, or tendons, lameness is a concomitant, and often irremediable, symptom."

It appears that various constitutional and local remedies have been tried for the prevention and dispersion of exostosis, viz., "the actual cautery, ammonia, cantharides, caustic, and setons." The constitutional remedies are of the same destructive nature, and have but too often aggravated that which they were intended to relieve; we do not believe that any specific treatment has ever had the honor of curing these forms of disease; that course of treatment we have ever found the most satisfactory that is calculated to promote the general health by sanative means; we cleanse the system, equalize the circulation, and excite healthy action to the parts by stimulants and counter-irritation (if the parts are inactive); poultices, fomentations, etc., if there is pain, or increased action. If this is done early, exostosis is easily arrested, unless an hereditary taint is manifest.

The removal of exostosis by an operation, we are told, has been performed with success, and no doubt there are cases in which it may safely be performed; yet it cannot be successful on spavined horses, the natural termination of spavin being anchylosis of the bones of the hock and inter-articular cartilage. A knowledge of this fact has led men to suppose that Nature has turned a summerset; and they endeavor to set her right with the firing iron and the implements of death; whereas, if her intentions were aided, the result would prove more satisfactory.

EXTRAVASATION. — The escape of blood or other fluids from their proper vessels.

EYE. — (See part first.)

FALLING OF THE YARD OR PENIS. — This disease sometimes happens to horses and bulls, in consequence of swelling, excrescence, and ulceration of the parts, sometimes of an obstinate or malignant nature. It may also be occasioned by too frequent sexual intercourse. It may also depend on weakness of the part; and, when this is the case, there is no ulceration nor excrescence about it. If it depend on debility, then tone up the whole animal, and wash the parts, first with castile soap, then with cold water. If it result from ulceration, wash with weak vinegar and water, afterwards with a mixture of powdered charcoal and water. The latter may be thrown up the sheath with a common syringe or injection pipe. When the ulcers show a disposition to heal, a little powdered bayberry bark will generally complete the cure. When excrescences form on the sheath or inside of it, they should be taken off by applying a ligature tight around their base.

FALSE QUARTER. — This can hardly be considered as a distinct complaint, but, more properly, as a consequence resulting from some one of the former diseases; in which, from the injury done to the coronary vascular ligament, it can never afterwards secrete horn; but the break or interruption, produced by the interposition of a portion of non-secreting substance, causes a part of the outer crust of the wall to be absent. Such a blemish is called a *false quarter;* and it is evident that it must greatly tend to weaken the hoof. It likewise sometimes produces the same unpleasant effects as a sand-crack, by the separation of the under layer of the wall admitting the vascular laminæ between the opening. The *treatment* can be only palliative. Keep the neighboring horn always thin; use a bar shoe, and "*lay off*" (as a smith calls it) the deficient quarter. This may be done either by paring the crust, or by an indentation in the shoe; the choice of which is left to the prudence of the operator, with this exception, that, in a weak, thin foot, the alteration should always be made in the shoe, and in a strong one, in the crust.

FARCY. — A disease of the lymphatics or absorbent vessels. Its most usual form is that of small tumors, or *buds*, as they are termed, which make their appearance in different parts of the surface, gradually become soft, or suppurate, and burst, and become a foul ulcer. Its cause may be found in anything that will derange the general system, or produce debility; its proximate cause is immoderate work, inattention to diet, hot unhealthy stables, sudden changes of temperature, standing on filthy litter, etc.

FAUCES. — That part of the throat which lies behind the tongue.

FEMORAL ARTERY. — The principal artery of the thigh.

FEMUR, or OS FEMORIS. — The thigh bone.

FETLOCK. — A lock of hair at the lower part of the fore and hind legs.

FEVER is a powerful effort of the vital principle to remove all obstructions to ordinary and proper action. The reason why veterinary practitioners have not ascertained this fact heretofore, is, because they have been guided by the false principle that *fever is disease.* Let them but receive the truth of the definition we have given, then the light will begin to shine, and medical darkness will be rendered more visible.

Fever, as we have said, is an effort of the vital power to regain its equilibrium of action through the system, and should never be subdued by the use of agents that deprive the organs of the power to produce it. Fever will be generally manifested in one or more of that combination of signs commonly given as a description of fever, viz: increased velocity of the pulse, heat, redness, pain and swelling, thirst, obstructed surface, etc., some of which will be present, local or general, in greater or less degree, in all forms of disease. In what is called acute attacks these signs are very manifest; in chronic cases, they are often faint; but still they exist. When an animal has taken cold, and there is power enough in the system to keep up a continual warfare against obstructions, the disturbance of vital action being unbroken, the fever is called pure, or unbroken. The powers of the system may become exhausted by efforts at relief, and the fever will be periodically reduced: this form of fever is called remittent. It would be as absurd to expect that the most accurate definition of fever would correspond, in all its details, with another case, as to expect all animals to be alike.

There are many agents that obstruct vital action, and many an organ to be obstructed, which some have classed as distinct fevers; for example, milk fever, puerperal fever, symptomatic, typhus, inflammatory, etc. Our system teaches us that there is but one cause of fever, viz., the natural motive power of the system, and but one fever itself, viz., accumulated vital action; hence the treatment must be physiological.

Veterinary Surgeon Percivall, in an article on fever, says: " We have no more reason, nor not near so much, to give fever a habitation in the abdomen, as we have to enthrone it in the head; but it would appear, from the full range of observation, that no part of the body can be said to be insusceptible of inflammation [local fever] in human fever, though, at the same time, no organ is invariably or exclusively affected. All I wish to contend for is, that both idiopathic and symptomatic fevers exhibit the same form, character, and species, and the same general means of cure; and that, were it not for the local affection, it would be difficult or impossible to distinguish them. When we come to examine the accounts of different authors on fever, and compare them one with another, we can hardly refrain from coming to the conclusion that their descriptions were originally derived from human medicine, and have been but variously modified to suit the prevailing doctrines of the day; they have gone through a system of imaginary fevers, and regularly transferred the observations and language of ancient authors upon diseases of the human species to the constitution of quadrupeds."

In the treatment of disease, and when fever is present, manifested by a determination of blood to the head, the object is to invite the blood downward and outward; or, in other words, equalize the circulation by warmth and moisture externally, as in lockjaw.

In neat cattle, should fever be present, the eyes appear dull and watery, the muzzle dry, and rumination has ceased; then the blood, for want of room in the nutritive tissues, is forced upon the lungs, liver, spleen, brain, or other glandular tissues, and men have named the disease congestive fever. The author advises the reader not to feel alarmed about the fever, but set to work and relieve the congestion. Disease of the bowels, garget of the teats and udder, will require fomentation and stimulants to the parts.

FILLY. — A name given to a mare until she is two or three years old.

FILM. — Opacity of the cornea.

FILTRATION. — Straining liquids through unsized paper; also through sand or porous stone.

FIRING. — A severe operation, often performed on horses, for spavins, curbs, ringbones, etc. Such barbarity should never be practised: it is a disgrace to this age of improvement. When discoveries are leaping on discoveries, and medical reform has germinated, shall we not permit the poor dumb brute to share the benefits of our investigation? Every man who loves a horse, or wishes well to the cause of horse-manity, will say that a more safe and effectual system of veterinary practice is necessary to rescue from the torture of the firing iron one of the noblest and most valuable quadrupeds in the world.

"The rage of firing is very generally, and much too frequently, adopted, and no doubt upon most occasions, hurried on by the pecuniary propensity and dictation of the interested operator, anxious to display his dexterity, or, as Scrub says, 'his newest flourish' in the operation; and when performed, and the horse is turned out to grass, if taken up sound, I shall ever attribute much more of the cure to that grand specific, rest, than to the effects of his fire." (See Taplin's *Farriery*, p. 83.) Hence the firing iron, like all other destructive agents, excites the system to rally her powers and resist the encroachments of disease; yet the process is like taking a citadel by storm; the breaches that are made by the weapons of warfare (such as the firing iron, scalpel, lancet, and poison) can be traced, and leave unmistakable evidences of their encroachments. Instead of provoking the vital powers to action by such destructive enginery, we should afford Nature all the aid we can, but never interfere with her operations.

FISTULA OF THE WITHERS. — "An obstinate disease of the horse's withers, or points of the shoulder, commonly produced by a bruise of the saddle." No wonder Dr. White calls it "obstinate," when the following treatment is recommended by him: "The scalding mixture — it consists of any fixed oil (as lamp or train oil), spirit of turpentine, verdigris, and corrosive sublimate. These are put into an iron ladle, and made nearly boiling hot; and in this state the mixture is to be applied to the diseased parts, by means of a little tow fastened to the end of a stick! It is necessary to prevent the mixture from flowing over the sound parts, as it would not only take off the hair, but cause ulceration of the skin." If this mixture will produce diseased action in the sound parts, we need not ask what will be the result when applied to parts already diseased. The author has cured many cases of fistula, by treating them as common abscesses, with the application of stimulating antiseptic and tonic poultices (see POULTICES), and by a purifying course of treatment, with proper attention to diet, etc.

FLEAM. — An instrument with which horses and cattle are bled.

FLESH. — A common name for the muscles of the body.

FLEXOR. — The flexors are those that bend one bone upon another. The tendons that serve to bend the leg, for example, are named flexors.

FOALING. — The bringing forth young in mares is not so often attended with difficulty as in cows, and they have seldom occasion for assistance. They should be placed in a situation where they may have shelter, and where they are free from danger.

FOMENTATIONS. — Fomentations are generally made by pouring boiling water on camomiles, burdock, poplar bark, etc. For an emollient fomentation, ground slippery elm is preferred. In inflammation of the bowels, for example, the parts may be fomented with flannels wrung out in a thin mixture of slippery elm.

FOOT. — (See part first.)

FOOT ROT. — This name is applied to a disease in the feet of sheep. This disease often happens to such as are fed in low meadows, or where the grass holds the frost or cold dews for a considerable time. Probably a foul habit of body may be a predisposing cause. In the treatment of foot rot, we should endeavor to find out the cause, or causes, of the disorder, and change the food or location of the sheep. If the disease has spread under the horny covering, all the superfluous horn should be carefully pared away, so that the dressing may be applied to the whole of the affected parts. The dressing is composed of powdered lobelia, formed to the consistence of paste, with honey.

FOUL FEEDERS. — Horses are so named that have depraved or vitiated appetites, eating foul litter and earth from the ground.

FOUNDER. — A term expressive of the different forms of rheumatism and ruin in the horse. Veterinary writers describe three different forms of this disease, viz., founder of the body, chest, and feet. This is one and the same disease, only located in different parts, and may arise from the same general causes; which consist in chilling the animal when exhausted, by which means the perspiration is obstructed, by much fatigue, and by violent and long-continued exertion: exposing the animal to cold wind or rain, or washing his legs and thighs, and sometimes his body, is often the cause of founder. Dr. White calls "founder, a term expressive of the *ruined state* of the horse." And well he might call it "ruined." How many thousand animals have been ruined, not by the disease, but by the *treatment!* Here is a specimen of it. Dr. White says: "The horse was bled before I saw him: five quarts of blood were taken off. I desired he might be bled again, when half a pailful more was abstracted. In less than an hour I saw him again, and, finding that he was not relieved, took another half-pailful, amounting in all to four gallons! The horse was sent home, and seemed to be doing well in a straw yard, though very weak and thin. [No wonder, after such a loss sustained by the vital powers.] At the end of three or four months he began to lose his fore hoofs, and, after declining some time longer, he died" — a victim to science. It is evident, from experience and facts, that the above treatment renders the disease incurable, and is the true cause of death; therefore, not suitable to the true ends to be accomplished. What, then, are the true ends to be accomplished? To relax muscular structure, determine action to the surface, improve the secretions, and remove obstructions which disturb or repel vital action. These will equalize the circulation of the blood, when *it will be* found that there will be no necessity for diminishing its quantity. The inflammation, as it is termed, is always sufficiently controlled when the circulation of the blood is free and universal. Therefore, instead of withdrawing vital action, promote its equal and universal diffusion.

FROG. — The posterior part of the horse's foot.

GALBANUM. — A gum resin, sometimes employed as an expectorant and anti-spasmodic; the dose, three or four drachms. It is used, also, in the composition of warm adhesive plasters, such as gum and diachylon plaster.

GALL. — A common name for bile.

GALL. — A sore produced by pressure, or chafing, of the saddle or harness.

GALL BLADDER. — The horse has no gall bladder, or reservoir for bile. A considerable quantity of bile, however, is formed by the horse's liver, and is conveyed by the hepatic duct into the first intestine, or duodenum. In the cow and sheep, the gall bladder is of considerable size.

GANGLION. — This term is applied to a natural enlargement, or knot, in the course of some of the nerves.

GANGRENE. — An incipient mortification. In this stage of the disease, there is generally absence of pain; the part is deprived of vital force, by causes inducing a loss of tone.

GARLIC. — It operates upon the horse as a diffusible stimulant and expectorant; possessing, also, diuretic properties. The author considers garlic a valuable remedial agent in the treatment of any disease where the constitution has suffered through hard work, or ill usage. It is a general custom, on the eastern coast of China, to allow cattle to eat as much as they choose. They are never known to suffer any inconvenience from it; on the contrary, they appear to thrive, and are scarcely if ever sick. The only objection to its long-continued use in cattle is, that it imparts an unpleasant flavor to the meat. It is considered by Gibson to be a valuable remedy in coughs. He advises two or three of the cloves or kernels, cut small, to be given in each feed, and observes that, by continuing this practice, with right and well-timed exercise and careful feeding, he has known many horses to recover, even when there has been a suspicion of their wind.

GASTRIC JUICE. — A juice formed in the stomach for the purpose of digestion.

GASTRITIS. — Inflammation of the stomach.

GAUNT-BELLIED. — A term applied to a horse when he is drawn up in the flank.

GELATINE. — A component part of animal matter.

GELDING. — A castrated horse. Such horses are not so vigorous as stallions; the latter are freer from disease than geldings, and will do more work, and keep a better appearance, as to coat and flesh, upon the same quantity of food.

GENTIAN ROOT. — A good tonic for a horse; the dose is two or three drachms.

GESTATION. — Being with young. The time of gestation in the mare is eleven months; in the cow, nine months.

GINGER. — An aromatic root, possessing stimulant and carminative properties.

GLANDERS. — A contagious disease peculiar to the horse, the ass, and the mule. Many persons suppose that glanders and farcy are the effect of a specific poison in the blood; but this theory is exploded. The following will throw some light on the subject, for which we are indebted to R. Vines, V. S.: "All the symptoms of disease which constitute glanders and farcy invariably depend on the unhealthy state of the system into which it is reduced or brought, and not, as is supposed, from a specific poison contained in the blood; and these symptoms of disease are found to depend on, and arise from, a variety of causes; whether they occur at the latter states or stages of common inflammatory diseases, such as strangles, common cold, distemper, disease of the lungs, dropsy, etc., or whether they arise independently of such causes; for, when the system is brought into an unhealthy state, and is more or less debilitated from neglect, or by the improper of any of these diseases, [many of them are improperly treated], farcy, or glanders, is the result. The disease of every animal will, therefore, assume a character according to the state of the system." Mr. Percivall, V. S., says: "The state of the body, or constitution, will always have considerable influence on the character and tendency of disease. In horses whose bodies are, and have long been, in an unthriving and unhealthy condition, a common swollen leg will occasionally run into farcy; and a common cold or strangles, or an attack of influenza, be followed by glanders. In other cases, such unfortunate sequels will supervene without any ostensible or discoverable cause." The great fault of those who have employed their talents in the investigation of the subject (glanders) is, that they take hold of the wrong end of it; they are engaged in attempting to discover the "specific poison," where none exists, when their time would be more profitably engaged in studying the principles of a system of medication that would rid the system of these early exciting causes, viz., common colds, etc., and thus prevent this great bugbear, glanders. The author can at any time, within a period of a few months, and without the assistance of "specific poison," manufacture a case of genuine glanders out of the following materials: A horse would be selected — and many such could be found in the city of Boston — whose general health shall be impaired; let the surface be obstructed by standing in a shower of rain, without anything to protect the animal from the pelting storm; then put him into a stall near the door, where a current of cold air will pass the hind extremities: he remains in this situation during the night. On the following morning, the animal appears dull, and is off his feed. It is soon ascertained that he has taken cold; now treat him according to the kill-or-cure practice: "If there is difficulty of breathing, and the throat is sore, — or, in other words, the usual symptoms, — the first thing to be done is, to bleed largely, until the horse faints. He should then be put into a cool place. It is often necessary to repeat the bleeding two or three times. If the throat is very sore, blister the part." (See CANTHARIDES.) The secretions now become impaired, there is loss of appetite, the coat stares; there is a dull, sleepy appearance about the animal; the discharge from the nostrils now assumes an acrimonious and putrid character, which, acting chemically on the membrane of the nose, constitutes ulceration: the latter corrode the cartilage and bones, and glanders is the result. Now we will view it in another form. The animal has taken cold (see CATARRH); the lungs — from previous disease, and the subsequent inhalation of impure air in a hot and crowded stable — are incapacitated, and their power to purify and vitalize the blood is destroyed; hence we have deposits of morbific matter on the mucous membrane, which corrode, ulcerate, and finally attack the substance of the lungs, and tubercle is the result, which may terminate in glanders. The expectoration, or passage of acrimonious humors through the nostril of the horse from the lungs, does, in its passage, irritate the schneiderian membrane at a point where it is in immediate contact with ossific or cartilaginous structure, and sufficiently accounts for the ulcers found in the nostrils in the above case. *We do not hesitate to say that glanders can be produced without infection, or contagion, and that a common cold or catarrh, neglected or improperly treated, will often terminate in glanders.* Mr. Vines, V. S., states "that the practice of physicing horses, and exposing them to wet and cold, when they have common catarrh, will produce confirmed glanders."

According to the testimony of Mr. G. Fenwick, V. S., of London, "Glanders is a symptom of tubercles in the lungs in nine times out of ten;" hence, when a horse has taken cold, and the surface is obstructed, the prudent owner will endeavor to force a crisis; that is, to open the pores of the skin, and promote perspiration. This can be done by the use of warmth and moisture externally, and the administration of warm, anti-spasmodic drinks. This will relieve the stricture of the surface, and permit the egress of morbific matter, which would otherwise be thrown on the lungs, or kidneys. If there is not sufficient power in the system to determine action to the surface, then administer diffusible stimulants. Mr. Youatt remarks: "Improper stable management is a more frequent cause of glanders than contagion. The air which is necessary to respiration is changed and empoisoned in its passage through the lungs; and a fresh supply is necessary for the support of life. That supply may be sufficient barely to support life, but not to prevent the vitiated air from again and again passing to the lungs, and producing irritation and disease. The membrane of the nose, possessed of extreme sensibility, is easily irritated by this poison. Professor Coleman relates a case which proves to demonstration the rapid and fatal agency of this cause. 'In the expedition to Quiberon, the horses had not been long on board the transports, before it became necessary to shut down the hatchways: the consequence of this was, that some of them were suffocated, and that all the rest were disembarked either glandered or farcied. In a close stable, the air is not only vitiated by breathing, but there are other and more powerful sources of mischief. The dung and the urine are suffered to remain, fermenting and giving out injurious gases.'"

GLANDS. — Soft, spongy substances in various parts of the body, which serve to secrete particular humors from the blood. They are vulgarly called kernels.

GLEET. — A discharge of a mucous fluid from the urethra, vagina, or nostrils.

GLOTTIS. — The upper part of the larynx, or top of the windpipe. The sensibility of this part is so great, that, if any substance happen to fall into the larynx, the most painful and distressing symptoms are produced; and, unless the extraneous matter be expelled by coughing, or removed by an operation (bronchotomy), a fatal termination will be the consequence.

GORED — A term applied to cattle with an overloaded stomach. When they are in this state, they are said to be blasted, blown, or hoven; probably from the quantity of carbonic acid gas that is generated, and by which the stomach is so distended that cattle often die in consequence of it. When cattle are put into a pasture, which abounds in nutritious food, to which they have been unaccustomed, or have an improper quantity given them, they frequently fill the paunch to such an extent that they are incapable of ruminating; hence, the food remaining in a warm situation, the combined action of heat and moisture generates the gas.

GRANULATIONS. — A term applied to the little, red, grain-like, fleshy bodies, which arise on the surface of ulcers and suppurating sores. Their use is to fill up cavities, and approximate the sides.

GREASE. — A swelling of the horse's heels, and discharge of stinking matter.

GRIPES.* — (See COLIC.)

GRISTLE. — A name commonly given to cartilage. (See CARTILAGE.)

GROGGINESS. — A horse is said to be groggy, when he has a tenderness, or stiffness, about the feet, which causes him to go in an uneasy, hobbling manner.

---

* SCIENTIFIC TREATMENT OF COLIC, OR GRIPES. — "On the 5th Sept., 1824, a young bay mare was admitted into the infirmary, with symptoms of colic, for which she lost eight pounds of blood before she came in. The following drench was prescribed to be given immediately: laudanum and oil of turpentine, of each, three ounces, with the addition of six ounces of decoction of aloes. In the course of half an hour this was repeated! But, shortly after, she vomited the greater part by the mouth and nostrils. No relief having been obtained, twelve pounds of blood were taken from her, and the same drink was given. In another hour this drench was repeated; and for the fourth time, during the succeeding hour; both of which, before death, she rejected, as she had done the second drink. Notwithstanding these active measures were promptly taken, she died about three hours after her admission." (See Clarke's Essay on Gripes.) It appears that the doctors made short work of it. Twelve ounces of laudanum, and the same of turpentine, in three hours! But this is secundum artem. This is called skilful treatment, and justifiable in every case where the symptoms are urgent.

Had the relaxing and stimulating plan, practised by us, been resorted to, and in a proper time and manner, it would probably have saved the poor brute. We have attended a large number of the same sort of cases, and have not yet lost the first.

GULLET, OR ŒSOPHAGUS. — A muscular and membranous tube, by which the food, etc., is conveyed from the mouth to the stomach. The upper part, or funnel-like cavity, is named pharynx. The gullet passes down the neck behind the windpipe, along the neck, through the diaphragm, and terminating in the stomach.

GUMS. — The fleshy parts of the sockets of the teeth.

HABIT. — By this term is meant the disposition, or temperament, of the body or constitution, whether natural or acquired. The term *habit* is also applied to any vice, as starting, kicking, rearing, etc. All bad habits, whether of the body, constitution, temper, or disposition of animals, may be in some measure corrected, if not entirely put a stop to, by proper attention to breaking, breeding, and stable management.

HALTER CAST. — Owing to the improper length of the halter, the horse is apt to get his fore leg across it, falls down, and sometimes injures himself considerably.

HAM. — This is the name given to the muscular part of the hind leg, terminating in the great tendo Achillis, or hamstring.

HAND. — The division in the standard for measuring horses is thus named. A hand is four inches.

HAW. — (See EYE, part first.)

HEART. — (See part first.)

HEEL. — A term applied to the back part of the termination of the hoof.

HEMLOCK. — A narcotic vegetable poison, deriving its deleterious properties from an alkaline principle, called conia. It has been known to kill many horses who have partaken of it.

HEMORRHAGE. — A flow of blood from any part of the body, in consequence of the rupture of an artery or vein. Hemorrhage, from external injury, is most readily stopped by taking up the bleeding vessel, and tying it with saddler's silk; but, when this cannot be done, the bleeding may generally be stopped by pressure, or styptics.

HEPATITIS. — Inflammation of the liver.

HIDE-BOUND. — When horses are out of condition, and have harsh dry coats, the skin will be contracted, and found tight about the ribs. It is a symptom of disease, and shows that the general health is impaired.

HIP-SHOT. — This is known one of the hip bones being lower than other. It generally depends on a fracture of the os innominatum, or part of the pelvis; the part having formed an irregular kind of union, so that the bone on that side is shorter than the other.

HOCK. — The horse's hock is composed of six bones. These bones are all connected together by very strong ligaments, which prevent dislocation, but allow a slight degree of motion among them. The surfaces that are opposed to each other are thickly covered by elastic cartilage, and by a membrane secreting the synovia, or oily fluid, which guards against friction. These bones are so strongly bound together as almost to defy dislocation.

HOOF-BOUND. — A dry, brittle, and morbid state of the foot. A want of vital action, occasioned, says Dr. White, "by inflammation," which he calls disease. Now it is evident that no vital action, as that of fever and inflammation, can be properly termed disease. The only action that can be properly termed disease, is the chemical action manifested in suppuration and gangrene. This is the great popular error that we are laboring to overcome. It is that of attributing disease and death to the action of the powers of life. When a part has become diseased, especially the foot (for from it the blood has a kind of up-hill work to perform, in returning to the heart by the veins), there is a low state of vitality; very little can be accomplished by the vital powers, amounting only to a low form of inflammation. And, of course, the chemical power of decomposition, always present and never tired, gets the advantage and decomposes the part; we then have thrush, which, if improperly treated, the hoof falls off by the process of decomposition, or, in other words, mortification. It becomes separated from the living parts, for want of inflammation, or vital supremacy, over chemical agency; and then the loss of the hoof is strangely attributed to inflammation, or the vital power, which did all it could to prevent such a termination.

HOOF-CASTING. — A partial or complete separation of the horse's hoof from the sensitive foot.

HOOSE. — A term used by cow doctors. It signifies a cough, either chronic or acute, with which cattle are affected, from exposure to cold winds or rain.

HOREHOUND. — A bitter vegetable, used in horse practice as a tonic and expectorant.

HOVEN. — (See BLASTED.)

HYDATID. — A thin bladder, containing a fluid resembling water, and nearly transparent. It is found in different animals. In sheep, it occasions a disease named gid, or giddiness; the hydatid being found in one of the ventricles of the brain, or in its convolutions. On account of the pressure it makes on the brain, it disturbs the functions of that important organ, especially when the sheep are hurried or driven.

HYDROCELE. — Dropsy of the testicle and its appendages.

HYDROPHOBIA. — Canine madness.

HYDROTHORAX, or DROPSY OF THE CHEST. — Mr. Percivall informs us, "that the objects to be pursued in the treatment of hydrothorax are twofold: first, we are to diminish any excess of action that may show itself in the sanguineous system,* and

* In plain English, abstract blood. This not only diminishes the sanguineous system, but every other function or system. The regulars have tried blood-letting to their hearts' content; their patients have been rowelled, blistered, calomelized, turpentinized, and hellebored, yet they have never been able to preserve life, "except two solitary cases in Mr. Sewell's practice;" for Mr. Percival tells us, in his lectures, that "he never saw a case terminate favorably." Is not this a proof that our brethren are on the wrong track? We are told that the proximate causes of dropsy are "debility and an obstructed perspiration;" and that it may result from "loss of blood, diarrhœa, diabetes, and other circumstances that rapidly exhaust the system." Hence the processes of cure are just the means calculated to produce the disease. The true indications in the treatment are, to warm and relax the surface, and promote perspiration; for whatever checks it stops the egress of morbific matter from the system, and, of course, determines it upon the internal surfaces. Diffusible stimulants may be given, to keep up the action on the surface. The general health must be improved.

thereby lessen the effusion of fluid into the chest; and secondly, by increasing the action of the absorbent system, effect the removal of what is already accumulated."* Most surgeons recommend early tapping in dropsy. The operation may be performed with the common trocar and canula. The best place for the introduction of the instrument is the space between the eighth and ninth ribs, close to their cartilages; not between the latter, lest the pericardium be punctured. Here, making the skin tense with the fingers of the left hand, the instrument, with its point directed upwards and inwards, may, with a little rotary movement, gradually be thrust in, until the resistance to its entry suddenly ceases; when the trocar should be withdrawn, and the canula at the same time pushed onwards, lest it slip out. If the flow of water suddenly ceases, a small whalebone probe should be introduced through the pipe.

ICHOR. — A thin, acrimonious discharge from ulcers, or diseased parts.

ICTERUS. — Jaundice, or yellows.

IDIOSYNCRASY. — A peculiar constitution, or temperament.

ILEUM. — The last portion of the small intestine. It terminates in the large intestine, or blind gut, named cæcum.

IMPOSTHUME. — A collection of matter, or pus, in any part.

INCONTINENCE OF URINE. — A continual dripping of the urine from a horse's sheath.

INFLAMMATION. — Inflammation and fever are one and the same thing. When fever is confined to a small part, it is called inflammation. (See INFLAMMATION, part first.) Dr. White, although an advocate of the popular error, viz., blood-letting, makes some very sensible remarks on the subject, if men generally would carry out these

* The action of the absorbent system never was, nor never can be, excited when the lancet is coöperative. Absorption is a physiological result, and cannot be excited by agents that act pathologically. The balance between exhalation and absorption is lost, in consequence of which, more fluid is poured out than is taken up; hence, if we excite the exhalents to throw off the morbid fluids from the surface, there will be less for the absorbents to take up, and the chances of success will be greater.

principles, they would prevent a great loss of property. "It must be obvious, that when an animal is laboring under general inflammation, or fever, in consequence of a suppression of the natural discharges, whether it be perspiration, urine, or dung, he cannot be cured merely by the abstraction of blood; for, however large the quantity abstracted, that which remains will be impure, or acrimonious, and unfit for carrying on a healthy action. It is absolutely necessary to restore the natural discharges by means of suitable medicines, unless that be effected by an effort of nature, which is not an uncommon occurrence, especially when the animal is supplied with some bland fluid, such as bran water, or thin bran mashes. The morbid matter sometimes runs off by the nostrils, sometimes by the kidneys or bowels, and sometimes by a general relaxation of the skin, and the body is thus restored to health." From the above we are led to the conclusion that, after all, Nature is the most efficient doctor, and that man should be her servant, to procure what she wants, merely to be used in her own way.

INFLUENZA. — Epidemic catarrh. Catarrhs, or violent colds, attended with sore throat, and a thin, watery discharge from the eyes and nose. It appears to be infectious, seldom making its appearance without attacking several horses in the same stable. The horse should subsist on warm gruel, and have a blanket thrown over him, and a drink of hyssop tea. As soon as the surface of the body is relaxed, and becomes moist, the catarrh will disappear.

INJECTION. — A term sometimes applied to clysters.

INOSCULATION. — The running of arteries and veins into one another, or the interunion of the extremities of arteries and veins.

INSPIRATION. — The act of drawing air into the lungs.

INTEGUMENT. — Any common covering of the body: it generally includes skin, muscle, and membrane.

INTERCOSTAL. — A term given to parts situated between the ribs: thus, we have intercostal muscles, etc.

INTERMITTENT. — A name given to disorders that appear to go off at certain periods, and return after some interval.

INTESTINES. — The horse's intestines are about ninety feet in length.

INTUS-SUSCEPTION. — This is occasioned by one portion of the bowels being drawn within the other.

IRIS. — That part of the eye by which the light admitted to the retina is regulated.

IRRITABILITY. — All muscular parts possess the property of contracting, or shrinking, when irritated, and are therefore endowed with irritability.

ISSUES. — (See ROWELS.)

ITCHING. — Itching in horses is generally a consequence of foul feeding, and may be occasioned by mange.

JAUNDICE. — In jaundice, the natural course of the bile is perverted, and re-absorbed into the circulation.

JAW-LOCKED. — (See LOCKJAW.)

JEJUNUM. — Part of the small intestine is thus named, from its being generally found empty.

JOINTS. — A joint is formed, generally speaking, by the heads of two or more bones. These ends are covered by a layer of cartilage or gristle, which is of a yielding nature. There is formed within the joint a slippery fluid, called synovia, or joint oil. The ends of the bones, thus covered with a smooth, yielding surface, so slippery that they move freely on each other without suffering from friction, are then firmly tied together by a strong substance, named ligament, which completely surrounds the head of the bones: this is termed capsular ligament. In some joints we find an additional ligament within the capsular ligament, or cavity: thus, in the hip joint, a strong ligament connects the head of the thigh bone with the socket that receives it.

Joints are subject to disease, either from external injury, or from long-continued exertion of them. In the former, the capsular ligament is penetrated, and a discharge of synovia ensues. Mr. Percival remarks "that, in many cases of open joint (commonly called so), there is no division nor injury whatever of the capsular ligament; but merely the exposure of some bursa mucosa placed between the joint and the external wound: the discharge is of the same kind as in the former case, and we can only determine which it is by carefully probing the wound. Most of all, we are likely to make this mistake in the shoulder joint and hock, when heat and swelling are present. From the acute sensibility of ligamentous parts when inflamed, the system quickly, and almost invariably, sympathizes; so that, in all severe cases of this nature, symptomatic fever supervenes, the pulse becomes accelerated, the horse heaves at the flanks, refuses his food, and shows symptoms of the most affecting suffering. It must be borne in mind that, although a joint be not open in the first instance, subsequent sloughing may expose its cavity. Now, the ordinary effects of disease in the synovial membrane are, first, a preternatural secretion of synovia, — hence the profuse discharge observed in these cases; second, an effusion of adhesive matter into the cavity of the joint; third, a thickening of the synovial membrane, a conversion of it into a substance resembling gristle, and an effusion of adhesive matter, and probably serum, into the cellular substance around, by which the external parts and those of the joints are firmly cemented together. In the latter stage the disease commonly extends itself to the cartilaginous surfaces; they exfoliate, leaving the extremities of the bones denuded, to grate on each other as often as the joint is moved. The bones, in their turn, throw out deposits from their ends around the joint, — a process that ultimately ensues, and anchylosis is the result."

The indication to be fulfilled is to promote adhesion by bringing the edges together and confining them in contact, either by taking a few stitches, or shaving the hair off around the parts and applying strips of adhesive plaster. The parts may have a pledget of lint bound on, moistened with healing balsam; and, if the limb will admit of it, a splint may be bound to the back part of it, so as to prevent all possibility of flexion. If union cannot be produced by this means, the parts may be poulticed with astringents. The object is to close the joint, and promote granulation. If the parts are inactive, sprinkle the surface of the poultice with charcoal and capsicum. In a case that came under the author's care in this city, and one in which there was no hope of its healing by the first intention, the tincture of capsicum was daily injected: this, together with tonic, stimulating, astringent poultices and fomentations, completed the cure. In cases where the external wound is large, and there is much heat, pain, and loss of motion, poultices of a relaxing and lubricating nature should be used; such are lobelia and slippery elm. A severe injury of this kind may be converted into a simple wound by the combined influence of these remedies. The horse should be kept at rest, on a light diet of scalded food, and an occasional dose of alterative medicine.

When lameness is manifest without heat or swelling, and there is reason to suppose that the animal has been overworked, rest and proper attention to diet will be all that is necessary. When the case is one of long standing, a run at grass may effect a cure, unless there is reason to suppose that the articulatory surfaces of the bones are diseased; we are not supposed to do more for these subjects than alleviate their sufferings, or, what amounts to the same thing, diminish their lameness.

JUGULAR VEINS. — The large veins of the neck, where a horse is bled.

KERNELS. — A common name for glands: thus, the parotid glands, situated beneath the ear, are termed the kernels under the ear.

KINO. — An astringent gum resin.

LACTEALS. — Absorbent vessels, which convey the chyle from the bowels into the thoracic duct.

LAMENESS. — The cause of lameness in horses is often very obscure, and can only be discovered by a patient and careful examination. A slight degree of lameness often passes unnoticed; or, if it be observed, the owner too often persuades himself that it will pass off. It is always the most prudent plan to lay up a horse the moment he is observed to be lame, and submit to the inconvenience of doing without his services until he is cured. When lameness is caused by wounds or bruises, the injured part is generally discovered without difficulty, though pricking, in shoeing, is not always so easily seen. All lameness from injuries within the hoof is often detected with difficulty. Slight lameness is most readily seen by making the horse trot gently, without giving any support to the head by the bridle or halter, and without urging him with the whip; the lameness is then seen by his dropping harder and dwelling longer on the sound leg than on the lame one, in order to favor the latter; and this, when the lameness is at all considerable, is attended with a corresponding motion of the head, which drops a little whenever he steps on the sound limb. An experienced observer can at any time distinguish lameness merely by seeing a horse walk out of the stable. It often happens, in very severe lameness of one or both fore feet, that the horse, when led out, will appear to be lame in the hind feet also: this is occasioned by the animal endeavoring to favor the fore foot or feet by throwing the bulk of his weight on the hind legs. In all cases of lameness, unless the cause is so evident as to render it unnecessary, it is proper to examine the foot carefully in the first place; and it should never be forgotten that swelling, heat, and tenderness of the fetlock joint, or even the leg, may arise from an injury to the foot. In lameness of the foot, the affected foot will be warmer than the other. Considerable relief may almost always be afforded in foot lameness by keeping the feet moist, or pasturing the animal in soft meadow land, or by stopping the bottoms of the feet with a wet sponge, by paring them when necessary. We sometimes find, on examining a lame foot, that there is an enlargement immediately above the coronet, at the heels and quarters, and that this enlargement feels hard and bony. This is termed ossification of the lateral cartilages; it is more distinctly seen by comparing it with a sound foot. In lameness of the foot, there is sometimes a crack in the horn towards the heels, extending from the coronet a little way down the hoof: this happens sometimes after a horse has been travelling. This is named a sand-crack. When the seat of lameness is in the fetlock joint, some degree of heat or swelling will be perceived. As the horse stands, he will be observed to favor the joint. Lameness of the back, sinews, or flexor tendons of the leg, is easily perceived by the heat and tenderness of the part.

LAMPAS. — A swelling and sometimes tenderness of the roof of the mouth, adjoining the front teeth. When the part is tender, and prevents the horse from feeding, he should be fed on scalded shorts for a few days; during that time, the mouth may be washed twice a day with an infusion of powdered bayberry bark. Two ounces of bark may be infused in one quart of boiling water: after macerating for one hour, it will be fit for use.

LARYNX. — The upper part of the trachea or windpipe.

LAX. — (See Scouring.)

LAXATIVE. — Medicines that purge gently; the most simple and safe is aloes.

LIGAMENTS are strong, elastic membranes, connecting the extremities of the movable bones.

LIGATURE. — Twine, thread, or silk, waxed, for the purpose of tying arteries, veins, or other parts.

LIGHTS. — A common name for lungs.

LILY. — The root of the white lily is frequently used for poulticing.

LINSEED, or FLAXSEED. — An excellent emollient drink is made by pouring two quarts of boiling water on four ounces of linseed, and suffering it to stand in a warm place for a short time. It is useful in cold, catarrh, and in diseases of the kidneys or bladder.

LIQUORICE. — The root, dried and powdered, is used for the same purpose as the last article.

LOINS. — A portion of the lungs and liver is thus named.

LOCKJAW. — This disease is too well known to require a particular description. It is evidently a disease of the spinal system, — other parts becoming sympathetically affected, — and often arises from a wound of a tendon, or nerve: it occasionally follows nicking, or docking. Mr. Youatt tells us, " This is one of the most fatal diseases to which the horse is subject." For the information of our readers, we will detail the treatment recommended by the above author. We presume that every man of common sense will come to the conclusion that the disease could not be otherwise than fatal under such unwarrantable barbarity. We have no personal disrespect for Mr. Youatt. It is the system of treatment recommended by him that we war against; a system that has killed more than it ever cured. Mr. Youatt observes : " The rational method of cure would seem to be, first to remove the local cause; but this will seldom avail much. The irritation has become general, and the spasmodic action constitutional. The habit is formed and will continue. It will, however, be prudent to endeavor to discover the local cause. If it be a wound in the foot, let it be touched with the hot iron, or caustic, and kept open with digestive ointment. If it follows nicking, let the incision be made deeper, and stimulated by digestive ointment; and, if it arise from docking, let the operation be repeated higher.* In treating the constitutional disease, efforts must be made to tranquillize the system; and the most powerful agent is bleeding. [Yes, most powerful to kill.] Twenty pounds of blood may be taken away with manifest advantage. There is not a more powerful means of allaying general irritation; the next thing is to resort to physic. Here again that physic is best which is speediest in its operation; the Croton nut has no rival in this respect; the first dose should be half a drachm, and the medicine repeated every six hours, in doses of ten grains, until it operates.* The bowels, in all these nervous affections, are very torpid.

" Then, as it is a diseased action of the nerves, proceeding from the spinal marrow, the whole of the spine should be blistered three or four inches wide. (See CANTHARIDES.) Having bled largely, and physiced, and blistered, we seek for other means to lull the irritation; and we have one at hand, small in bulk and potent in energy, — opium ! † Give at once a quarter of an ounce, and an additional drachm every six hours."

The best method we know of, in the treatment of lockjaw, is, first, to apply a poultice to the foot (if it has been wounded), consisting of about six ounces of lobelia, four ounces of slippery elm, two ounces

---

* " First, to remove the local cause; but this will seldom avail much." Then why torture the poor brute? We need not trouble ourselves about the particular nerve affected, to enable us to relieve a sympathetic disease, when we have a medicine — lobelia and milkweed, or Indian hemp — that will relax every nerve in the animal. " If it be a wound in the foot, let it be touched with the hot iron." This is a means better calculated to injure than relieve. We should apply, at once, the means that are known to act on the whole nervous structure. " If it follows nicking, let the incision be made deeper; and if it arise from docking, let the operation be repeated higher." What beautiful philosophy this is ! — make one disease to cure another. Is it strange that " this is one of the most fatal diseases?" Is it not a wonder that any live ? Must not their escape be attributed to the conservative power of the system, in spite of the violence done ? When Mr. Youatt recommends cutting the tail a little higher, to cure a disease that was produced by the same operation, — viz., docking, — he puts the author in mind of the man who filed the edge of his razor to sharpen it.

* In the first part of this paragraph, Mr. Youatt observes, "the most powerful agent to tranquillize the system is bleeding." So say the butchers when they bleed the ox, and conduct the process till no blood remains.

† This is a narcotic, vegetable poison, and, although large quantities have been occasionally given to the horse without apparent injury, experience teaches us that poisons in general — notwithstanding the various modes of their action, and the difference in their symptoms — all agree in the abstraction of vitality from the system. Dr. Eberle says, " Opiates never fail to operate perniciously on the whole organization." Dr. Gallop says : " The practice of using opiates to mitigate pain is greatly to be deprecated. It is probable that opium and its preparations have done seven times the injury that they have rendered benefit on the great scale of the civilized world. Opium is the most destructive of all narcotics."

of capsicum, powdered; mix them with a suitable quantity of meal sufficient for two poultices, which should be renewed every twelve hours. After the second application examine the foot, and, if suppuration has taken place, and the matter can be felt, or seen, a small puncture may be made, taking care not to let the instrument penetrate beyond the bony part of the hoof. Next stimulate the surface to action, by warmth and moisture, as follows: take about two quarts of vinegar, into which stir a handful of lobelia; have a hot brick ready (*the animal having a large cloth, or blanket, thrown around him*), pour the mixture gradually on the brick, which is held over a bucket to prevent waste; the steam arising will relax the surface. After repeating the operation, apply the following mixture around the jaws, back, and extremities: chloroform, and olive oil, equal parts; rub the mixture well in with a coarse sponge; this will relax the jaws a trifle, so that the animal can manage to suck up thin gruel, which may be given warm, in any quantity. This process must be persevered in; although it may not succeed in every case, yet it will be more satisfactory than the blood-letting and poisoning system. No medicine is necessary; the gruel will soften the fæces sufficiently; if the rectum is loaded with fæces, give injections of an infusion of lobelia.

LUMBAR MUSCLES. — Muscles of the loins within the body, and in the region of the kidneys. These muscles are sometimes injured by violent exertions, and the kidneys often participate in the injury.

LUNGS, or LIGHTS. — The organ of respiration. (See RESPIRATION, part first.)

LUXATION. — A partial displacement of the bones forming a joint.

LYMPH. — (See BLOOD.)

LYMPHATICS. — (See part first.)

MACERATION implies soaking or steeping any substance in water, or other fluids, so as to soften, dissolve, or separate it from some other parts with which it is combined.

MALLENDERS. — A scurvy kind of eruption on the back part or bend of the knee joint.

MANGE. — A disease which manifests itself in the skin, and causes a horse to be perpetually rubbing himself. Cattle, sheep, and dogs are also subject to mange. It is a well known fact, that horses are very apt to become mangy, if kept long in the stable without grooming; yet the disease may arise from causes independent of a neglected skin, though it seldom attacks a well-cleansed animal. Mr. Percivall observes: " It seems that mange may be generated either from immediate excitement to the skin itself, or through the medium of that sympathetic influence which is known to exist between the skin and the organs of digestion. We have, it appears to me, an excellent illustration of this in the case of mange supervening upon poverty, — a fact too notorious to be disputed, though there may be different ways of theorizing upon it."

Mr. Blaine says, " Mange has three origins — filth, debility, and contagion."

Owners of horses must bear in mind, that mange can be communicated by the brush or comb used about a mangy subject; the pustules on the surface contain acari or mites. The author has been very successful in the treatment of this disease, by the daily use of sulphur and soda.

MARASMUS. — A decay or wasting of the whole body.

MARSH MALLOWS. — A plant used for making emollient drinks and fomentations.

MASH. — A mash is made by pouring boiling water on bran, or shorts, then covering the bucket until sufficiently cool for use. Mashes are excellent for sick and convalescent horses, and such as have not sufficient exercise to keep them in health.

MASSETER. — The name of a muscle of the cheek, by which mastication is performed.

MASTICATION. — (See part first.)

MATERIA MEDICA. — A catalogue and description of the various articles used in medicine.

MAXILLA. — The jaw.

MAXILLARY. — Belonging to the jaw; as the maxillary arteries and glands. The glands under the jaw are named sub-maxillary glands.

MEDIASTINUM.— A duplicature of the membrane, named pleura, by which the cavity of the chest is divided into two parts.

MEDICINE.— Mr. Clark, veterinary surgeon of Edinburgh, says: "Medicine is often given to the poor brutes unnecessarily, and, of course, mischievously. If a man, or horse, be in a state of health, what more is required, or how can they be rendered better? Health is the more proper state of the animal body, *and it is not in the power of medicine to make it better, or to preserve it in the same state.*"

Dr. White says: "The custom of giving medicines too frequently, is a bad one; the constitution adapts itself to it, which circumstance renders medicine inefficacious when necessary, or, at least, it greatly reduces the effects."

If a horse is in health, the proper way to promote it is to proportion the food to the labor.

Dr. White continues: "Medicines are given to the horse under the title of alteratives. These alteratives are composed of *antimony, mercury, sulphur, nitre, aloes, salts*" (*generally altering bad for worse*).

Mr. Clark says, "That sulphur not only opens the body, but the skin also, and therefore should be used with caution, as horses are very apt to catch cold on too liberal a use of it."

Salts bring on great sickness, and sometimes violent purging, and, instead of promoting the secretions, occasion great dryness of the skin.

"Aloes given in small quantities, by way of alteratives, and too frequently repeated, weaken the stomach, so as to bring on a lax, or what is called a washy, habit of body.

"Antimony should always be rejected, if coarse and black, like gunpowder." (See White's *Farriery*, p. 559.)

The above author says: "It is amazing what different kinds of trash are forced down horses' throats. The following is a striking instance: A gentleman, in London, was greatly prejudiced in favor of vinegar, as a cure in many diseases. His favorite horse was taken ill in very warm weather, and, as he thought vinegar was a cooling article, he ordered a pint to be given to his horse at once. It was no sooner given, than the horse lay down, stretched himself out, and died."

MEDULLA OBLONGATA.— The commencement of the spinal marrow, within the cranium.

MEMBRANE, MUCOUS.— This membrane is folded into all the orifices of the animal, as the mouth, eyes, nose, ears, lungs, intestines, bladder, etc.; in fact, into every cavity that has a direct communication with the external surface. Its structure of arterial capillaries, venous radicles, nervous projections, etc., is similar to the skin, and is considered a duplicate of the external surface. Its most extensive surfaces are those of the lungs and intestines. This membrane furnishes from the blood a fluid called mucus, to lubricate its own surface, and protect it from the action of materials taken into the system. The skin and mucous membrane are a counterpart of each other. If the action of the skin is suppressed, the mucous membrane performs a part of its office; thus, a cold, which closes the pores of the skin, stops perspiration, which is now forced through the membrane, producing discharges at the nose, eyes, etc.

SEROUS MEMBRANE.—Of this kind are the pleura and peritoneum: they are distributed in all parts of the system, lining muscles, tendons, and tendinous sheaths, the ends of movable bones, etc.; in short, wherever there is need of the protection of parts against friction. They secrete from the blood a fluid called serum, for the purpose of affording this protection. The excessive discharge of fluids into cavities lined by serous membrane, constitutes the different forms of dropsy. There are other membranes, viz., adipose, which secrete the fat of the body; synovial, which secrete synovia, or joint oil; and cellular membrane, or tissue, is the common connecting substance of most parts of the body.

MESENTERY.— A thin membrane by which the bowels are held together, and over which the lacteals, or chyle vessels, pass. Besides the chyle vessels, there are considerable veins and arteries passing over the mesentery. The arteries are distributed to the bowels, and the veins terminate in the vena porta, or great vein of the liver.

METACARPUS.— The metacarpus of the horse consists of one great bone, commonly named the canon, shank bone, or fore leg, and two small bones, or splent bones, attached by ligaments to the back part of the canon bone, rather towards the sides. The suspensory ligament passes down on the back part of the canon bone, and between the two splents. The flexor tendons, or back sinews, pass down over the suspensory ligament. When the bones only of the fore leg are spoken of, they are termed metacarpus. They begin at the knee, and end at the fetlock joint.

METATARSUS.— The hind leg, between the hock and fetlock joints.

METATARSAL BONES.— The hind canon, or shank bone, with the two small splent bones attached to it. The large blood-vessels and nerves, in this situation, are also named metatarsal.

MIASMATA.— Poisonous effluvia.

MIDRIFF.— (See DIAPHRAGM.)

MOLARES.— The name of the grinding teeth.

MOLTEN GREASE.— A name which Mr. Blaine has given to dysentery.

MORBID DISPLACEMENTS OF THE INTESTINES, ETC.— *Rupture, or Intestinal Hernia.** Hernia, in its strict sense, is a protrusion of any viscus out of its natural cavity; hence we have hernia of the brain, of the lungs, and of the various viscera of the abdominal regions. Hernia, as we propose to consider it, is a displacement of the intestines from the abdominal cavity, either through some of the natural openings, or through artificial ones, the effects of accident. When such protrusion takes place through a moderate opening, and the portion of gut can be readily returned, it is called a *reducible* hernia; but when it occurs through a small opening, and the intestine cannot be replaced, it is termed an *irreducible* hernia. If the mouth of the opening, round the intestine, constringe, and prevent the return of the bowel, it then forms a *strangulated* hernia, and usually proves fatal, unless relief be promptly obtained.

The hernia, by far the most common in the horse, is the *inguinal*, of which the scrotal, or when the bowel descends into the scrotum, is most frequently observed in the stallion. *Bubonocele*, or that of the groin, is a very rare form of disease, but it is occasionally witnessed in geldings. In the former, the intestine accompanies the spermatic cord by the inguinal canal through the abdominal rings into the scrotum: in the latter, the bowel alone lodges in the groin. The ruptures we have named may be considered as the only ones common to the horse. Some of them are very rarely seen: hernia is more frequently on the right than on the left side; and scarcely ever appears in mares. However, ventral hernia, or rupture of the muscles of the abdominal sides, and protrusion beneath the skin of a portion of intestine, is sometimes beheld in either sex, and perhaps, of the two, is more frequently witnessed in the female.

The *causes* which produce hernia are various, but all arise from violence of exertion, or the effects consequent upon external injuries. With us the efforts used in racing, and the leaps taken in hunting, are causes, as we may readily suppose; when we consider that the dilatation of the abdomen, restrained as it is by weight and tight girthings, must press backwards the intestinal mass. Rearing and kicking also, and being cast for operations, particularly the rising up after castration, have all brought it on. Blows with a thick stick, or from the horn of a cow, may likewise induce it.

---

* "Blaine's Outlines," by Mahew.

The *symptoms* of strangulated hernia are very similar to those of acute enteritis; there is the same uneasiness, shifting of position, getting up and lying down again. The horse rolls in the same manner, and in turning on his back sometimes seems to get a momentary respite from pain; yet it is but momentary, for the suffering is not one of remission; it is constant; this will serve as one distinguishing mark between it and spasmodic colic, with which it has been confounded. In stallions, a pathognomonic symptom is, that the testicle on the hernial side is drawn up to the abdomen, and is retained there, with only momentary fits of relaxation; toward the last, the pulse is quick and wiry; the horse paws, looks at his flanks, but seldom kicks at his belly. We assure ourselves of hernia by an oblong tumor in the groin, of larger or smaller bulk; hard or soft, as it may contain either fæces or gas, in which latter case it will also be elastic. When the tumor is raised by the hand, or pressed, a gurgling sound is emitted: or, if the horse be coughed, it will be sensibly increased in dimensions.

The *treatment of strangulated hernia.*— The horse suffering under the affection we will suppose to be a stallion, and then describe the various manipulations for his relief: firstly, the examination into the state of the hernia; secondly, the application of means preparatory to the application of pressure; thirdly, the application of pressure itself; also, the operation of removing the stricture; and, likewise, the application of the various processes to hernia in the horse.

The *treatment of hernia in a stallion.*— First, *the examination of the hernial sac.* In this manipulation both hands are employed; one is introduced into the rectum, the other into the sheath. The one within the rectum must seek the internal ring; while the other, pursuing the course of the cord on the side affected, is to be pushed up to the external ring; and thus, in the natural state, the opposed fingers may be made nearly to meet, and so estimate the size of the opening. However small the protruded portion of gut, the practitioner will be able to detect, and even to reduce it. *This exploration may be made in the standing posture; but it will be conducted with more facility and certainty if the animal be cast, which is the preferable mode of proceeding.*

Secondly, the *application of means preparatory to the taxis:* these are said to be bleeding, and partially paralyzing the parts by administration of chloroform; or lessening the volume of distention by dashing the part with cold water; or, if the horse be already cast, by spreading ice over the belly.

Thirdly, *the manual efforts to return the displaced gut.* To fulfil this indication, we are, with the same hope, at once to proceed thus: The horse is to be thrown upon the opposite side to that disordered; and, after one hind leg has been drawn and fixed forward, as for castration, he is to be turned upon his back, and in that position maintained by trusses of straw, while other trusses are placed under him to raise the croup. With both arms well oiled, or covered with some mucilaginous decoction, the operator will now commence his exploration, taking the precaution of emptying the rectum as he proceeds. As soon as he shall have ascertained that it is a case of hernia,—have assured himself the gut protruded through the ring is undergoing neither stricture nor strangulation,—he may endeavor to disengage the hernial part, by softly drawing it inward within the cavity, at the same time pushing it in the like direction with the hand within the sheath. Should he experience much difficulty in these attempts, he is to desist; violence being too often the forerunner of strangulation and gangrene. He must bear in mind, also, that, although the reduction is effected, unless it be followed by immediate castration, it does not always prove to be a cure: the protrusion recurs after a time, and occasionally even the moment the animal has risen. If the taxis should be fortunate enough to reduce the hernia, and it be not intended to castrate the horse, apply a well-wadded pledget, or folded cloth, to the part; this may be retained with a bandage crossed between the legs from side to side, and fastened by one part under the belly to a girth; and also passing between the legs, it may be again made fast to the back portion of the same girth; the intention of this is, to prevent the protrusion of the gut by the exertion of rising, and consequently it should be removed as soon as that danger is over. If a radical cure were attempted, of course the clams would supersede this, either in the stallion or gelding.

And concerning the *treatment of strangulated hernia in geldings.* Inguinal hernia, taking the same course, is susceptible of the same terminations, and requires the same treatment as in stallions. The taxis is to be employed, and will be used with most effect, the operator (the horse lying upon his back) extending the hernial sheath with one hand, while he manipulates with the other; or, should this fail, by instructing his assistant to hold up the hernial mass from the belly, so as to take its pressure off the ring, and thus give him an opportunity to renew his efforts with more effect. In some cases, the introduction of one hand into the rectum becomes necessary. The reduction of the hernia should be followed up immediately by the application of the clams, if we unite with the reduction an attempt at permanent cure of the hernia; taking care, at the time, to draw out the part of the scrotum to which the vaginal sheath is adherent, and to push up the clams as close as possible to the belly; they are then to be closed, as for castration.

Of *congenital hernia,* our limits allow of little more than the mention; nor need more be detailed, as its consequences are seldom injurious. It appears that inguinal hernia commonly exists in the fœtus in utero. M. Lineguard, V. S., of Normandy, where breeding is very extensively pursued, has ascertained that enterocele is invariably present at birth; even in abortions, and in subjects still-born. The *congenital enterocele* is an attendant on birth, increasing up to the third or sixth month, but afterwards diminishing, and ultimately vanishing. Should it continue beyond a year or eighteen months, it is to be regarded as a chronic or permanent hernia. *Chronic* or *permanent hernia,* it may be remarked, our observations being so much limited to geldings, we see little of. Castration, however, with the armed clams, is the evident cure.

Strangulation of the Intestines, or Morbid Displacement of the Intestines. The intestines, in consequence of their peristaltic motion, become sometimes entangled together, and a fatal strangulation takes place; this happens, occasionally, from some of the mesenteric folds entwining them; sometimes by their rupturing the mesentery, and becoming strangulated by passing through the opening they have made: but it is much oftener the consequence of *spasmodic* action, and during colic these *inversions, involutions, invaginations,* and *introsusceptions* occur. When thus affected, it is not unusual for the ileum to become reversed in its usual course; in which case, a portion, then contracted by spasm, becomes forced into a part less constringed, and an impenetrable obstruction thence is formed. We may draw a practical inference from these cases,—that in spasm we should attempt an early relief; and likewise that we should endeavor, in all cases of failure in bowel affections, invariably to make a *post-mortem examination:* and this we may do on the ground that repeated cases may enable us accurately to interpret symptoms; then, although we cannot relieve, we may offer such an opinion as will convince our employers it is not our ignorance of the signs, but our circumscribed means, which is the cause of our inability to afford assistance.*

MORTIFICATION.—A part deprived of vital force, by causes inducing a loss of tone.

MOULTING.—Casting the coat. In spring the old coat is shed, or thrown off, and the horse gradually improves in spirit and in appearance; but, during the change, he is more liable to take cold. In the latter part of the year, the coat becomes longer and

* Blaines' "Outlines."

coarser, and loses its healthy gloss; at the same time, the horse often becomes weak, sweats readily upon moderate exercise, and is often incapable of performing his usual labor. This is more especially the case with horses that have been hard worked and badly fed. At both these periods it is necessary to take particular care of horses, and work them moderately. A horse, when moulting, should not be exposed in the stable to a current of air, but kept in a ventilated stable. Warm clothing is improper.

MUCILAGE. — A solution of gum, or anything that partakes of the nature of gum. Gummy or mucilaginous drinks are useful in internal disease; the cheapest is an infusion of linseed or marsh-mallows; but the best, perhaps, is a solution of gum arabic.

MUCOUS MEMBRANES. — (See MEMBRANE.)

MUCOUS. — Many of the secretions of the body are of a mucous nature.

MUCUS. — A fluid secreted by mucous surfaces.

MUSCLE. — The parts that are usually included under this name consist of distinct portions of flesh, susceptible of contraction and relaxation.

MUSK. — A powerful odorous substance, whose medical virtues are chiefly anti-spasmodic.

MYRRH. — A gum resin of a fragrant smell and bitter taste. It is given internally, as a tonic, in doses of one or two drachms. Tincture of myrrh is sometimes applied to wounds, ulcers, and sinuses.

NAG. — A name sometimes applied to road horses, and such as have been docked, in contradistinction to those that have long tails, or are used in harness.

NARCOTICS. — Medicines which stupefy, relieve pain, and promote sleep. There are, however, two different ways to effect these objects, and, of course, two different characters of remedies to be used for the purpose. The popular method is to administer opium, whose natural tendency is to depress the vital powers, and deprive them of sensibility. All mixtures, in any form, that contain opium, though soothing for the present, are ultimately and surely pernicious. The true plan is to give antispasmodics. (See ANTI-SPASMODICS.)

NARES. — The nostrils.

NECROSIS. — The mortification and separation of a portion of dead bone from the other parts of the bone.

NEPHRITICS. — Medicines that act on the kidneys.

NERVING, NERVE OPERATION. — It consists of cutting out a portion of the nerve which supplies the foot, either just above the fetlock joint, which is named the high operation, or in the pastern, which is called the low operation. In the former the sensibility of the foot is supposed to be entirely destroyed, and in the latter only partly so. Dr. White observes, serious mischief, such as the loss of the hoof, has sometimes followed the higher nerve operation.

"After the division of a nerve, the extremities of the divided portion retract, become enlarged and more vascular; but especially the upper portion; and coagulable lymph is effused, which soon becomes vascular. In a few days the coagulable lymph from each portion becomes united, and anastomosis forms between the blood-vessels; the lymph gradually assumes a firmer texture, and the number of the blood-vessels diminishes, and the newly-formed substance appears to contract, like all other cicatrices, so as to bring the extremities of the divided portions nearer and nearer to each other. It is difficult to determine, from an experiment on the limb of an animal, the exact time at which the nerve again performs its functions after being divided. In eight weeks after the division of the sciatic nerve, I have observed a rabbit to be in some degree improved in the use of its leg; but at the end of eighteen weeks it was not perfect. When the nerves of the leg of a horse are divided just above the foot, they are sufficiently restored to perform their functions, in some degree, in six or eight weeks; but it must be observed *that these nerves are only formed for sensation,* and it is very different with the nerves of nutrition, voluntary motion, etc.; the re-union is sometimes accomplished by granulations. Secondly, I would observe, that punctures and partial divisions of nerves heal in the same way as when there has been a total division; and that, even on the first infliction of the wounds, the function of the nerves is very little impaired." (See Swan's work on morbid local affections.)

Mr. Sewell finds "that, in cases of entire section of a nerve, sensation returns in about two months; but in others, in which a portion of nerve has been exercised, that the period of restoring feeling can by no means be foretold: in one of his own horses, he ascertained that there was no sensibility in the foot, even at the expiration of three years; and in some others, after a longer interval, the organ appeared to be wholly destitute of feeling."

NICKING.— An operation often performed on horses, to raise the tail, and make them carry it more gracefully, or rather to suit the taste of man.

NIPPERS.— The front teeth, above and below, have been thus named.

NITRE.— Mr. Morton writes: "Nitre given internally is a febrifuge and diuretic. The dose is from two to four drachms. In order to obtain its full effect as a febrifuge, it should be exhibited in the form of ball, so that it may undergo solution in the stomach: but as a diuretic, it is best given in solution. It passes to the kidneys unchanged, and its presence may be readily detected in the urine by means of bibulous paper immersed in it, which, on being dried, deflagrates; or, if the quantity given be great, it may be procured in crystals from the urine. Very large doses of this salt act as an irritating poison. Two pounds being given in six pints of water to a horse, apparently in health, within half an hour irritation of the mucous lining of the alimentary canal began, evidenced by the fæces being voided frequently and in small quantities. The kidneys were soon after excited into increased action, the urine being forcibly expelled, and the act accompanied with uneasiness. In about four hours after, the pulse had risen to nearly double the number of beats, and the visible mucous membranes were highly injected. Blood being withdrawn from the jugular vein, it presented all the appearances of arterial blood. In the serum the existence of the salt could be detected, but it was obtained in abundance from the urine. From this period the symptoms became less urgent, and the pulse gradually regained its healthy standard; but the dung and urine continued to be passed more frequently than natural throughout the day.

Externally applied, nitrate of potassa is a valuable stimulant to wounds, and it may be employed with much benefit when gangrene has taken place. For this purpose, a saturated solution is ordered to be kept in the pharmacy.

OATS.— According to Sir H. Davy's analysis, oats contain 742 parts of nutritive matter out of 1000, which is composed of 641 mucilage, or starch, 15 saccharine matter, and 87 gluten, or albumen. New oats are difficult of digestion.

OBLIQUE MUSCLES.— The muscles of the abdomen, or belly, are thus named. There are four of them; two external and two internal. Some of the muscles of the eye are also named oblique muscles.

OCCIPUT.— The back part of the head.

ŒDEMA.— A watery or dropsical swelling.

ŒSOPHAGUS, or ESOPHAGUS.— The tube passing from the mouth to the stomach.

OINTMENTS.— Unctuous substances of the consistence of butter; when made considerably thinner by the addition of oil, they are termed liniments; but when their solidity is increased by wax, rosin, etc., they are termed plasters.

OLECRANON.— The head of the bone named ulnar (see cut), in the horse; it affords a powerful lever for the triceps extensor cubiti muscle to act upon, in straightening the fore arm upon the humerus. (See SKELETON.)

OLFACTORY NERVES are spread over all the interior of the nostril, and constitute the sense of smell.

OMENTUM.— The omentum, or caul, is a

double membrane, containing within its folds a considerable quantity of fat, in the human body and many animals. But in the horse this is never seen; nor does the omentum contain much fat; what there is lies in the region of the stomach.

OPACITY. — A want of transparency in those parts of the eye named pupil, or cornea.

OPERATIONS.

OF SURGICAL OPERATIONS,* AND THE VARIOUS RESTRAINTS IT IS SOMETIMES NECESSARY TO PLACE THE HORSE UNDER FOR THEIR PERFORMANCE. — " When it is necessary to perform any painful operation on so powerful an animal as the horse, it is of consequence to subject him to a *restraint* equal to the occasion. Horses are very dissimilar in their tempers, and bear pain very differently; but it is always prudent to prepare for the worst, and few important operations should be attempted without casting. Humanity should be the fundamental principle of every proceeding, and we ought always to subject this noble animal to pain with reluctance; but when circumstances absolutely call for it, we should joyfully close our hearts to all necessary suffering. The resistance of the horse is terrible, and it is but common prudence to guard against the effects of it. The *lesser restraints* are various: among them may be first noticed the *twitch*. The *twitch* is a very necessary instrument in a stable, though, when frequently and officiously used, it may have the ill effect of rendering some horses violent to resist its application. In many instances blindfolding will do more than the twitch; and some horses may be quieted, when the pain is not excessive, by holding the ear in one hand, and rubbing the point of the nose with the other. A soothing manner will often engage the attention and prevent violence; but it is seldom that either threats or punishment render an unruly horse more calm. Inexperienced persons guard themselves only against the hind legs; but they should be aware that some horses strike terribly with their fore feet: it is prudent, therefore, in all operations, to blindfold the animal, as by this he becomes particularly intimidated, and if he strikes he cannot aim. When one of the fore extremities requires a very minute examination, it is prudent to have the opposite leg held up; it may, in some cases, be tied: and when one of the hinder feet is the object of attention, the fore one of the same side should be held up, as by this means the animal is commonly prevented from striking. If this precaution be not taken, still observe to keep one hand on the hock, while the other is employed in what is necessary; by which means, if the foot become elevated to kick, sufficient warning is given, and the very action of the horse throws the operator away from the stroke. Without the use of these arts the practitioner will expose himself to much risk. * The *trevis* is the very utmost limit of restraint, and is seldom used save by smiths, to shoe very violent and powerful horses: whenever recourse is had to it, the greatest caution is necessary to bed and bolster all the parts that are likely to come in contact with the body. On the Continent we have seen horses shod in this machine, and apparently put into it from no necessity greater than to prevent the clothes of the smith from being dirtied. Horses have been destroyed by the trevis, as well as by casting; or their aversion to the restraint has been such, they have died from the consequences of their own resistance. The *side-line* is now very generally used, not only in minor operations, but also in those more important. Many veterinarians do not use any other restraint than this, in which they consider there is safety both to the horse and to the operator. It is applicable to such horses as are disposed to strike behind; and consists in placing a hobble strap around the pastern of one hind leg, and then carrying from a web collar passed over the head the end of a rope through the D of the hobble, and back again under the webbing round the neck. A man is then set to pull at the free end of the rope, by which the hinder leg is drawn forward without elevating it from the ground. By this displace-

* Blaines' " Outlines."

ment of one leg the horse is effectually secured from kicking with either. Occasionally it is thus applied: hobbles are put on both hind legs, and the rope is passed through each of the rings. According to this last method, the horse is actually cast, as he must fall when the ropes are pulled. Take a long rope, and tie a loop in the middle, which is to be of such a size as it may serve for a collar; pass the loop over the head, letting the knot rest upon the withers; then take the free ends, pass them through the hobbles, and bring it under the loop. Let two men pull at the ropes, and the hind legs will be drawn forward.

"*Casting.*— The objections to this practice arise from the dangers incurred by forcing the horse to the ground. Mr. Bracy Clark simplified casting, by inventing some patent hobbles, having a running chain instead of rope, and which, by a shifting D, made the loosening of all the hobbles, for the purpose of getting at a particular leg, unnecessary. These were still further improved by Mr. Budd, so as to render a release from all the hobbles at once practicable. Hobble leathers and ropes should be kept supple and pliant with oil, and ought to be always examined previous to using; nor should the D or ring of the strap be of any other metal than iron. Brass, however thick, is brittle, and not to be depended on. To the D or ring of one pastern hobble, a chain of about four feet long is attached; to this a strong rope is well fastened, and, according to the way the horse is to be thrown, this hobble is to be fixed on the fore foot of the contrary side: the rope is then passed from the hobble on the fore foot to the D of the hind foot of that side, then to the other hind foot, and, lastly, through the D of the other fore foot. After this, much of the ease and safety of the *throw* depend on bringing the legs as near together as possible. This should be done by gradually moving them nearer to each other, without alarming the horse; which will very much facilitate the business, and is really of more moment than is generally imagined. A space sufficiently large should be chosen for the purpose of casting, as some horses struggle much, and throw themselves with great violence a considerable way to one side or the other; and *they are able to do this if the feet have not been brought near together previous to attempting the cast.* The place should be also very well littered down. The legs having been brought together, the assistants must act in concert; one particularly should be at the head, which must be carefully held throughout by means of a strong snaffle bridle; another should be at the hind part to direct the fall, and to force the body of the horse to the side which is requisite. Pursuing these instructions, the animal may be at once rather *let* down than *thrown*, by a dexterous and quick drawing of the rope; the whole assistants acting in concert. The moment the horse is down, the person at the head must throw himself upon that member, and keep it secure; for all the efforts of the animal to disengage himself are begun by elevating the head and fore parts. The rope is tightened. The chain is fixed by inserting a hook through one of the links, of sufficient size not to pass the hobbles. When the operation is over, the screw which fastens the chain to the hobble, first put upon one fore leg, is withdrawn. The chain then flies through the D's of the other hobbles, and all the legs are free, save the fore leg first alluded to; the strap of this has to be afterwards unbuckled. There are also other apparatus used in casting, as a strong leathern case to pass over the head, serving as a blind when the animal is being thrown; and as a protection against his rubbing the skin off his eyes when down. Then a surcingle is also used. This is fastened round the horse's body, and from the back hangs a broad strap and a rope: the strap is fastened to the fore leg of that side which it is desired should be uppermost; the line is given to a man who stands on the opposite side to the generality of the pullers. On the signal being given, the men having hold of the hobble rope pull the legs one way, while he who has hold of the rope attached to the surcingle pulls the back in a contrary direction, and the horse is immediately cast

"*Slinging* is a restraint which horses submit to with great impatience, and not without much inconvenience, from the violent excoriations occasioned by the friction and pressure of the bandaging around his body. Graver evils are also brought about by the abdominal pressure: some horses stale and dung with difficulty when suspended; and inflammation of the bowels has not unfrequently come on during slinging. The slings are, however, forced on us in some cases, as in fractured bones, the treatment of open joints, and some other wounds where motion would be most unfavorable to the curative treatment. *Suspension* may be partial or complete. Suspension of any kind will require the application of pulleys and ropes affixed to the beams, that the whole body of the horse may be supported. A sling may be formed of a piece of strong sacking, which is to pass under the belly, the two ends being fastened firmly to pieces of wood; each of about three feet long, and which are to reach a little higher than the horse's back: to the pieces of wood, cords and pulleys are to be firmly attached, by which means the sacking can be lowered or raised at pleasure. To the sacking, also, are to be sewn strong straps, both before and behind, to prevent the horse sliding in either direction, without carrying the sacking with him. Upon this so-formed cradle he is to recline. If horses when they are fresh should be placed in this machine, most of them would either injure themselves, or break through all restraint. However, by tying up their heads for three or four nights, their spirit is destroyed. The slings may then be applied without the fear of resistance: it is the best method not to pull the canvas firm up, but to leave about an inch between the horse's belly and the cloth, so that the animal may stand free, or throw his weight into the slings when he pleases. In this fashion a horse may remain for months in the slings, and at the end of the time display none of the wear and tear so feelingly described by old authors.

"*Castration.*— This practice is of very ancient origin; and is as extensive as ancient. It is founded on the superior placidity of temper it gives. The castrated horse no longer evinces the superiorities of his masculine character, but approaches the softer form and, milder character of the mare. Losing his ungovernable desires, he submits to discipline and confinement without resistance; and, if he be less worthy of the painter's delineation and the poet's song, he is valuable to his possessor in a tenfold degree. In England, where length in the arms and of the wide spread angles of the limbs is absolutely necessary in the horse to accomplish the rapid travelling so much in vogue among us, the exchange of the lofty carriage and high action of the stallion is absolutely necessary; and, when we have added the lessened tendency of the gelding to some diseases, as hernia, founder, cutaneous affections, etc., we may be content to leave the sexual type with the racer for his breed; also with the drayhorse for his weight, and the fancy of his owner. Supposing it, therefore, eligible to castrate our horses, what is the proper age for the operation? What are the relative advantages and disadvantages of the different methods of performing it? The proper age to castrate the young horse must depend on circumstances; as on his present appearance, his growth, and the future purposes we intend him for; observing, generally, that the more early it is done, the safer is the operation: for, until these organs begin to secrete, they are purely structural parts, and as such are not so intimately connected with the sympathies of the constitution. Some breeders of horses castrate at twelve months; others object to this period, because they think the animal has not sufficiently recovered the check experienced from weaning, before this new shock to the system occurs. In the more common sort of horses used for agricultural purposes, it is probably indifferent at what time the operation is performed; this consideration being kept in view, that the earlier it is done the lighter will the horse be in his fore-hand; and the longer it is protracted the heavier will be his crest, and the greater his weight before, which in heavy draught work is desirable. For carriage horses it would be less so, and the period of two years is not a bad one for their castration. The better sort of saddle-horses should be well examined every three or four months; particularly at the ages of twelve, eighteen, and twenty-four months; at either of which times, according to circumstances or to fancy, provided the forehand be sufficiently developed, it may be proceeded with. Waiting longer may make the horse heavy; but, if his neck appear too long and thin, and his shoulders spare, he will assuredly be improved by being allowed to remain entire for six or eight months later. Many of the Yorkshire breeders never *cut* till two years, and think their horses stronger and handsomer for it: some wait even longer, but the fear in this case is, that the stallion form will be too predominant, and a heavy crest and weighty fore-hand be the consequence; perhaps also the temper may suffer. Young colts require little preparation, provided they are healthy and not too full from high living; if so, they must be kept somewhat short for a few days; and in all, the choice of a mild season and moderate temperature is proper. When a full grown horse is operated on, some further preparation is necessary. He should not be in a state of debility, and certainly not in one of plethora: in the latter case, lower his diet, and it would be prudent to give him a purgative. It is also advisable that it be done when no influenza or strangles rage, as we have found the effects of castration render a horse very obnoxious to any prevalent disease. The advanced spring season, previous however to the flies becoming troublesome, is the proper time for the performance of the operation upon all valuable horses; and be careful that it be not done until *after* the winter coat has been shed, which will have a favorable effect on the future *coating* of the horse, independent of the circumstance, that at a period of change the constitution is not favorable to any unusual excitement.

"Castration is performed in various ways, but in all it expresses the removal of the testicles; there are methods of rendering the animal impotent without the actual destruction of these organs; for if by any other method the secretion of the spermatic glands is prevented, our end is answered.

"*Castration by cauterization* is the method which has been principally practised among us. But this by no means proves it the best; on the contrary, many of our most expert veterinarians do not castrate by this method. Mr. Goodwin, and many other practitioners of eminence, never castrate by cautery.

"A preliminary observation should be made previously to casting, to see that the horse is not suffering from a rupture: such cases have happened; and as in our method we open a direct communication with the abdomen, when the horse rises it is not improbable that his bowels protrude until they trail on the ground. Hernia as a consequence of castration may easily occur by the *uncovered* operation; for, as already observed, it makes the scrotal sac and abdominal cavity one continuous opening. It is not to be wondered at, therefore, if the violent struggles of the animal should force a quantity of intestine through the rings into the scrotal bag. Should we be called on to operate on a horse which already had hernia, it is evident we ought not to proceed with it, unless the owner be apprised of the risk, and willing to abide by it. In such case we would recommend that the method of Girard be practised, *i. e.* to inclose the tunica vaginalis within the clams (sufficiently tight to retain them, but not to produce death in the part) pushed high up against the abdominal ring, and then to remove the testicle, being very careful to avoid injuring any portion of intestine in the operation. When a discovery is made of the existence of hernia *after an opening has been already made* for the common purpose of castration, should the operator continue his process, and castrate? We should say, by no means; but, on the contrary, we would greatly prefer the method recommended by Mr. Percivall,—firmly to unite the lips of the external wound by suture, allowing the testicle itself to assist in blocking up the passage; with a hope also that the inflammation caused by the excision might altogether stop up the scrotal communication with the abdomen. But, in the appalling case of immense protrusion of intestine, what is to be done? Mr. Coleman, in such a case, proposes to make an opening near the umbilicus, large enough to introduce the hand, and thus draw in the bowels. Mr. Percivall would prefer dilating the external ring; but the testicle must be very firmly retained, and even permanently fixed against the dilated ring, or the bowels would again descend. The intestines probably would become inflated in any such case.

"As unbroken young horses are the most usual subjects of this operation, and as such often have not yet been bridled, if a colt cannot be enticed with oats, etc., he must be driven into a corner between two steady horses; where, if a halter cannot be put on, at least a running hempen noose can be got round his neck; but, which ever is used, it should be flat, or the struggles, which are often long and violent, may bruise the neck, and produce abscess or injury. When his exertions have tired him, he may be then led to the operating spot; here his attention should be engaged while the hobbles are put on, if possible; if not, a long and strong cart-rope, having its middle portion formed into a noose sufficiently large to take in the head and neck, is to be slipped on, with the knotted part applied to the counter or breast; the long pendant ends are passed backward between the fore legs, then carried round the hind fetlocks; brought forward again on the outside, run under the collar-rope; a second time carried backward on the outer side of all, and extended to the full length in a direct line behind the animal. Thus fettered, Mr. Percivall says his hind feet may be drawn under him toward the elbows; it has been, however, often found that, at the moment the rope touches the legs, the colt either kicks and displaces the rope, or altogether displaces himself; but his attention can generally be engaged by one fore leg being held up, or by having his ear or muzzle rubbed, or even by the twitch; if not, the rope may be carried actually round each fetlock, which then acts like a hobble; and this rope may be gradually tightened: this last, however, is a very questionable method, and the others therefore ought to be long tried before it is resorted to; in this way people have succeeded with very refractory colts; but it requires very able assistants, and, if possible, the man who has been used to the individual colt should be present. In either way, as soon as the rope is fixed, with a man to each end of it behind the colt, let them, by a sudden and forcible effort in concert, approximate his hind legs to his fore, and thus throw him. Before the colt is cast, however, it should be endeavored to ascertain that he is free from strangles and hernia.

"Being satisfied that no hernia exists on either side, proceed to cast the colt, turning him, not directly on the left side, but principally inclining that way; and, if possible, let the croup be very slightly elevated; it is usual to place him directly flat on the left side, but the above is more convenient. Next secure the near hind leg with a piece of hempen tackle, having a running noose; or, in default of this not being at hand, make use of the flat part of a hempen halter, which should for safety be put on before the hobble of that leg is removed; as may be readily done, if the hobbles having shifting or screw D's, as described in casting, are made use of. Every requisite being at hand, the operator, having his scalpel ready, should place himself behind the horse, as the most convenient way to perform his manipulations; and, firmly grasping the left testicle with his left hand, and drawing it out so as to render the scrotum tense, he should make an incision lengthways, from the anterior to the posterior part of the bag. The resistance of the cremaster muscle has to be overcome before the testicle can be forced to the bottom of the scrotum; and this is the more readily accomplished if the animal's attention be engaged. The incision may be carried at once through the integuments, the thin dartos expansion, and the vaginal coat of the testicles, with a sweep of the scalpel: but with one less dexterous at the operation, it will be more prudent to make the first incision through the scrotum and dartus only, to the required extent; and then to do the same by the vaginal coat, thus avoiding to wound the testicle, which would produce violent resistance, and give unnecessary pain. We, however, take this opportunity of noting, that cases have occurred, when the tunica vaginalis was divided, no testicle followed; firm adhesions between this tunic and the tunica albuginea having retained it fast. In such cases the scalpel must be employed to free the testicle, by dissecting it away from the vaginal sac. When no such obstruction occurs, the testicle, if the opening be sufficiently large, will slip out; but the operator must be prepared at the moment of so doing to expect some violent struggles, more particularly if he attempt to restrain the contractions of the cremaster, and by main force to draw out the testicle. Preparatory to this, therefore, the twitch should be tightened; the attendants, especially the man at the head, must be on the alert; and the testicle itself, at the time of this violent retraction of the cremaster, should be merely held, but not dragged in opposition to the contraction. If the clams have been put on over the whole, according to Mr. Percivall's method, they will assist in retaining the retracting parts; but they must not be used with too much pressure. The resistance having subsided, the clams must now be removed; or, if they have not been previously in use, they must now be taken in hand, and, having been prepared by some tow being wound round them, should be placed easily on the cord, while time is found to free from the grip of the pincers the *vas deferens*, or spermatic tube, which is seen continued from the epididymis. The Russians, Mr. Goodwin informs us, cut it through when they operate. Humanity is much concerned in its removal from pressure, because of the excess of pain felt when it is included. It is necessary, before the final fixing of the clams, to determine on the part where the division of the cord is to take place. To use Mr. Percivall's words, 'If it be left too long, it is apt to hang out of the wound afterward, and retard the process of union;' on the other hand, if it be cut very short, and the arteries happen to bleed afresh after it has been released from the clams, the operator will find it no easy task to recover it. The natural length of the cord, which will mainly depend on the degree of the descent of the gland, will be our best guide in this particular. The place of section determined on and marked, close the clams sufficiently tight to retain firm hold of the cord, and to effectually stop the circulation within it. There are now two modes of making the division: the one is to sever it with a scalpel, and then to sufficiently sear the end of it as to prevent a flow of blood. The other, and in some respects the preferable method, is to employ a blunt-edged iron, which is to divide by little crucial sawings, so that, when the cord is separated, it shall not present a uniform surface, but ragged edges, which will perfectly close the mouths of the vessels. This done, loosen the clams sufficiently to observe whether there be any flow of blood; gently wipe the end of the cord also with the finger, as sometimes an accidental small plug gets within the vessel; this had better be removed at the time. Retain a hold on the clams a few minutes longer; and, while loosening them gradually, observe to have an iron in readiness again to touch the end of the cord, if any blood makes its appearance. Satisfied on this point, sponge the parts with cold water; no sort of external application is necessary, still less any resin seared on the end of the cord, which can only irritate, and will never adhere. On the after-treatment much difference of opinion has existed, and even yet exists. The powerful evidence of accumulated facts has now convinced us of the necessity and propriety of some motion for the newly castrated horse, as a preventive of local congestion; such practice is common in most countries, and seems salutary in all. Hurtrel d'Arboval, thus impressed, recommends the horse, immediately after the operation, to be led out to walk for an hour; and it is a general plan in France to walk such horses in hand an hour night and morning. Mr. Goodwin, in proof of its not being hurtful, informs us that whole studs of horses, brought to St. Petersburgh to be operated on, are immediately travelled back a certain portion of the distance, night and morning, until they arrive at home. We have, therefore, no hesitation in recommending a moderate degree of motion in preference to absolute rest.

"*The French method of castration* is advocated by Mr. Goodwin; and it is sufficient that it receives his recommendation to entitle it to attention; it is rendered the more so, as he observes on the method in general use among us, 'that the operation performed by the actual cautery always induces, more or less, symptoms that often become alarming; and that it cannot be performed on the adult without incurring more swelling and severer consequences than attend other methods of operation. If I ever use the actual cautery, it is for the sake of expedition, and then only on a yearling, or a two-year old; but I am resolved never to employ it again on an adult.' These observations, as emanating from such a source, must be deemed important. Mr. Goodwin then offers the description of the French method of operating, from Hurtrel d'Arboval. 'Castration, by means of the clams, is the method in general use, if not the only one now employed; it is the most ancient, since it was recommended by Hieroclius among the Greeks. It is performed in two ways, the testicle being *covered* or *uncovered*. In the former, the exterior of the scrotum, formed by the skin and dartos muscle, is cut through, and the testicle is brought out by dissecting away the laminated tissue, the gland being covered by the tunica vaginalis; the clam is then placed above the epididymis, *outside* the external peritoneal covering, of the cord. In the uncovered operation, the incision is made through the servus capsule of the testicle; the tunica vaginalis being divided, the testicle presents itself, and the clam is placed well above the epididymis, on the cord. The operation, performed in either way, requires us to provide ourselves with a scalpel, a pair of clams, a pair of long pincers, made purposely to bring the ends of the clams together, and some waxed string. The clams may be formed of different kinds of wood; but the elder is considered the best, and generally made use of. To make a clam, we procure a branch of old and dry elder, whose diameter should be about an inch, and whose length should be from five to six inches: of course, the dimensions must at all times be proportioned to the size of the cord we have to operate on. At the distance of half an inch from each end, a small niche, sufficiently deep to hold the string, must be made, and then the wood should be sawed through the middle lengthways. Each divided surface should be planed, so as to facilitate the opening of the clams, either when about to place them on or take them off. The pith of the wood is then to be taken out, and the hollow should be filled with corrosive sublimate and flour, mixed with sufficient water to form it into a paste. Some persons are not in the habit of using any caustic whatever; then, of course, scooping out of the inside of the clam is not necessary: notwithstanding, the caustic, inasmuch as it produces a speedier dissolution of the parts, must be useful, and ought not to be neglected.' The addition of the caustic, however, Mr. Goodwin objects to with great reason, remarking, that unless it be a very strong one, and therefore dangerous to employ, it cannot be of any use to parts compressed and deprived of circulation and life. He further informs us that he has operated in six cases in succession with the same effect, without any escharotic matter whatever. An experimental case of Mr. Percivall's terminated fatally: by the use of caustic the cord was greatly inflamed, as high as the ring, and which unquestionably produced the unfortunate result. 'The covered operation,' continues Mr. Goodwin, 'is the one that I am about to advocate, and which differs only insomuch, that the scrotum and dartos muscle must be cautiously cut through, without dividing the tunica vaginalis. It was Monsieur Berger, who was accidentally at my house when I was about to castrate a horse, and who, on my saying that I should probably do it with the cautery, expressed his surprise that I should perform the operation in any other way than on the plan generally approved of in France. Being a stranger to it, he kindly consented to preside at the operation, and, after seeing him perform on the near testicle, I did the same on the right, but, of course, not with the same facility. After opening the scrotum, and dissecting through the dartos, which is very readily done by passing the knife lightly over its fibres; the testicle, and its covering, the tunica vaginalis, must be taken in the right hand, while the left should be employed in pushing back the scrotum from its attachments; and, having your assistant ready, as before, with the clam, it must be placed well above the epididymis, and greater pressure is, of course, necessary, as the vaginal covering is included in the clam.'

" Mr. Goodwin further observes, that in Russia he has seen hundreds of horses operated on, even after the human fashion, with safety; and, he remarks, it certainly produces less pain, the animal loses less flesh and condition, and is sooner recovered than when operated on by the actual cautery.

"*Castration by ligature* is a painful, barbarous, and very dangerous practice: and consists in inclosing the testicles and scrotum within ligatures, until mortification occurs, and they drop off. It is practised by some breeders on their young colts, but it is always hazardous, and disgracefully cruel. The substance of the testicle in some countries is also broken down either by rubbing, or otherwise by pressure between two hard bodies: this is practised in Algiers, instead of excision, and tetanus is a frequent consequence of it. In Portugal they twist round the testicle, and thus stop the circulation of the gland. Division of the *vas deferens* has been performed, it is said, with success, on many animals; and is proposed as a safe and less painful process than the emasculation of the horse. It consists in a longitudinal section through the scrotum, dartos, and vaginal sheath, so as to expose the cord, from which the vas deferens is to be separated and severed from the artery and vein. There is a certain consent of parts, by which the sympathy of an organ remains after its functional offices are *apparently* destroyed. There can be little doubt but the nervous excitement would continue, the vein and artery remaining entire. There are certain nice conditions of the organ necessary for propagation; thus, the horse who retains his testicles within his abdomen, possesses all the roguish qualities of him with one perfectly evolved: he is lustful, and can cover, but is seldom fruitful. Of the morbid consequences of castration we have little to say: by early evacuations, green food, a loose box, a cool air, moderate clothing, but particularly by walking exercise, swellings of the parts may be prevented: if not, bleed and foment; should suppuration follow, and sinuses form, treat as directed under those heads; and if tetanic symptoms start up, refer to that article. There has been lately practised in India a novel mode of castration, which is said to be the invention of a Boer settled at the Cape of Good Hope. The cord is exposed in the usual manner; from the cord the artery is singled out; this vessel is scraped through with a coarse-edged blunt knife, when the other constitutents of the cord are cut away, and the operation is finished. This method is much praised by those who have adopted it, and is said to be always attended with success.

" *Lithotomy.* — Hurtrel d'Arboval's account of the progress of lithotomy in veterinary practice commences in 1774. The second case was successfully operated on in 1794; and at later periods other veterinary surgeons have also performed it. 'In monodactyles there are two methods of operating for the stone; one through the rectum, the other through the bladder. The first, which consists in lying open the bladder by a longtitudinal incision made through the parietes of the part of the rectum adherent to it, by means of a straight bistoury, is easily practised; but in its consequences is dangerous in the extreme: in fact, it is an operation never to be adopted but in a case where the magnitude of the stone precludes its extraction through the neck of the bladder. In all other cases, lithotomy by the urethra is to be pursued. For its performance, are required a straight probe-pointed bistoury, a whalebone fluted staff, and a pair of forceps curved at the extremities. The animal should, if practicable, be maintained in the *erect* posture. The tail plaited and carried round on the right quarter, the operator feels for the end of the staff introduced up the urethra, and makes an incision directly upon it, from above downwards, an inch and a half or two inches in length. Next, he introduces the sound, and passes it onward into the bladder. Now, placing the back of the bistoury within the groove of the sound, by gliding the knife forwards, the pelvic portion of the urethra, and also the neck of the bladder, become slit open; the latter in two places, in consequence of a second cut being made in withdrawing the bistoury. The opening made being considered of sufficient dimensions, the operator introduces the forceps into the bladder, and seizes the calculus, one hand being up the rectum, to aid him in so doing. The forceps clasping the stone are now to be withdrawn, but with gentleness; and with a vacillating sort of movement of the hand from side to side, in order more easily to surmount any difficulties in the passage, and the more effectually to avoid contusion or laceration. M. Girard tells us, 'That the cut through the pelvic portion of the urethra ought always to be made obliquely to one side; the operator should hold his bistoury in such a direction that its cutting edge be turned toward the angle of the thigh. By this procedure we shall gain easier access to the bladder; and not only avoid wounding the rectum, but also the artery of the bulb, as well as the bulb itself, and suspensory ligaments of the penis.' The parts cut through in the operation are, 1st, the fine thin *skin* of the perineum, smooth externally and marked with a raphe; densely cellular internally: 2ndly, adhering to the tissue, the *faschial covering*, derived from the faschia superficialis abdominis, which has here become fibrous: it forms the common envelope to the parts underneath, and is closely connected with the corpus musculosum urethræ: 3rdly, *the corpus musculosum urethræ*, that penniform band of fleshy fibres which springs by two branches from the ischiatic tuberosities embracing the sphincter ani, and concealing the arteries of the bulb; whence they unite, and proceed to envelop the urethra: 4thly, *the corpus spongiosum urethræ*, the part immediately covered by the muscular envelope, and which here is bulbous. It is more particularly worthy our remark, from two arteries penetrating the bulb, which come from without the pelvis, ascending obliquely outward to reach the part: 5thly, *the suspensory ligaments* of the penis, pursuing the course of, and adhering to, the tendinous union of the erectores. An attention to the relative position of these parts will demonstrate the advantages of the lateral oblique incision over one made directly along the raphe: by pursuing the latter, we necessarily cut through the suspensory ligaments and into the bulb, wounding thereby the arteries; whereas, by the former, all this danger is avoided, besides that it renders the operation more simple and facile.

"*Tracheotomy*. — Cases occur when this operation is required; as in strangles, when the tumors threaten suffocation, or when any substance has remained unswallowed in the œsophagus, the pressure of which obstructs respiration. In a distressing case of gunpowder bursting immediately under a horse's nose, the effects of which tumefied his mouth and nostrils, so as to prevent free inspiration, the animal owed his life entirely to our excising a portion from the tracheal rings, about ten inches below the angle of the throat. The operation is a very simple one, and may consist either in a longtitudinal section made through two or three of the rings, or a portion, occupying about an inch round, may be excised from the anterior cartilaginous substance. The proper mode, when it can be done, however, is to make a circular opening with a very narrow knife, removing a portion of two cartilages, or taking a semicircular piece from each; and this last, although it is seldom performed, is by far the best method: the integuments should be first divided in the exact centre of the neck, three or four inches below the obstruction; then the skin and tissues should be sufficiently separated to allow a tube adapted to the size of the trachea to be introduced; the tube having an acute turn and a rim, which must be furnished with holes for the adaptation of tapes to secure it around the neck. There are several instruments of this sort in use, of which that adopted by the French, or the one invented by Mr. Gowing of Camden-Town, is to be preferred. The operation has been also performed in cases of roaring, under an idea of dividing the stricture which impeded respiration; but, unless the exact situation of this were discovered, it would be but an experimental attempt.

"*Œsophagotomy.*— It was long thought that a wound in the œsophagus must be necessarily fatal, but we have now sufficient proofs to the contrary on record; so that we are not deterred from cutting into the œsophageal tube when it is necessary; but it is an operation requiring skill and anatomical knowledge; and its future results are sometimes very serious. The cases that call for œsophagotomy are the lodgment of accidental substances within the tube. An apple once so lodged was removed by incision by a veterinary surgeon at Windsor. Carrots, parsneps, beets, etc., are liable to produce such obstruction when not sliced. Too large a medicinal mass also has lodged there; and a voracious eater has, by attempting to swallow too large a quantity of not salivated bran or chaff, produced an obstruction which pressed on the trachea and threatened suffocation. In all cases of obstruction of this kind we will suppose that a probang well oiled has been previously attempted to be passed, and has completely failed. The probang for the horse, however, differs materially from that used for the cow. It is formed after the fashion of the one adopted by the human practitioner, consisting of a pliable piece of whalebone, having a sponge tied to one end. The operation being determined on may be practised standing; if the swelling be large, no fear need be entertained about cutting important organs, as the enlargement will push them on one side. Cut down, therefore, directly upon the centre of the impacted substance. If the horse be cast, which is quite unnecessary, have him of course thrown with his left side uppermost. It will also be necessary to command a good light. The part of the neck chosen for the opening must of course be governed by the obstructing mass. A section should be made through the integuments and cellular tissue beneath them, right into the œsophagus, if possible with one cut, and into the centre of the pipe. If this be not done at once, and it requires some dexterity so as to effect it, mind to make all future incisions in a line with the first opening; as it is important that the cellular tissue should be little interfered with. The œsophagus, fairly cut into the impactment, should jump forth; should it not do so, do not manipulate, or attempt to force it out, but enlarge the opening, and the substance will come through when that is long enough; but no fingering could compel its exit while the opening is too small. The end gained for which the incision was made in the œsophagus, the wound may be then closed by the interrupted sutures, each holding a small piece of tow above the orifice, and having their ends hanging out of the external opening, which should also be brought together by sutures. The after-treatment should be, to interdict all dry food; the animal ought to subsist on very thick gruel for three, four, or five days. If the condition appears to suffer much, allow malt mashes, and when so doing watch the wound; and if the matters taken in are seen to ooze out, wash them away frequently with warm water, to prevent lodgment, which might encourage sinuses to form; and after each washing, syringe with some very mild stimulant, as a very weak solution of sulphate of zinc (*white vitriol*), etc. etc.

"*Neurotomy.— Division of the sentient nerves of the foot.*— Neurotomy has now stood the test of very extensive application: our writers offer innumerable proofs of its restoring almost useless animals to a state of much utility. And, if there are chances that it may occasion such injury as to hasten the end of some horses, it is usually in such as the disease would have done the same for at no distant period. Having stated thus much in its favor, it must not be supposed that we recommend it as an unqualified benefit, even where it succeeds best. No neurotomized horse ever after goes with the same freedom, nor with equal safety, as he did before the operation was performed: indifference to the nature of the ground gone over, is said to have fractured legs; it is quite common to batter the feet to pieces; and, although horses have hunted afterwards, and hackneys have carried their riders long distances, yet it is more calculated to prove beneficial to carriage than to saddle horses. This we believe to be a just statement of its merits; but there are benefits which it offers to the animal of a more extensive and constitutional kind. Those gained by the bodily system generally have been in some cases very marked: thus, an aged and crippled stallion, from the irritation constantly kept up, became so emaciated as to be unable to fecundate; but, being relieved from a constant state of suffering by neurotomy, improved in health and condition, and was again used to cover. It happened, also, that a mare similarly circumstanced ceased to feel œstrum; but after neurotomy it again returned, and she resumed her character of a brood mare. It appears to act with most certainty when a *portion* of the irritated nerve is *excised*. One case has actually occurred where the tetanus, occasioned by a wound in the foot, was arrested and removed by neurotomy. It also promises much in the painful state of some cankers, where the irritation has rendered the application of dressings almost impossible: here, by depriving the foot of sensibility, we deprive the horse of that which is injurious to him: the sore itself is often amended by it; but in every instance the dressings can be effectively applied, and the healthy processes cannot be at all suspended.

"With respect to whether the lower or upper incision ought to have the preference, the decision should be guided by the circumstances, as regards the intensity and the seat of the disease. The operation commonly leaves, for a considerable time, some enlargement around the spot, the effects of the adhesive matter interposed between the severed portions of nerve; and which can be remedied by no application of bandages. This bulging remains so long as life continues; and, however cunningly the incision be concealed, this can be felt with ease, and tells the truth ever after the operation has been performed. Such a circumstance has, however, led some practitioners, when it has been wished to make the upper section, and yet to avoid the chance of detection, to operate on the metacarpal nerve on the *outside*, and on the pastern or plantar nerve on the inside.

"*Mode of performing the operation.*— The situation of the section through the skin being determined on, a guide to which may be gained from the perforatus tendon, and having firmly secured the leg to be first operated on, cut the hair from the part. This being done, and the exact course of the artery being ascertained by its pulsation, make a section close to the edge of the flexor tendon. Let the cut be near, but rather behind, the artery, if below the fetlock joint. The cellular substance being cleared away will bring the vessels into view, and the nerve will be readily distinguished from them by its whiteness. Elevating it from the vessels, and its membranous attachments, by means of a crooked needle armed with thread, pass a bistoury under it, as near to the upper angle of the section as possible. The violent spasm the division of the nerve produces may be somewhat lessened by pressing the nerve between the finger and the thumb; when an opportunity may be taken, either with the scalpel or scissors, of dividing it; then, taking hold of the lowermost portion between a pair of forceps, excise about three-fourths of an inch of its trunk. Having finished, if both feet are affected, proceed to operate upon the contrary side of the other leg; after which turn the horse, and repeat the operations on the like parts of each leg as they come in succession. The integuments may be now drawn neatly together, and secured by a twisted suture, the whole being properly covered by a light compress. Tie up the head for a day or two, after which put on a cradle; keep the horse very quiet and low; give mashes to open the bowels; but we should avoid *physicking*, from the fear that griping might occur, which would make him restless, or probably require exercise.

"*Periosteotomy.*—This operation consists in having the horse thrown upon his side, and the leg to be operated upon released from the hobble, and extended upon a sack filled with refuse hay or straw: this is done by means of a piece of webbing passed round the hoof, and the end given to a man to hold, who pulls rather violently at the member. The operator then kneels down and feels for the exostosis he intends to perform periosteotomy upon. This may be a splint or a node, and commonly exists upon the metacarpal portion of the fore limb. The operator having found the excrescence, snips just below it with a pair of rowelling scissors. He then takes a blunt seton needle and drives it through the cellular tissue, and immediately over the enlargement. Next, another slit in the skin, above the exostosis, is made with the rowelling scissors, and through this last opening the point of the seton needle is forced and then withdrawn. Into the free space thus made a curved knife is introduced: the point of this knife is blunt, and the blade curves upward, the cutting part being below. Some persons use a very diminutive blade, but the editor prefers a rather large instrument, as being more under the command of the hand.

Having introduced this knife, he turns the cutting edge downward, and with it incises the enlargement, sending the blade right through the periosteum, and also through the substance of the exostosis, if it be not too solid for the knife to penetrate. This latter fact is only to be ascertained by actual experience, and no opinion formed after an external examination can be of any value; such being much more the guess of a pretender than the judgment of a surgeon. The age of the animal may be some guide, but even this it is better not to depend upon too entirely. It is true that young horses freely cast forth exostoses, which aged animals mostly absorb; but this rule, though very general, has exceptions, and by no means is to be absolutely depended upon.

"The enlargement being cut through, next take a seton needle armed with a tape, and draw it through the channel already made: tie a knot at either end of the tape, large enough to prevent its being pulled through the opening at either end, and the business is over. The affair is very simple, and the horse may be at once let up. It is, however, in some cases, and only in some, of so much benefit that the horse, being thrown 'dead lame,' gets up and trots off quite sound. However, ere you adopt the operation, apprise the owner of the risk incurred, and that it is by no means a certain cure. Leave the choice with him, but be sure and tell him the openings made for the entrance and exit of the seton commonly leave a blemish behind them; and where the seton travelled, often there remains a thickening, which it may require months to obliterate.

"The after part of the treatment consists in merely having the seton daily moved to and fro: though some persons apply an active blister all over the parts immediately in the neighborhood of the seton; under the idea that the vesicatory renders the operation of greater efficacy, which however is very questionable. When periosteotomy acts at all, it mostly does so at once; and when its benefits are not immediate, it is better to withdraw the seton to prevent after blemish, rather than hazard further and useless treatment by the application of a stimulant to skin already deprived of any connection with the deeper seated structures.

"*Division of the flexor tendon.* — There are so many accidents and diseases that may produce contraction of the flexor tendons, that we only wonder we do not more often meet with them; which we should certainly do, but that the attendants, despairing of being able to afford relief, advise their destruction. The operation consists in making a longitudinal incision of about three inches in length along the inner lateral edge of the tendon; dissecting each portion from its cellular attachments, so as to expose the nerve, artery, and tendons. This opening will allow the perforans to be freed from the perforatus, when a division should be made by a scalpel applied to its surface. It is evident that this should take place below any thickening, or adhesions which may have permanently connected the tendon with the neighboring parts; any lesser attachments will be broken through, by forcing back the foot to the just position. By Mr. Dick this was done 'by placing his knee against the front or projecting part of the pastern, at the same time laying hold of the foot with one hand and the upper part of the leg with the other, and using considerable force: and this appears to be necessary, in order to break any adhesions that may have formed.' The limb should now be placed in a poultice; and, if any fear of future contraction should arise during the cure, lengthen the toe of the shoe proper to the foot operated on. Some slow exercise after the first week may be allowed, but previously to that the horse should be confined to a stall, during which the bowels must be kept open with mashes.

"*Amputations.* — These have been hitherto confined principally to the tail, the ears, and other parts of minor importance in the animal frame; but veterinary surgery now takes a wider field, and the extremities are amputated with a certainty of making horned cattle still serviceable for the purpose of yielding milk; and, without doubt, the same might be done with the brood mare, or stallion, particularly in fractures of the fore extremities. Professor Dick, of the Edinburgh Veterinary College, furnishes a case sent to him by one of his pupils, to the following effect: 'I performed amputation upon the cow on the 7th of July; after having properly secured the animal, and applied a tourniquet above the carpus, I made a circular incision through the integuments round the leg, a little below the carpus; and, having separated the skin so as to allow of its being pushed up a little, I cut through the sinews, and lastly sawed off the stump: the parts are now completely whole, although she has been going at grass all the time; and, now that she has got the advantage of a cork stump, makes a wonderful shift for herself, and yields a good supply of milk to her owner.' Mr. Dick also notices another case of amputation of the fore leg of a two-year-old heifer; and of a third, where the hind leg was removed above the tarsus. Such operations have occasionally occurred from time immemorial, with a few enterprising characters. We have heard of them, but they were mostly regarded as mere matters of curiosity or wonder; and therefore were not imitated. We shall, however, probably ere long have them more common, in cows at least; for, occurring below the carpus and tarsus, they are as easily performed as nicking or docking: and there is no doubt but, were a hollow padded stump applied, such low operations might be prudent in many cases. Fractures, with great comminution of bone, considerable ravages of disease within the foot, or extensive gangrene, are the cases which might call for amputation. Of the method of amputation little need be added to the above. The principal practical points are, the fixing of a tourniquet of sufficient force, which should be padded to make its principal pressure on the leading arterial trunks, while its general circumference will act on the smaller vessels: a ring should be cut lightly below the intended place of operation, only through the integuments; which, when separated from their cellular adhesions for about six inches, should be turned back; and a circular incision may then be made through the muscles, etc., taking up by ligature such vessels, both venous and arterial, as display a disposition for much hæmorrhage. The section thus made, free the bone from the soft parts by the scalpel, where the adherences are very intimate; and having by means of a crucial bandage retracted the soft parts altogether above the bone, saw it through. Finally, effectually secure the principal vessels, when, bringing the soft parts and skin over the bone, retain them there by proper bandaging, which suffer to remain without disturbance until the third or fourth day.

"*Amputation of the Penis.*— Amputation of the penis is not unknown among us: it has been performed several times, and it is found that no canula is necessary to keep open the urethra: the force of the urinary flow breaking down any incipient cicatrization of its orifice. The sheath is first forced back, and the penis brought forward to its greatest possible extent: whatever portion it is intended to remove is now cut through by means of an amputating knife; when the remainder is retracted within the sheath, and little hæmorrhage has afterwards occurred, except at the time of passing the urine; but there appears to have been no alarming quantity of blood lost.

"*Amputation of the Tail, or Docking.*— We are most happy to state this filthy and unnecessary operation is now discarded. It never consisted of more than the cutting off a portion of the stump with brute force, and the cruel application afterwards of a hot iron to the small artery of the tail.

"*Nicking.*— We should be grateful that this barbarous and dangerous process is no longer numbered among the necessary operations. It is so beset with accidents which no skill or prudence can prevent, that no one who has a free will ought to mutilate a horse by nicking.

"*Firing.*— The practice of firing was not always confined to quadrupeds; on the contrary, it probably was first used on man; and to this day in many countries it is a very popular remedy among human surgeons. In India it is applied over the abdomen for the cure of scirrhosity of the liver. Firing in veterinary practice has, by Mr. Coleman's pupils, been justified as only men will justify a favorite operation, the virtues of which have been impressed upon their minds by an eloquent teacher. When Coleman was the chief of the veterinary profession, firing under his rule was used for any and every occasion. It was ridiculously supposed to act as a permanent bandage; as if a few strokes with a heated iron could destroy the elastic property inherent in the skin. It was the favorite styptic of these practitioners, and was applied to arteries (as of the tail), as though it possessed within itself some medicinal virtue. It was used to promote absorption, as in callus; and was likewise resorted to to check absorption, as in ulceration. It was called into action to promote granulation, in broken knees; and was also a favorite agent to check granulations, when they were too luxuriant. In short, there was no folly which a hot iron did not cover. It has now happily fallen into disuse. Most modern practitioners will now confess that their chief reason for exercising the iron is to satisfy the proprietor, not to benefit the animal. After such an acknowledgment, who would submit to have his patient servant's skin scored and burnt with red hot metal?

"The *mode of cauterization* differs according to circumstances. As a general rule it ought, of course, to be applied in the direction of the hair, by which the blemish is lessened: but this rule cannot be arbitrarily followed, although it ought to do away with all the false pride of displaying the taste in the figures scored upon a prostrate beast. The Veterinary College recommends that the limbs be *always* fired in perpendicular lines; others advocate all manner of fanciful marks. Some cast the horse; many surgeons perform standing. The irons used are of various shapes and dimensions. Some recommend the firing of all things to be very light; others persist there is no virtue in hot iron unless it burns very deep. The operation consists in having irons of some substance made red hot, and then drawing them mechanically along, or twisting them about upon the skin. The figures are various, so is the depth of the incision. Both must be decided by the taste, judgment, or heartlessness of the operator.

"*Blistering.*—This is an operation of very great utility, and is, perhaps, compared with its benefits and importance, the safest that is performed. When a *vesicatory* becomes absorbed through the pores of the skin, it inflames the sensible cutis underneath; the consequence of which is, an effusion of serum through the part, which, in the human subject, elevates the cuticle into a bladder equal to the surface inflamed, but in the horse, from the greater tenacity of the cuticular connections, it becomes separated in the form of small distinct vesicles only. If the irritating cause be quickly removed, the serum may be re-absorbed, and the surface restored by a slight effort of adhesive inflammation. If the irritant act in a still minor degree, it simply irritates the vessels of the cutis to an infiltration of fluid through the sensible pores, but produces no desquamation of cuticle: such has been called, a *sweating blister*. But when, by continued irritation, the cutis is exposed, suppuration succeeds, and the part is fully blistered. The salutary action of blisters is exerted in several ways; in promoting absorption; in combating deep-seated inflammations, and in aiding others. As a stimulus to the absorbents, they act beneficially in the removal of injurious deposits, as the coagula remaining after inflammatory lesions. But it is to be remarked, that when any existing deposit is of long continuance, or is osseous, it requires that the action of the vesicatory be kept up. Mercurial blisters have been thought to have a superior influence in accelerating absorption. Mercurials, *rubbed in* some weeks or days previously to blistering, are certainly great assistants, and should always be employed in the treatment of obstinate osseous or ligamentry enlargements. Blisters are very important aids, in inflammatory affections, as counter-irritants, derived from a law in the animal economy, that two inflammations seldom exist in the vicinity of each other; therefore, when such an affection has taken place in any part, and we wish to remove it, we attempt to raise an artificial inflammation in the neighborhood by means of blisters; which, if persevered in, destroy, or at least lessen, the original one. Occasionally also we blister the immediate inflamed part, with an intention to hasten the suppurative process by increasing the activity of the vessels; as in deep-seated abscesses and also in those which attack glandular parts. We therefore employ blisters, to hasten the maturation of the tumors in strangles. When the flagging powers vacillate between resolution and suppuration, as they often do in the phlegmonous inflammations of glandular or of deep-seated parts, blisters may either hasten the resolution, or they may add their influence to the attempted suppuration, and thus bring it to maturity. But we carefully avoid, in other cases, applying a vesicant to a part immediately in a state of active inflammation: particularly we should avoid what is too often done, that of blistering over the tendons, ligaments, and articulatory surfaces of a tumid limb, laboring under a congested state of the parts from excess of vascular action. Here we should do great injury were we to blister, by causing a greater deposit of lymph, and by hastening its organization into an injurious bond of union between the inflamed parts. The *vesicatory* or *blister*, for general use in veterinary medicine, as a simple stimulant, should be principally composed of Spanish flies. Cheaper substitutes are used, but they irritate violently: in extensive inflammatory affections, they are on this account perfectly inadmissible; and wherever a case requires anything more it will be noted. The *mode of blistering* with the Spanish fly is sufficiently known. The hair should be cut or shorn as close as possible from around the part; the blistering matter should then be well rubbed in for ten or fifteen minutes. If the pasterns and fetlocks are the parts to be blistered, previous to rubbing in the ointment, smear some lard, tallow, or melted suet, over the heels, and within the hollow at the back of the small pastern. This will often prevent some troublesome sores forming, from the blistering ointment falling on these sensitive parts. While a blister is acting, the litter should be removed from under the feet, or it will tickle the legs, and irritate; but, above all, the head ought to be most carefully secured, for two days and nights, to oppose lying down, more especially to prevent the horse biting the blistered part. On the third evening he may be permitted to repose; but a prevention should even then be continued, by means of what is called a cradle. This apparatus may be bought at every turning shop; or may be made of eight or ten pieces of round wood, an inch and a half in diameter, and two feet long: these are strung at each end on a rope, and fastened around the neck. When it is intended to blister repeatedly, the effects of the first should have subsided before the second is applied; the scurf and scabs also be cleared away, and the part well washed with soap and water. In all cases, the third or fourth day after the application the part should be thoroughly painted over by means of a long haired brush (such as are in use with pastry-cooks to glaze their crusts) with lead liniment, which should be repeated every day; and when it is proposed to turn a horse out, it should never be done until the whole blistered surface be quite healed; otherwise dirt, flies, etc., may prove hurtful. It remains to observe, that, instead of repeated active blistering, it is in some cases preferable to keep up a continual slight irritation on the original blister by means of stimulants, as iodine ointment, mild blistering applications, etc.; caution is however necessary to avoid forming an eschar, and thereby a permanent blemish; but when a blemish is not of consequence this plan will be found often more efficacious than firing, as in splints, spavins, etc. Some practitioners blister mildly one day, and on the next wash off the blistering matter, thereby saving loss of hair. But there is more of apparent than real good in this plan. If a blister be necessary, it requires all its activity.

"*Ammoniacal blister.*— Spanish flies are only efficacious when the animal can afford to wait their action, which is rather slow. In most of the acute diseases, the horse would perish before the blister began to rise, wherefore resort has been had to boiling water and red-hot iron. The action of these last coarse and brutal measures was alone controlled by the violence of the internal inflammation; and, if the practitioner was mistaken in his estimate of the immediate danger, extensive and lasting blemish was the consequence. We have in the liquor ammonia an agent quite as formidable as boiling water or heated iron, but it is rather longer in displaying its force; wherefore, it allows time for watching its action, and of checking it the instant it has sufficiently blistered the skin. It is true the liquor ammonia upon the skin cannot be removed, neither need it be counteracted. Ammonia is like steam, only powerful when confined. The ordinary soap liniment, if covered over, would, because of the ammonia it contains, produce a lasting blemish; but every veterinary surgeon knows how very harmless a preparation that is when simply rubbed upon the surface. So, when we desire the active effects of liquor ammonia, we double a blanket or rug four or five times and hold it over the liquid. It takes from ten to twenty minutes to raise a blister, and it consequently can from time to time be observed; and, when its action has reached the wished-for point, all we have to do, effectually to stop it, is to take away the rug or blanket. That removed, the free surface and the heat of the body occasions the ammoniacal vapor to be dispersed, and the animal is safe.

"*Rowelling.*— Rowels act as foreign substances within the body; they cause irritation and suppuration, whereby more deep-seated inflammations are supposed to be removed; they are, however, often very convenient, because they stand as signboards to show the proprietor that something has been done. The common mode of making a rowel is after the following manner: a slit is first made by means of the rowel scissors, on any part of the integuments held between the finger and thumb; with the handle of the scissors separate from its cellular connections a circle of two or three inches in diameter, into which introduce something to prevent the reunion of the skin. A piece of circular leather, tolerably stiff, with a central hole, is a very common substance used, but is objected to by some on account of the difficulty of changing it without injuring the skin; tow, as more pliant, is frequently introduced into this cavity. If the rowel runs freely, it should be dressed every day, by changing the plug, if of tow, and by cleansing it, if of leather. No rowel should go undressed beyond the second day, for the comfort of the horse. They are very favorite applications with farriers, and therefore are frequently abused, by being employed in all cases indiscriminately; they are now, however, falling into disuse, setons having almost superseded them.

"*Setons*, in their action, resemble a very mild form of rowel, but are more convenient in their application. There is hardly a part of the body where a seton may not be conveniently placed: they have been put around the eye; they have also been entered at the withers, and brought out between the humerus and the sternum,—so extensive or so diminutive can they be made. In sinuous ulcers of the withers and of the neck, they may be placed; through the heels, in foot diseases, they have been inserted. In cavernous sores, they are entered at the superior part, and are brought out at an inferior, so as to form a depending orifice. The formation of a seton is very simple: a skein of thread, or a piece of tape of a convenient size, may be used: at the one end place a large knot; arm the eye of a corresponding-sized seton needle with the free end of the tape; introduce this into any proposed part, and, bringing it out at some other, either make a second knot, or tie the two ends of the tape together; which last method of fastening is, however, often objected to, from the danger of its catching in something, and being torn out, to the disfigurement of the horse. When a seton is placed in a sinuous track, for the purpose of inflaming, it is moved twice a day frequently, and moistened each time with some stimulant, as oil of turpentine, tincture of aloes and of benjamin. All setons require daily cleaning and moving. When they are required to act more quickly, the tape is infused in terebinthinate of cantharides, or small pieces of black hellebore are sewn within it. An old material, composed of woollen, flax, or cotton, and hair, is also used instead of tape. Setons, however, are of small service in acute cases. They are chiefly in use for disturbances of a chronic description.

"*Abstraction of Blood, or Bleeding.—Arteriotomy.*—Blood is abstracted by opening the conducting vessels, arterial and venous. When taken from arteries, the process is called *arteriotomy*; when by the latter, *phlebotomy*. Some bleedings include both these operations, as general scarifications of the soft parts; bleeding at the toe point; divisions of the vessels of the cornea, etc., etc. Blood-letting is called *local* when it is practised on or very near the affected part; and it is supposed to act more immediately than general bleeding, because it produces more effect with the loss of less blood. Local bleeding is therefore usually practised on the minor branches of the arteries and veins, as on the temporal artery, the plate vein, the vena saphena, etc. *Leeches* are a means of local bleeding not often used by us in veterinary practice; but there is no reason whatever why they should not be employed; when applied to the eye, and occasionally to other parts also, they adhere readily, abstracting blood rapidly, and therefore might be valuable aids in violent local inflammation. *Cupping* is also practised in France and other parts of the Continent with very large glasses, and it is there supposed to act remedially in many local inflammations. By *general bleeding* we understand the depletion of the system at large; and this we practice in extensive inflammations.

"*Division of the temporal artery.*—The proper spot for either its puncture or division is directly where the vessel leaves the parotid gland, to curve upward and forward around the jaw, a little below its condyle. When it is *punctured*, it usually affords much blood; and in such case, enough having been obtained, *divide* the trunk; when, the receding portions becoming pressed by the integuments, and lessening by their own contractility, the hemorrhage is stopped. It should be punctured by a lancet; a fleam may fix itself in the bone. Its division can be readily made also either by a lancet or scalpel.

"*Bleeding by the palate* is also a species of arterio-phlebotomy, and is a very favorite spot for abstracting blood with most ignorant persons, who vehemently recommend it in spasmodic colic or gripes, and in megrims. In such cases, however, a want of knowledge of the anatomy of the parts has occasioned a serious hemorrhage to occur; it may prove a fatal one, if the artery proper to the part be divided incompletely. The palatine artery and nerve run near each other, on each side of the roof of the mouth, so as to divide the inner surface of the hard palate into three nearly equal portions. No other than a direct division of the vein should ever be made; therefore, when bleeding is determined on at this place, do it by plunging a lancet or even a penknife in a *direct* line *across* the ruga, *one inch within the mouth, exactly between the middle and second nippers*; there these vessels form a curve, which curve will then be divided, and will then yield three or four pints of blood. If the instrument enter too much on one side, as about the middle of the second nipper, then a *partial* and *longitudinal* division of the artery may be made, and an alarming hemorrhage may follow. In this case, the section must be enlarged and deepened inwardly, that is, away from the teeth, which *completely severs* the vessel, and its retraction will stop the hemorrhage. A moderate or slight flow of blood from the palate may be obtained by light scarifications of the rugæ: but all bleedings here, except under circumstances of the most urgent necessity, had better be avoided.

"*Bleeding by the toe* is also arterior-phlebotomy. By no means cut out a portion of the sole at the point of the frog, which frequently occasions abscess; but with a very fine drawing-knife cut down exactly in the line of union between the crust and the sole; then, by puncturing the part with a lancet, a vast flow of blood may be obtained, the benefits of which in some cases are very marked, particularly in acute founder. If the blood should not flow with sufficient freedom, place the foot in warm water: the bleeding finished, cover the puncture with some tow and a little tar, and lightly tack on the shoe. There are, however, other methods of bleeding from the toe. Mr. Mayer uses a drawing-knife with a long curve, so that one sweep of the blade may cut a piece out of the foot. This appears to us bad practice as it leaves nature a space to fill up, instead of a simple incised wound to heal. Others take away none of the horn, but merely make a slit through the outer covering on to the vascular portion of the foot. The flap of horn they hold up so long as they desire blood, by the insertion of a piece of wood; and when they have obtained blood enough, they take out the wood so as to let the horny flap down. This last method, of all others, appears to us the easiest and the best.

"Sometimes the *plantar vein* is opened as a substitute. *Scarifications* are also occasionally practised, which, of course, divide both venous and arterial branches. In France extensive scarifications used to be made into indurations before the suppurative process had commenced, which in some cases prevented that from going on; and the *remedial wounds* made were healed by adhesive inflammation, or by healthy granulation: the same method has also been occasionally practised here, but it is not now often attempted.

"*Phlebotomy*, or the puncture of a venous branch, is the most usual mode of drawing blood in veterinary practice, and may be employed on any point of the body; but some vessels are much more frequently opened than others, and most of all the jugular.— *Bleeding by the thigh vein.* The *saphena* is a prominent vein, continued from the inner part of the hock, and may be opened by the fleam; but with much greater safety and propriety by a lancet. The opposite leg being held up, the operator placing himself in front of the thigh, and steadying himself and the horse by placing one hand on the hock, may fix the vein with the little finger of the other; while the lancet held between the thumb and fore finger punctures it. This vein should never be opened, save upon absolute necessity, as it is often troublesome to pin up. A horse has been cast for the trivial matter of stopping the hæmorrhage.— *Bleeding from the plate vein.* This vein is frequently opened to abstract blood after injuries of the fore extremities.— *The superficial brachial vein* is a continuation of the superficial division of the metacarpal veins, and in the passage upwards receives more than one branch; its principal trunk ascends along the inner side of the radius. It may also be well to remark, that, when taking blood from the superficial veins of the arm or fore-arm, if any difficulty is experienced in obtaining a sufficient flow, the lifting up of the other leg, by throwing the muscles of the punctured one into action, will force the blood from the inner to the outer set; and an increased quantity may be obtained. The plate vein, or external thoracic, is often opened, as it emerges from behind the arm, and is pinned up without any difficulty.

"*Bleeding by the jugular vein.*— The situation of this important vessel is well known, but its internal connections are not so familiar, though such knowledge is essential to the *uniform safety* of the operation. The horse has only external jugular veins, a right and a left one: as each emerges from the chest, it is found deep-seated, and approaching the trachea; it then passes forwards in company with the external carotid artery: toward the middle of the neck it becomes more superficial, and is now distinctly seen progressing rather *above* and *without* the carotid artery and trachea, or windpipe. The carotid, therefore, in the future course of the jugular, is situated a little *below* and more deep-seated than the vein. The jugular is also separated from the carotid by a slight muscular band, derived from the levator humeri. Its further track is marked in the hollow formed by the inferior edge of the levator humeri, where it is covered by the panniculus carnosus and integuments only; when, having nearly reached the jaw, it makes its well-known *division* into two portions. Bleeding by the jugular is usually practised with a lancet or with a fleam. The proper spot for the puncture may be found anywhere between two inches and six from the division of the vein: this latitude is here mentioned, because it is prudent to avoid puncturing directly over a former bleeding-place, known by the scar and enlargement: it should also be avoided where a little knot in the course of the vein will sometimes denote the existence of one of the venous valves. But in all ordinary cases, where these hindrances do not appear, operate at two or three inches from the division of the vein: which will be sufficiently evident when it is pressed on below the place punctured. Avoid operating low down in the neck, as there the vessel is deeper seated, and near to important parts.

"First moisten the hair and smooth it down; then, steadying and enlarging the vessel with one hand, with the other plunge the point of the lancet into the integuments, so as just to puncture them and the vein; then, by a slight turn of the wrist, carry the instrument obliquely forward to finish the cut. For opening the smaller veins, the *lancet* should always be used. In all but the practised hand, the *fleam* is the safest for bleeding from the jugular; it is always prudent to have the eye of the horse covered: unless the eye be covered, the horse will be likely to flinch at the moment of the stroke, and the puncture may be made in any place but where we wish. The hair being first wetted and smoothed, and the fleam being retained in the left hand, the unemployed fingers pressing on the vein so as to fix and swell the vessel; let the point rest exactly in the middle of the swelling; strike the fleam sufficiently hard to penetrate the skin and vein. A blood-stick is preferable for the purpose of striking the fleam: there is a vibration between two hard bodies when they meet, which, in this instance, is favorable to a quick and moderate puncture of the vein. After the vein has been opened, moderate pressure with the edge of the can which catches the blood is sufficient to keep up the flow: it may also be encouraged by putting a finger within the horse's mouth. The requisite quantity of blood being drawn, remove the can. The remaining process of securing the vessel is of equal importance. The sides of the orifice are first to be brought in apposition, without pinching them, and without drawing them from the vein: the same cautions should also be observed when the pin is introduced: let it be small, with an irregular point, and when inserted wrap round it a few hairs or a little tow.

"Common, however, as this operation is, and qualified as every one thinks himself to perform it, yet there are very serious accidents which do arise occasionally. It has occurred that the carotid artery has become penetrated. When the puncture has been made through the vein, the accident is known immediately by the forcible and pulsatory gush of florid arterial and dark venous blood together. In one instance of this kind, which occurred to a French practitioner, he immediately thrust his finger into the opening through the vein, and thus plugged up the artery, intending to wait for assistance. In this state he remained, we believe, an hour or more; when, removing his finger, to his surprise, he found the hæmorrhage had ceased, and did not again return. In another case, where an English practitioner accidentally opened the carotid, he placed a compress on the orifice, and had relays of men to hold it there for eight-and-forty hours; when it was found the bleeding had stopped. The admission of air is also another serious accident that now and then attends bleeding: it sometimes happens from the sudden removal of the fingers or blood-can, or whatever was used to distend the vessel by obstructing the return of the blood: this being suddenly taken away, allows the escape of the blood toward the heart, and occasions a momentary vacuum, the air being heard to rush with a gurgling noise into the vein through the orifice; it then mixes with the blood, and occasions, in some instances, almost immediate death. The animal begins to tremble; he next staggers, and finally falls in a state of convulsion: if the quantity of air taken in has been considerable, death ensues. The remedy must, therefore, be instantaneous, and consists in again opening the orifice, or making a new one, to gain an immediate renewed flow of blood, which will, in most cases, renovate the horse, who has been found afterwards to be tormented with an intolerable itching.

OPIATES. — (See NARCOTIC.)

OPIUM. — A narcotic vegetable poison. Mr. Coleman "thought, from some experiments made at the veterinary college, that opium has no apparent influence over the nervous system of the horse, and that it does not alleviate pain." Dr. White says, "I think that opium, as to its effect on the horse, does not possess that soothing anodyne and soporific quality for which it is justly distinguished in human medicine." Opium always tends to depress the vital organs in proportion to its quantity.

OPODELDOC, or SOAP LINIMENT. — A solution of soap and camphor in spirits of rosemary.

OPTIC NERVE. — The nerve on which sight depends.

ORBIT. — The socket of the eye is thus named.

ORGANIC. — A disease is said to be organic when any particular organ of the body is affected.

OSSIFICATION. — Ligaments and cartilages sometimes become bony, especially those ligaments which unite the splent to the canon bones, and the lateral cartilages of the foot.

OVARIES.—Two appendages to the womb, or uterus, which are cut out in the operation of spaying.

OVERREACH.—A horse is said to overreach, or overlash, when he wounds the fore heel with the hind foot.

OVERWORK.—Many of the diseases of horses originate in overwork.

PACE.—The peculiar manner of motion, or progression. The natural paces of the horse are, the walk, trot, and gallop, to which some add the amble.

PALATE.—The upper part or roof of the mouth.

PALLIATIVE.—Medicines and operations by which diseases appear to be relieved, but not cured. However desirable palliatives may be in the diseases of the human body, they are seldom satisfactory in the diseases and lameness of horses.

PALPITATION.—Beating of the heart against the breast-bone or ribs.

PALSY, or PARALYSIS.—A loss of muscular power, or an inability to move any part of the body.

PANCREAS, or SWEETBREAD.—A glandular substance situated in the abdomen, near the stomach. It secretes the pancreatic juice.

PANNICULUS, OR FLESHY PANNICLE.—A thin muscular covering attached to the skin of brute animals, by means of which they are enabled to shake it, and get rid of flies, etc.

PAXTON SHOE.—A contrivance for expanding contracted feet; but, like all other mechanical contrivances, they are useless or pernicious.

PAPS.—When young horses are cutting their teeth, and sometimes after that period, the excretory ducts of some of the salivary glands under the tongue become enlarged. These are named paps. They should be touched with a solution of alum, and the animal fed on mashes. (See MASH.)

PAPILLARY.—Pap-like; or rather like small or minute paps. A term applied to small elevations on different parts of the body, whether morbid or natural. Those little eminences on the internal surface of the leaves, or laminæ, of the cow's third stomach, or manyplus, are termed papillary.

PAR VAGUM.—The eighth pair of nerves are thus named.

PARACENTESIS.—The operation of tapping, for the purpose of giving vent to water collected in the chest, abdomen, etc. It has frequently been performed on animals without any benefit.

PARIETAL.—The bones that form the sides of the skull are thus named.

PARING.—Cutting the hoof in order to prepare it for the shoe.

PAROTID GLANDS.—Two large glands situated under the ears; they secrete saliva, which is conveyed by a duct into the mouth.

PAROXYSM.—The periodical accession, or the periodical increase, of a disorder.

PARTURITION.—The act of bringing forth young.

PASTERN.—The part between the fetlock joint and the hoof. (See cut of the foot, part first.)

PASTERN NERVE.—The nerve from which a portion is cut out in the operation of nerving.

PASTURE.—Pastures in elevated situations are the best for horses.

PATELLA.—The knee-pan of the human body, and the stifle of the horse. (See SKELETON.)

PATHOLOGY.—The doctrine of diseases.

PAUNCH.—The common name for the first stomach of the cow.

PECTORALS.—Medicines that relieve cough, and other diseases of the lungs.

PECTORAL MUSCLES.—The muscles of the breast.

PELVIS.—The basin, or that cavity wherein is lodged the bladder, uterus, and the rectum.

PENIS.—The yard or male genital organ.

PERFORANS TENDON.—The innermost

of the back sinews, or that which goes to the os pedis.

PERICARDIUM. — The heart bag. (See HEART, part first.)

PERICRANIUM. — The membrane that is closely connected with the bones of the head.

PERIOSTEUM. — The investing membrane of the bones.

PERISTALTIC MOTION. — That motion of the muscular coat of the bowels which causes the food and excrement to pass through them.

PERITONEUM. — The membrane which forms the external coat of the bowels, and some other of the viscera of the abdomen; it is, therefore, named the peritoneal coat of the bowels. It lines, also, the internal surface of the belly.

PERITONITIS. — Inflammation of the peritoneum. Diseases of the peritoneum are very rare in horses, and, when treated on the depleting, antiphlogistic principles of allopathy, generally terminate fatally. When the physiological equilibrium is interrupted, and inflammation of the peritoneum ensues, the available vital force is concentrated upon a small region of the body. The true indication is, to invite this force away from that region, and to distribute it over the general system, that it shall not be excessive anywhere. This mode of relief we call equalizing the circulation; the allopaths term it counter-irritation; they concentrate it in one spot, in the form of rowel and blister, their own works will show with what success.* Our principles teach us to accomplish the object by the stimulating influence of medicated vapor, enemas, nervines, and a mucilaginous diet. Whenever the disease is treated by the abstraction of blood, it generally terminates in dropsy.

PERSPIRATION. — The fluid which is secreted by the vessels of the skin. Perspiration is a highly important discharge in horses and other animals. The best medicine to promote sweating in the horse is a tea composed of lobelia, capsicum, and pennyroyal.

PHARYNX. — The upper part of the œsophagus, or gullet.

PHLEGM. — A mucous liquid thrown up from the lungs.

PHRENIC NERVE. — A nerve that passes through the thorax, over the heart, to the diaphragm.

PHRENITIS. — Inflammation of the brain.

PHYSIC. — In stable language, the term is applied to purgative medicines.

PHYSIOLOGY. — That branch of medical science which describes the functions of every part of the body.

PIA MATER. — A delicate membrane, that closely invests the brain.

PITCH, BURGUNDY. — A resin obtained from fir; it is used in the composition of plasters and charges.

PLACENTA. — The afterbirth.

PLATE VEIN. — A large vein that runs from the inside of the fore leg to the chest.

PLETHORA. — A fulness of vessels. Horses are often brought into this state from over-feeding, and want of sufficient exercise. It is known by heaviness, dulness, unwillingness to work. The urine is high-colored, and the dung generally hard and slimy. The cure consists in the reduction of the quantity of food, warm mashes, and regular exercise.

---

* Mr. Percivall details a case of peritonitis, after the usual symptoms in the early stage had subsided. "The horse's bowels became much relaxed: suspecting that there was some disorder in the alimentary canal, and that this was an effort of nature to get rid of it, I promoted the diarrhœa by giving mild doses of cathartic medicine, in combination with calomel!

"On the third day from this, prolapsus ani (falling of the fundament) made its appearance. After the return of the gut, the animal grew daily duller and more dejected, manifesting evident signs of considerable inward disorder, though he showed none of acute pain. The diarrhœa continued; swelling of the belly and tumefaction of the legs speedily followed. Eight pounds of blood were drawn, and two ounces of oil of turpentine were given internally, and in spite of another bleeding, and some subordinate measures, carried him off [the treatment, we presume] in the course of a few hours.

"Dissection: a slight blush pervaded the peritoneum, at least the parietal portion of it, for the coats of the stomach and intestines preserved their natural whiteness. About eight gallons of water were measured out of the belly. The abdominal viscera, as well as the thoracic, showed no marks of disease."

PLEURA. — The membrane which covers the lungs so closely as to appear a part of their substance.

PLEURISY, PLEURITIS. — Inflammation of the pleura.

PLEXUS. — A network of blood-vessels or nerves.

PNEUMONIA. — A general term for inflammation of the lungs.

POISONS. — Articles which impede or destroy the vital operations. Some people proclaim that all food is poison; that the difference in the effect produced lies in the quantity given. We deny this: good corn, oats, and hay, whose nature is to nourish and support the animal, can never be a legitimate cause of disease. Its excess in quantity, and its chemical decomposition for want of digestive power, are all of true food that can prove injurious. On the other hand, experience teaches us that opium, arsenic, corrosive sublimate, tobacco, and calomel are inimical to the animal organization, and will never change their chemical equivalents. A grain of arsenic will always be a grain of that poison, and can be detected after death: the same applies to opium. A very few grains of opium injected into the carotid artery of a dog killed him in four minutes; when the same quantity was injected into a vein, the animal lived twenty-five minutes. When injected into the bladder, it required a larger quantity to destroy life.

Again: one drop of the oil of tobacco applied on the tongue of a rabbit killed him instantly; one drop applied to the same organ of a cat, threw her into convulsions; two drops placed on the tongue of a squirrel killed it instantly. Hence, it does not require much penetration in order to decide what is and what is not poison. Animals often get, apparently, well, although large quantities of the above poisons have been given. This is no proof that the poisons cured them. In the early stage of the disease, the constitution can bear more violence — blood-letting and poisoning — than when it becomes debilitated. This explains the reason why large quantities of opium may be given to a horse at a certain time, without any perceptible effect; at another time, one-half the quantity will destroy him.

POLL-EVIL. — An obstinate disease, which often happens to horses. It generally proceeds from a blow received upon the poll or back part of the head.

PORTA. — The name of the great vein of the liver.

POULTICE, or CATAPLASM. — The emollient poultice may be composed of equal parts of slippery elm and flaxseed. The intentions to be answered by poultices are relaxation and stimulation. To relax a part, add to the above emollient, lobelia; when it is necessary to stimulate, use cayenne. Poultices that are designed for foul ulcers, in addition to the above articles, should contain at least one-third powdered charcoal.

PREVENTION OF DISEASE. — It is an old, but true saying, that prevention is better than cure, and, we may safely add, less expensive.

PRICKS, or PRICKING. — In shoeing a horse, the nail is sometimes driven in a wrong direction, and the sensible parts are wounded; he is then said to be pricked. When a horse has been slightly pricked, and the nail immediately withdrawn, it may not be followed by lameness; but, when the wound is considerable, matter will form; if the matter is not let out by paring away the horn, it quickly spreads under the horny sole, and upwards through the laminated substance of the foot, and breaks out at the coronet. (See CORONET.) To prevent this, the parts, as soon as the accident has happened, should be bathed with healing balsam. If the horse goes lame for several days, a poultice must be applied to promote suppuration.

PRIMÆ VIÆ. — The first passages, or stomach, and first intestines.

PROBANG. — An instrument for removing any obstruction in the œsophagus or gullet. It consists of a rather flexible rod, covered with leather, with a round, smooth knob at one end.

PROBE.— An instrument for examining wounds.

PROLAPSUS.— The falling down of a part, as of the uterus or fundament.

PSOAS MUSCLES.— The muscles that lie under the loins. These muscles are sometimes injured in strains of the loins.

PULMONARY DISEASES.— Diseases of the lungs.

PULMONARY VESSELS.— The blood-vessels and air-vessels of the lungs, which consist of the pulmonary artery and vein, and the bronchia, or branches of the windpipe.

PULSE.— The beating of the arteries. The horse's pulse is most conveniently felt in that part of the carotid artery which passes under the angle of the lower jaw.

PUNCTA LACHRYMALIA.— Two orifices near the inner corner of the eye, through which the tears pass.

PUPIL.— A part of the eye.

PUS.— The white matter formed by the process of suppuration.

PYLORUS.— The inferior portion of the stomach.

QUARTER ILL, or QUARTER EVIL.— There is a variety of names given to this disorder, such as joint murrain, or garget, black quarter, quarter evil, black leg, etc. The true causes of this disease are generally too liberal an allowance of food, or a sudden transition from poor keep to luxurious and nutritious diet. In some cases the energy of the body is lessened by exposure to cold and wet; hence the quarters and feet swell, and it is this circumstance which has given rise to the name by which the disorder is commonly known. The approach of this complaint is generally indicated by the animal separating himself from his companions; by his appearing dull, listless, and heavy, and by his refusing food. The more immediate symptoms are lameness and swelling of the hind quarters, and occasionally of the shoulders or back. These swellings, when pressed, make a crackling noise. (See EMPHYSEMA.) The mouth and tongue are frequently found blistered in this disease. A spare diet, and keeping the animal in a dry barn, are strictly to be attended to, with an occasional dose of nitrate of potassa, and clysters of thin gruel and common salt. By this means the disease may be subdued. If the disease first appears in the foot, a charcoal poultice must be applied.

RACK BONES.— The vertebræ of the back.

RADIUS.— The bone of the fore-arm.

RAKING.— (See BACK RAKING.)

RECTUM.— (See INTESTINES.)

RED-WATER.— This disease often attacks cows, and is named from the red appearance of the urine.

RESPIRATION.— The act of breathing; which includes inspiration, or the taking in of air by the lungs, and expiration, or the act of discharging it.

RINGBONE.— A bony excrescence on the lower part of the pastern, generally, but not always, causing lameness.

ROARING.— A disease which takes its name from the wheezing noise the horse makes in breathing, when put into quick motion. It is supposed by most veterinary writers to be caused by an effusion of lymph in the windpipe. Our own view of the subject is, that it is owing to a contraction of the respiratory tubes.

ROSEMARY.— The essential oil of this shrub is a useful ingredient in stimulating liniments.

ROT.— A disease of sheep, resembling pulmonary consumption, complicated with dropsy. Its causes are flooded lands and unsubstantial food.

ROWELS.— These are considered as artificial abscesses, or drains. They act on the principle of making one disease to cure another.

RUMINATION.— Chewing the cud.

RUPTURE.— A swelling caused by the protrusion of some parts of the bowels out of the cavity of the abdomen into a kind of sac, formed by that portion of the peritoneum (which see) which is pushed before it.

SACRUM.— That part of the back bone from which the tail proceeds.

SALIVA.— Spittle.

SALIVATION. — A profuse and continued flow of saliva.

SAND CRACK. — A perpendicular crack on the side or quarter of the hoof.

SANIES. — A bloody or greenish matter, which is sometimes discharged from foul ulcers.

SARSAPARILLA. — An infusion of equal parts of sarsaparilla and sassafras is useful for animals when the blood is loaded with morbific agents.

SCAPULA. — The shoulder blade.

SCARF SKIN. — (See CUTICLE.)

SCARIFICATION. — An incision of the skin with a lancet.

SCIRRHUS. — An indolent, hard tumor.

SCLEROTIC COAT. — (See EYE.)

SCOURING. — A scouring, or purging, is common among all our domestic animals. It is not a disease, but only a symptom of a loss of equilibrium, which may proceed from improper food, exposure to the cold and rain, which, of course, includes a loss of caloric, or heat. There is no general remedy, or one more speedy and effectual in the onset than mucilaginous drinks composed of slippery elm, combined with injections of the same. Warmth and moisture to the surface, and antispasmodics (which see), combined with astringents (bayberry bark is the best, in doses of half a tablespoonful every six hours), will seldom fail to effect a cure.

SCRATCHES. — Troublesome sores about the heels, depending on morbific agents in the system; for the cure of which, see "Modern Horse Doctor."

SCROTUM. — The bag or covering of the testicles.

SECRETION. — The word secretion is used to express that function.

SERUM. — The watery part of the blood.

SESSAMOID BONES. — Two small bones on the back part of the fetlock joint.

SINEW-SPRUNG. — A term sometimes applied to strains in the back sinews.

SITFAST. — A horny kind of scab, which forms on the skin in consequence of a saddle-gall.

SKIN. — (See CUTIS.)

SLIPPING. — (See ABORTION.)

SOLE. — (See FOOT, part first.)

SPASM. — An involuntary and continued contraction of muscles; thus, lockjaw depends on a spasmodic contraction of muscles.

SPAVIN. — A disease of the horse's hock, which generally causes lameness. Spavins are of two kinds: the bone, and the bog or blood, spavin. The former consists of a bony enlargement of the inside of the hock joint, towards the lower part; the latter, of a soft but elastic tumor, towards the bend of the joint. Mr. Percivall remarks: "Notwithstanding our confessed inability to cure this disease, we are often called on to treat it, as to the removal of it by means of a chisel, file, or saw. Although the practice is exceedingly commendable in cases of common exostosis, it is not so well adapted to spavin; those who employ such means seldom fail to leave the parts ultimately in a worse state than they found them. Our most successful remedies are such as come under the denomination of counter-irritants."

SPERMATIC CORD. — The vessels, etc., by which the testicles are suspended, consisting of the spermatic artery and vein, the vas deferens, or seed duct, cremaster muscle, and cellular membrane.

SPHINCTER. — A name given to muscles whose fibres are arranged in a circular direction, and whose office is to shut up the parts to which they are attached; such are the sphincter of the neck of the bladder, and the muscles which close the fundament.

SPINE. — The spine of the neck and back is composed of many small bones named vertebræ. Seven belong to the neck, eighteen to the back, six to the loins, five to the sacrum, and in the tail there are about thirteen.

SPLEEN, or MILT. — A soft substance, of a long, oval form, and purple color. It seems to be a reservoir for the blood that may be designed for the secretion of bile in the liver.

SPLENTS. — These are bony excrescences,

which grow on the inside of the shank bone.

STAGGERS. — This is named from the staggering gait of the animal. It may be brought on by the horse eating too greedily, swallowing his food when imperfectly chewed, or eating freely of food that is difficult of digestion. Horses of rather an advanced age, and weak digestive organs, when improperly fed, or when a large quantity of meal is allowed, are very liable to apoplexy, or staggers. The disease is generally symptomatic of derangement of the stomach, indigestion, and over-distention of the digestive organs.

STERNUM. — The breast bone.

STIFLE JOINT. — This joint is composed of the bones called os femoris, tibia, and patella. (See SKELETON.)

STOPPING. — A mixture of clay and cow-dung is employed for the purpose of stopping horses' feet, and keeping them moist.

STRAINS. — For all kinds of strains, rest is the best remedy; sometimes they require poultices, fomentations, etc. The latter will be indicated by pain and swelling.

STUBS. — When a horse is wounded by a splinter of wood, about the foot or leg, he is said to be stubbed.

STYPTICS. — Medicines which stop bleeding. The most effectual method of stopping bleeding is to tie the wounded vessel.

SUDORIFICS. — Medicines which excite sweating. It is very difficult to sweat a horse, except the process be assisted by warmth and vapor externally. Lobelia, pennyroyal, and capsicum, promote the insensible perspiration; they must be given in infusion to the amount of half a gallon or more.

SULPHUR. — Used in cutaneous diseases, as an alterative.

SWEETBREAD. — (See PANCREAS.)

SYNOVIA, JOINT OIL. — A mucilaginous fluid formed within joints, to render motion easy, or diminish friction.

TANSY. — A medicine used to expel worms.

TAR. — Common tar is used as an astringent for horses' feet.

TARTAR, CREAM OF. — Used on horses to promote the secretion of urine.

TENACULUM. — A kind of hook, for taking up an artery.

TENDO ACHILLIS. — The great tendon, which is fixed or inserted into the calcaneum, or projecting bone of the hock.

TENDON. — The white shining extremity of a muscle.

TENESMUS. — Continual efforts to void dung, without any discharge.

TENT. — A piece of lint, or tow, smeared with ointment, and thrust into a sore, in order to prevent a too hasty and superficial healing.

THORACIC DUCT. — The trunk of the absorbents. (See ABSORBENTS, part first.)

THRUSH. — A disease has lately prevailed to a great extent in the New England States, which deserves some consideration. It is called thrush, and is supposed to be a disease of the horse's frog, consisting in a discharge of matter from its cleft, or division; sometimes the other parts of the frog are also affected, — become soft, ragged, and incapable of affording protection to the sensitive frog, which it covers. We cannot agree with many writers, that thrush is a strictly local disease; for, after it has passed through the different stages, viz., inflammation, suppuration, etc., the whole system takes up the diseased action, either by sympathy or irritation. Hence the reader will see the folly of depending on local agents, in the form of ointments, for the cure of the disease, in which all the organs are more or less concerned.

The internal remedies we recommend, are alteratives; remove the cause, if any exist, in the form of bad ventilation, poor diet, hard work, partial grooming, or the sluicing of cold water on the legs. Let the animal have bran mashes, with a few boiled carrots, every night.

The local remedies consist in paring away the ragged or uneven parts of the frog; then wash the surface with castile soap and lukewarm water; afterwards with a solution of common salt, in the following proportions: one tablespoonful Liverpool salt to a pint of rain water; then apply linseed oil, spirits of turpentine, pyroligneous acid, — equal parts, — in the cleft of the frog; let the whole surface be covered with tow, then upon the tow place a flat piece of wood, about the width of the frog, — one of the ends passing under the toe of the shoe, the other extending to the back part of the frog, and bound down by transverse slips of wood, the ends of which are to be placed under the shoe. The moderate pressure thus applied will contribute materially to the cure and to the production of solid horn. This dressing must be repeated daily. If, after this treatment, matter should discharge, the heel contract, and the horn soften, then apply a poultice of Indian meal, with half a tablespoonful of cayenne pepper on the surface, washing the foot, as above, every night.

TIBIA. — The bone of the horse's thigh; that is, the bone between the hock and the stifle.

TICKS. — Insects that infest sheep and other animals. A strong infusion of lobelia will destroy them.

TONGUE. — The tongue is a muscular substance, composed of fibres variously arranged, by which it is rendered capable of that diversity of action which we observe; it has also several muscles attached to it. The small bone, to which it has a muscular attachment, is named os hyoides.

TONICS. — Medicines that augment the strength of the body, such as gentian, wild cherry, poplar bark, etc.

TRAINING. — By the word training is meant, putting a horse in that state in which all the functions of the body are in equilibrium. In order to bring a horse into this desirable state, we refer the reader to the "Modern Horse Doctor."

TUBERCLES. — Small tumors that sometimes suppurate and discharge pus; they are often found in the lungs.

TUMOR. — A swelling on any part of the body. Tumors are of various kinds: sometimes caused by bruises, or other accidents; at others, arising without any visible cause.

TUNIC. — A coat, or membrane, investing a part; such as the tunica vaginalis of the testicles.

TURGESCENCE. — An over-fulness of the vessels in any part.

TUMERIC. — Tumeric root, an aromatic stimulant, sometimes used in jaundice or yellows.

TWITCH. — An instrument made by fixing a noose, or cord, to the end of a stick; this is put on the horse's upper lip and twisted rather tight, which makes him stand quiet during an operation.

TYMPANY. — A distention of the abdomen by air.

TYPHUS. — Putrid fever.

UDDER. — The udder is a glandular body, whose office is to secrete milk. It is divided, in the cow, into four quarters; each of which has an excretory duct, or teat, whose office is to facilitate the extraction of milk. At the extremity of each teat is a contrivance for the purpose of retaining the fluid contained in the udder, until it becomes much distended; when, if not drawn off, it flows spontaneously, and the animal is thereby partly relieved of her burden. Sometimes the udder swells and becomes sore, as is often caused by improper feeding. As there is great sympathy existing between the stomach and udder, whatever deranges the former will also affect the latter, through the medium of sympathetic action. In this case, the cow should be drenched with a tea of pennyroyal and thoroughwort, and fed on gruel. The udder should be fomented with an infusion of mullen leaves. Should the swelling continue, and appear painful, the following embrocation may be used: linseed oil and lime-water, equal parts, mix. If an abscess forms, and matter can be felt, it should be opened at its most depending part, so that the matter may run freely off.

ULCERS. — There are quite a variety of ulcers to be found in animals; the most of them will heal by the application of a mild astringent, or tonic, such as an infusion of bayberry bark, or the tincture of capsicum. If it be foul or callous in any part, then powdered bloodroot will be proper.

URETERS.— Two small tubes by which the urine is conveyed from the kidneys to the bladder.

URETHRA.— A membranous and muscular tube by which the urine is conveyed from the bladder; it is of considerable length in the horse.

URINE, EXCESSIVE DISCHARGE OF.— (See DIABETES.)

URINE, INCONTINENCE OF.— (See INCONTINENCE.)

UTERUS.— The womb. The uterus of the mare is very unlike that of the human subject, in whom it consists of one bag, rather of an oval shape, somewhat resembling a pear; but in the mare and other quadrupeds it has a body and two branches, called its horns. The uterus terminates in the vagina by a narrow portion, called the neck or mouth of the womb. The extremities of these horns have tubes attached to them, which, from the name of the discoverer, are called Fallopian tubes; one end of each is expanded, and has a fringed kind of edge: this is named the fimbria of the Fallopian tube. The Fallopian tube is very tortuous in its form; and that end which proceeds from the horn of the uterus is extremely small; but the other, which is slightly attached to the ovarium, is considerably larger. The ovarium is an oblong body, about the size of a small hen's egg. The ovaria — for there are two of them — are composed of a number of transparent vesiculæ, called ova (eggs); each ovum is surrounded with cellular membrane; and when the ovum is impregnated and passes into the uterus, it leaves a mark which is named corpus luteum.

UVULA.— In the human subject, the small flesh-like substance hanging in the middle and back part of the throat, is thus named. In the horse, this is of a very different form. The uvula completely closes the opening to the pharynx, though it readily yields to the passage of food, or any liquid, toward the gullet; it prevents, also, the return of anything to the mouth, even the air which is expired from the lungs, unless it be thrown aside by a violent effort, as in coughing. It is on this account that, when the horse is affected with nausea, or has the action of the stomach inverted, — which sometimes happens, though very rarely, — the contents of the stomach will be discharged through the nostrils; but if the horse happens to cough during the process, some part will be discharged by the mouth.

VAGINA.— The passage from the external pudendum, or shape, to the mouth of the womb.

VALERIAN.— The root of valerian is used as an antispasmodic; its virtues have been underrated by writers on veterinary medicine.

VEINS.— The motion of the heart is known to communicate momentum to the blood through the veins. Mr. Percivall says: " We are not to reject the power of the heart altogether, merely because the blood flows with a uniform stream in the veins; for the absence of pulsation in them is no proof that the motion of the blood is not influenced by the contractions of the heart; the extreme division which this fluid undergoes in its circulation through the capillaries, and the tortuosity and complication of the numberless small veins, account for the regular and uninterrupted stream which we meet with in the larger branches. To prove that this is the explanation of the fact, if you open a vein that has free and direct communication with the extremity of an artery (its capillary structure), the blood will flow from it with the same pulsatory motion as if the artery itself had been penetrated: but if the vein be one of large size, remotely situated from any arterial communication, or if it be one that springs from the union of numerous capillaries, that smooth and even stream, with which the blood circulates in the trunks, will be observable here. These facts, then, lead us to conclude that the force of the heart is not sufficient of itself to propel blood through the venous system."

From the collected accounts of writers on this subject, it seems highly probable, that the blood flowing in the veins receives additional momentum from the reaction of the capillaries, and that it is further urged on by some contractile force resident in these vessels themselves. That the blood is advanced in its course by the action of those muscles contiguous to veins furnished with valves, is, without doubt, well founded, as far as an occasional auxiliary is concerned, as the common operation of bleeding demonstrates; for it is in consequence of muscular pressure upon the veins about the head, that the motion of a horse's jaw accelerates the flow of blood through the jugular vein; as such, however, it cannot be ranked among the essential causes of the blood's motion in them.

VENTRICLE. — One of the cavities of the heart. (See HEART.)

VERMIFUGE. — Medicines that destroy or expel worms.

VERTEBRÆ. — The bones of the neck and spine.

VERTIGO. — A slight degree of apoplexy.

VISCERA. — The plural of *viscus*, a term applied to the internal organs, as the lungs, bowels, etc.

VIVES. — A swelling of the parotid gland, which is situated between the ear and the angle of the jaw.

VULVA. — A name given to the external parts of generation in females.

WALL EYES. — A horse is said to have a wall eye, when the iris is of a light or white color.

WARBLES. — Small, hard swellings on the horse's back, caused by the pressure, or heat, of the saddle.

WARTS. — Spongy excrescences which arise in various parts of the body.

WENS. — Hard tumors, of various sizes, in different parts of the body. The most effectual method of removing them is to dissect them out, together with the cyst, or bag, in which they are formed. The skin is then to be sutured, and treated as a common wound.

WHIRL BONE, or ROUND BONE. — The hip joint is thus named.

WIND. — The most effectual method of bringing a horse to his wind, is to give him regular exercise.

WINDGALLS. — Elastic tumors on each side of the back sinews, immediately above the fetlock joint; they are often caused by hard work, or trotting on hard roads, at too early an age. There are various operations recommended, such as firing, blistering, etc.; but the remedy is generally worse than the disease. Rest, bandaging, and the occasional use of liniment, is all that can be done with safety.

WITHERS. — The part where the mane ends is thus named in the horse.

WORMS. — The stomach and bowels of horses are liable to be infested with different kinds of worms; but as the same treatment is proper, of whatever kind they may be, it is needless to enter into a particular description of them. Many articles are recommended by veterinary writers, for the purpose of ridding the animal of these pests, viz., antimony, calomel, turpentine, either of which would be just as likely to kill the horse as the worms. The true indications to be fulfilled are to tone up the stomach and digestive organs.*

YARD, FALLEN. — (See FALLING OF THE YARD.)

YARD, FOUL. — The horse's penis sometimes requires to be washed with soap and water, in order to free it from mucous matter and dirt.

YELLOWS. — This disease is indicated by a yellowness of the membranes that line the eyelid, and the inner parts of the lips and mouth. In this disease, the natural course of the bile is perverted; it becomes absorbed into the circulation, and thus tinges the membranes and fluids of a yellow color. The excrement is generally of a lighter color than usual. The disease may be produced by a want of tone in the liver, caused by obstructing the surface.

* Dr. J. Hinds says, " Since the worms are not always to be killed, even by strong poisons (calomel), nor brought away by brisk purgatives, reason dictates and nature beckons us to follow her course in affording to the horse a run at grass; if that is impossible, adopt the means nearest thereto that lie within our reach." If calomel is a poison, — and thousands declare it is, — then it must entail a disease more formidable than the one it is intended to cure.

APPENDIX TO PART FIRST.

LIGAMENTARY MECHANISM OF ARTICULATIONS.

ARTICULATIONS OF THE TRUNK.

Ligaments of the spine. — Those between the head and first and second vertebræ are:

Lateral ligaments, one on each side, that run from the coronoid processes of the occipital bone to the fore part of the atlas, and are fixed in the roots of the transverse processes.

Suspensory ligament of the head is a broad ligament enclosed within the capsular. It proceeds from the body of the atlas to the occipital bone.

Capsular ligament is attached to the occipital bone, around the roots of the condyloid process, and to the anterior articular processes of the atlas.

Superior ligament runs from the long ring of the atlas to the spine of the vertebra dentata.

Odontoid ligaments are three in number: the two long pass from the sides of the process dentata to the occipital condyles; the last runs from the point of that process to the anterior and inferior parts of the atlas.

Inferior ligament runs from the inferior spinous process of the first to the second vertebra.

The ligaments common to the spine are:

Intervertebral ligaments. — They are the chief bond of union by which one vertebra is bound to another.

The common inferior and superior ligaments. — The former passes obliquely along the inferior parts of the vertebræ, and the latter runs within the spinal canal.

Capsular ligaments surround the smooth cartilaginous surfaces of the articulatory processes.

Intertransverse ligaments fix the transverse processes of the dorsal vertebræ together.

Interspinous ligaments are found between the spinous processes of the back and loins.

Ligamentum subflavum (or nuchæ) extends from the occipital bone to the tail. It covers and connects the spinous processes of the back, loins, sacrum, and coccyx. This ligament forms a strong connecting medium between the spines of the vertebræ.

Ligaments of the pelvis. — Two superior transverse ligaments are fixed to the transverse processes of these bones above; two inferior, below, run from the fourth and fifth transverse processes of the loins to the crest of the ileum. Sacro-iliac symphysis consists of a cartilago-ligamentous substance interposed between, and firmly adherent to, the transverse processes of the sacrum and the inward part of the ileum. This union is strengthened by ligamentary bands, which run from the posterior spine, and border of the ileum, to the transverse process of the sacrum.

Sacro-sciatic ligaments are broad expansions stretched across the sacro-sciatic notch. They arise from the transverse processes of the sacrum, and those of two or three uppermost bones of the coccyx, and are extended to the posterior parts of the ileum and ischium, and to the tuberosity of the latter.

Obturator ligament is an expansion, thinner than the last, which passes across the foramen magnum ischii.

Ligament of the symphysis is the cartilago-ligamentous substance which unites the pubic bones.

Ligaments of the ribs. — Every rib is connected to two vertebræ by four ligaments, viz., two capsular, internal and external ligaments.

Capsular ligament of the head invests and holds it within the vertebral socket. Two articular cavities are found within it, one with each vertebra, which have separate synovial linings.

Capsular ligament of the tubercle surrounds it at its articulation with the transverse process of the vertebra.

External and internal ligaments consist of strong fibres, which connect the neck of the rib, above and below, to the spine.

Intercostal ligaments are broad fibrous bands which run obliquely across the intercostal spaces, and hold the ribs and their cartilage firmly together.

Sternal ligaments. — These several pieces of the breast bone are united to each other by intervening cartilaginous substance; in addition to which they are connected by ligamentary bands, both inwardly and outwardly. The fore part of it is surmounted by a broad portion of cartilage, which runs along its under part.

ARTICULATIONS OF THE FORE EXTREMITY.

Shoulder joint. — The capsular ligament around this joint is strengthened in many places by additional fibres dispersed upon its exterior. It is fixed to the rough margin of the glenoid cavity, and to the neck of the os humeri. A synovial membrane lines it, which may be followed upon the cartilaginous surfaces of the bones. Externally, this ligamentous capsule is clothed on every side by muscles, and to them is attributed the main strength of the joint.

Elbow joint. — The ligaments of it are two lateral and a capsular.

Knee joint. — In the knee there are five distinct articulations; one between the radius and the three small bones of the upper row; a second between the small bones, above and below; a third between those of the lower row and the metacarpal bones; a fourth between the os trapezium and the os cuneiforme; and a fifth between the os pisiforme and os trapezoides; they have all separate capsular ligaments and synovial linings.

The ligaments of the knee, and the tendons passing over it, are girt by, broad, glistening, ligamentous bands, which retain the latter in their places, and render the joint stronger and more compact. Between these ligaments, fascia, and the extensor tendons, are some small bursæ.

External lateral ligament runs from a tubercle upon the radius to the head of the external metacarpal bone.

Internal lateral ligament consists of two parts, which proceed from a similar tubercle upon the inside, and from the body of the radius. The longer is fixed to the inner head of the metacarpal bone, and the shorter to the fore part of the metacarpal.

Ligamentum annulare passes from the os trapezium to the ossa scaphoides and cuneiforme; it confines the flexor tendons.

Fetlock joint. — Capsular ligament is attached to the articulatory surfaces of these bones; and the synovial membrane, after having lined it, is reflected upon their cartilages; it is guarded in front by the extensor tendon.

Long lateral ligament is fixed to a projection upon the side of the metacarpal bone, and to the os suffraginis.

Short lateral ligament runs underneath the latter. These ligaments prevent motion sideways.

The ligaments of the sessamoid bones are seven, viz.: superior suspensory, the long inferior, the short inferior, the two lateral, and the two crucial.

Pastern joint is formed by the adaptation of the ossa suffraginis and corona. It has a capsular, and two pairs of lateral ligaments.

The capsular ligament is inserted into the smooth cartilaginous ends of these bones; it is blended with the extensor tendon in front, and behind with the long inferior ligaments of the sessamoids.

The lateral ligaments are inserted on the sides of the os coronæ and suffraginis.

Coffin joint is made up of three bones: the os corona, pedis, and naviculare.

Capsular ligament envelopes the articulatory surfaces, and is inserted beyond their limits; in front it is united with the extensor tendon; behind, it is strengthened by the tendo perforans. In addition to the capsular, there are three pairs of ligaments.

The first pair passes from the superior edges of the os pedis to the lateral parts of the os corona, and are inserted about its middle.

The second pair is stretched from the extremities of the os pedis to the os corona, and are fixed below and behind the first.

Third pair arise from the sides of the coronal process, and terminate in the cartilages.

The ligaments of the os naviculare are four, viz.: two single, and one pair.

Superior ligament runs from its upper and posterior part to the tendo-perforans.

Inferior is a very broad ligament, arising from the whole of the lower edge of the bone, and thence extending to the os pedis, above the long extensor tendon.

Lateral ligaments fix the os naviculare, by its two ends, to the sides of the os corona.

ARTICULATIONS OF THE HIND EXTREMITY.

The thigh joint is formed by the reception of the head of the os femoris into its socket.

Capsular ligament is attached around the cervix of the os femoris and the margin of the acetabulum; it is thickly clothed on every side by muscle, which assists to maintain its position.

The *acetabulum* is surrounded by the circular ligament, whose border turns inward to embrace the cartilaginous head of the os femoris.

The *notch* in this cavity, to its inward side, is crossed by the transverse ligament, which here makes up for the deficiency in the bone.

Ligamentum teres consists of a bundle of ligamentous fibres inclosed in a sheath, which proceed from a pit in the inner and upper part of the ball to a similar one in the roof of the socket. Another portion of it leaves the cavity under the transverse ligament, and is implanted in the pubes.

The synovial membrane lines the socket, and is reflected over these parts.

Stifle joint is composed of the os femoris, the tibia, and patella.

Ligamenta patella are composed of four strong cords, which descend over the condyles of the os femoris, and are inserted into the tubercle of the tibia. The external one passes upon the outer and anterior part of the external condyle; the internal, upon the inward part of the internal condyle; and the middle one, between them. They approach each other in their descent. Concealed by the external one is the fourth ligament of the patella; it runs to the outward part of the tibia.

The *patella*, with its articulatory surface of the condyles in front, forms a joint of its own, perfectly distinct from that between the tibia and os femoris.

Its capsular ligament is fixed to its surrounding border.

Internal lateral ligament descends from the internal condyle to the inner and upper part of the tibia.

External lateral ligament — stronger than the internal — runs from the external condyle to the upper end of the fibula.

Crucial ligaments, short and strong, and deeply buried within the joint, run from the space within the condyles to the tibia.

The *synovial membrane*, after having lined the capsule, is reflected upon the cartilages and ligaments included within it.

Hock joint has four lateral ligaments, two on each side, called internal and external.

Capsular ligament includes the lower end of the tibia, and the pully-like part of the astragalus; to both of which, and the lateral ligaments, and to the os calcis, it is firmly attached.

The *os calcis* forms a joint with the os cuboides, and the ossa cuneiforme are also a joint, and the middle and small bones make joints with the cuboid above, and the metatarsi below; hence, there are six articulations in addition to what we commonly understand by the hock joint, that between the tibia and astragalus.

www.ingramcontent.com/pod-product-compliance
Lightning Source LLC
Chambersburg PA
CBHW020311240426
43673CB00039B/772